FAO Fisheries and Aquaculture Report No. 990
FAO, Rapport sur les pêches et l'aquaculture nº 990

FIPM/R990 (Bi)

Report and papers presented at the

THIRD WORKSHOP ON FISH TECHNOLOGY, UTILIZATION AND QUALITY ASSURANCE
IN AFRICA

Victoria, Mahe, Seychelles, 22–25 November 2011


Rapport et documents présentés au

TROISIÈME ATELIER SUR LA TECHNOLOGIE, L'UTILISATION ET L'ASSURANCE DE QUALITÉ
DU POISSON EN AFRIQUE

Victoria, Mahe, Seychelles, 22-25 novembre 2011


FOOD AND AGRICULTURE ORGANIZATION OF THE UNITED NATIONS
ORGANISATION DES NATIONS UNIES POUR L'ALIMENTATION ET L'AGRICULTURE
Rome, 2011

## PREPARATION OF THIS DOCUMENT

This document contains the report and papers of the third Workshop on Fish Technology, Utilization and Quality Assurance in Africa. The workshop was held at the International Conference Centre (ICC), Victoria, Mahe, Seychelles, from 22 to 25 November 2011. It was attended by 30 experts who reviewed the progress in post-harvest fish utilization, with particular attention to fresh fish handling, fish processing, quality assurance, marketing and socio-economic issues. The meeting included:

- presentation by the secretariat of a report on the progress and events since the second FAO Workshop on Fish Technology, Utilization and Quality Assurance in Africa held in Agadir, Morocco, in 2008;
- presentation of 26 papers selected by the screening panel. This panel was established by the technical secretariat following a call for contributions launched a year before the meeting which recorded 38 papers of experts; and
- a field trip to the Indian Ocean Tuna factory.

During the discussions a number of recommendations were made to FAO, its member countries and to all institutes involved in fish utilization in Africa.

The workshop was organized by the Products, Trade and Marketing Service (FIPM) of FAO's Fisheries and Aquaculture, Policy and Economics Division (FIP), in collaboration with the Seychelles Bureau of Standards (SBS) in Victoria, Mahe, Seychelles. The workshop was funded by the Regular Programme of FAO.

The views expressed in this publication are those of the authors and do not necessarily reflect the views of FAO. The papers do not follow the FAO editorial guidelines: they appear as presented by their authors.

## PRÉPARATION DE CE DOCUMENT

Ce document contient le rapport et les contributions du troisième Atelier sur la technologie, l'utilisation et l'assurance de qualité du poisson en Afrique. L'atelier, auquel assistaient 30 experts en technologie du poisson, s'est tenu du 22 au 25 novembre 2011 aux Centre international des conférences (ICC) (Victoria, Mahe, Seychelles). Il a passé en revue les progrès dans l'utilisation du poisson post-capture avec une attention particulière à la manutention du poisson frais, à la transformation du poisson, l'assurance de qualité, la commercialisation et les aspects socioéconomiques. Ces questions étaient adressées à travers:

- la présentation par le secrétariat du rapport sur les progrès et événements depuis le deuxième Atelier de la FAO sur la technologie, l'utilisation et l'assurance de qualité du poisson en Afrique qui s'est tenu à Agadir, (Maroc) en 2008;
- la présentation de 26 communications sélectionnées par le groupe d'experts chargé de les examiner. Ce groupe d'experts a été établi par le secrétariat technique suite à l'appel à contributions lancé un an avant la réunion et qui a enregistré 38 communications d'experts; et
- une visite de terrain à la conserverie de thon «Indian Ocean Tuna factory».

Lors des discussions un certain nombre de recommandations ont été faites à la FAO, à ses pays membres et aux instituts impliqués dans l'utilisation du poisson en Afrique.

Cet atelier a été organisé par le Service des Produits, échanges et commercialisation (FIPM) de la Division des politiques et de l'économie de la pêche et de l'aquaculture de la FAO (FIP), en collaboration avec le Seychelles Bureau of Standards (SBS) de Victoria, Mahe, Seychelles. L'atelier a été financé par le Programme régulier de la FAO.

Les opinions exprimées dans cette publication sont celles des auteurs et ne reflètent pas nécessairement celles de la FAO. Les contributions ne suivent pas les Directives éditoriales de la FAO: elles figurent telles que présentées par leur auteurs.

FAO.
*Report and papers presented at the third Workshop on Fish Technology, Utilization and Quality Assurance in Africa. Victoria, Mahe, Seychelles, 22–25 November 2011.*
*Rapport et documents présentés au troisième Atelier sur la technologie, l'utilisation et l'assurance de qualité du poisson en Afrique. Victoria, Mahe, Seychelles, 22-25 novembre 2011.*
FAO Fisheries and Aquaculture Report/FAO Rapport sur les pêches et l'aquaculture. No. 990. Rome, FAO. 2012. 263 pp.

## ABSTRACT

The third Workshop on Fish Technology, Utilization and Quality Assurance in Africa was organized by the Products, Trade and Marketing Service of FAO's Fisheries and Aquaculture, Policy and Economics Division (FIP), in collaboration with the Seychelles Bureau of Standards (SBS) in Victoria, Mahe, Seychelles. The workshop reviewed progress in post-harvest fish utilization in Africa and made recommendations to FAO, its member countries and institutes interested in fish utilization in Africa. The experts reviewed in particular fresh or live fish handling, fish processing, post-harvest loss assessment, quality and safety, and marketing and socio-economic issues. The meeting included: a presentation by the secretariat of a report on progress and events since the workshop held in 2008, presentation of 26 papers and a field trip to the Indian Ocean Tuna factory. The report includes the recommendations as well as the papers that were made available to the experts.

## RÉSUMÉ

Le troisième Atelier sur la technologie, l'utilisation et l'assurance de qualité du poisson en Afrique a été organisé par le Service des produits, échanges et commercialisation de la Division des politiques et de l'économie de la pêche et de l'aquaculture de la FAO, en collaboration avec le Seychelles Bureau of Standards (SBS) de Victoria, Mahe, aux Seychelles. L'atelier a passé en revue les progrès dans l'utilisation du poisson post-capture en Afrique et fait des recommandations à la FAO, à ses pays membres et aux instituts intéressés par l'utilisation du poisson en Afrique. Les experts ont passé en revue notamment la manutention du poisson frais ou vivant, la transformation du poisson, l'évaluation des pertes post-captures, la sécurité sanitaire et la qualité, la commercialisation et les questions socioéconomiques. Cette révision s'est effectuée à travers la présentation, par le secrétariat, du rapport sur les progrès et événements depuis l'atelier qui s'est tenu en 2008, des présentations de 26 communications et une visite de terrain à la conserverie de thon «Indian Ocean Tuna factory». Le rapport inclut les recommandations de même que les communications qui ont été mises à la disposition des experts.

# CONTENTS/TABLE DES MATIÈRES

## ORGANIZATION

1.     The third meeting on Fish Technology, Utilization and Quality Assurance in Africa was organized by the Products, Trade and Marketing Service, Fisheries and Aquaculture Policy and Economics Division of the Food and Agriculture Organization (FAO), in collaboration with the Seychelles Bureau of Standards (SBS).

## OPENING

2.     The workshop was held at the International Conference Centre (ICC), Mahé, Seychelles, from 22 to 25 November 2011.

3.     Mrs Amy Quatre, Chief Executive Officer (CEO) of SBS, expressed the delight and honour of her institution at being asked to coorganize the Expert Meeting on Fish Technology, Utilization and Quality Assurance. She then extended her warm welcome to the participants before thanking FAO for pursuing the initiative to give African countries the opportunity to discuss their experiences and exchange information. She recounted the significant role of fish and fishery products in the economy of Seychelles. The issue of the magnitude of post-harvest losses in small-scale fisheries was raised, as was the need to reduce them so as to optimize the use of fisheries resources, as entrenched in the FAO Code of Conduct for Responsible Fisheries. Activities implemented by FAO in post-harvest fisheries were also underscored. She then wished the meeting fruitful deliberations.

4.     Mr Alejandro Anganuzzi, Executive Secretary of the Indian Ocean Tuna Commission (IOTC), addressed the meeting on behalf of FAO, and thanked the audience for having taken the time to attend the workshop. He also thanked the Government of Seychelles for hosting the meeting and for having given their full support to its organization. He then elaborated on the steady increase in the numbers of people from developing countries deriving their livelihoods from fishing amid deficient post-harvest technologies and quality assurance systems and marked by high losses, which cumulatively had significant impacts on the economies of households and fishing communities. He emphasized the timeliness of the meeting, which would be addressing important socio-economic issues to improve the livelihoods and food security situation of the fishing communities and the larger rural populations, and hence contribute towards the realization of the Millennium Development Goal on reducing food insecurity and alleviating poverty. He also mentioned past and present interventions conducted in this regard by FAO, including the thematic workshops, such as this event, that the UN agency has been sustaining for the past 30 years.

5.     The meeting was then addressed by H.E. Peter Sinon, Minister for Investment, Natural Resources and Industry, who expressed the pride of Seychelles in hosting the meeting. He underscored the significance of fisheries to the country, the second pillar of its economy after tourism and, in terms of food security, reminded participants that Seychelles is one of the major per capita fish-consuming nations. He outlined the Government's sectoral policies to diversify further, giving emphasis to value-added product development from bycatch, which represent about 5 percent of the industrial tuna fishing catch volumes. Regarding small enterprises in particular, H.E. Sinon brought up the challenges faced in meeting international standards and in reducing losses in earnings, wastage of valuable resources and production of low-quality products, therefore the importance of the deliberations.

6.     The detailed speeches are provided in Appendixes D, E and F and the programme is presented in Appendix A.

## PARTICIPANTS

7.     The meeting was attended by 29 experts from: France (1), Ghana (2), Italy (3), Kenya (1), Mali (1), Morocco (1), Mozambique (1), Senegal (5), Seychelles (7), Somalia (1), South Africa (1), Tunisia (1), Uganda (1), United Kingdom of Great Britain and Northern Ireland (2), and United Republic of Tanzania (2). Twelve participants were women. The list of attending experts is presented in Appendix B.

## NOMINATION OF MEETING OFFICERS

8.     Mr Christopher Hoareau (SBS), Ms Ottah-Atikpo (Food Research Institute), Mr Alain Le Bail (International Institute for Refrigeration), Mr Richard Abila (KMFRI), Mr Alphonse Tine (Université Cheick Anta Diop) and Mr Mike Dillon (Grimsby Institute, Secretary of the International Association of Fish Inspectors [IAFI]) acted as the facilitators during the meeting sessions. Mr Christopher Hoareau facilitated the final adoption of the recommendations. The names of the selected rapporteurs of the different sessions are listed in

Appendix A. Mrs Yvette Diei-Ouadi, Mr Jogeir Toppe and Mr John Ryder, from FAO, served as the technical secretariat of the meeting.

9. The experts reviewed the progress in the area of post-harvest fish utilization, technology and quality assurance and made a series of recommendations. Emphasis was placed on fresh and live fish handling, fish processing, safety and quality, marketing and socio-economic issues, and the regional networking mechanism. The meeting included:

- Presentation by the Secretariat of a report on the progress and events since the workshop on Fish Technology and Quality Assurance in Africa, held in Agadir, Morocco, in 2008. The participants accepted the report as presented.
- Presentation of 26 papers selected by the screening panel. This panel was established by the Technical Secretariat following a call for contributions launched 6 months prior to the meeting, which resulted in the receipt of 38 papers from experts; and
- A field trip to the Indian Ocean Tuna factory.

10. The papers are provided in Appendix C.

11. On the basis of the presentations and discussions, draft highlights and recommendations were prepared by the Secretariat. These were discussed, amended and adopted by the experts in the final session of the meeting.

## HIGHLIGHTS AND RECOMMENDATIONS

12. Below are the highlights of the meeting and the recommendations made to FAO, its African Member Countries and institutes involved in fish utilization in Africa:

### Regional networking

1. The progress on the recommendations made at the second meeting held in Agadir in 2008 was presented and discussed. It was noted that the African Network on Fish Technology and Safety (ANFTS) had been established, and a logo had been designed. However, the ANFTS was not as operational as it should be.

2. After some long and constructive deliberations, aided by presentations on two national networking initiatives and a presentation on the state of the regional network, some meeting officers were nominated to come up with concrete and sound proposals to make the network operational. Based on these discussions, an elected Executive Committee made up of six members (Mr Christopher Hoareau [president]; Mr Yahya Mgawe, Mr Khalifa Sylla and Mrs Margaret Adzoa Ottah Atipko [strategic planning]; Mr Anass Karzazi [communication and focal point]; and Mr Clifford Edmund Clifford [operations and action plan]), was confirmed with the challenge to take the organization forward with the strategic support of FAO.

3. Twelve seats were made available on the wider board and, ideally, these should be filled by members reflecting the regions of the network. This body now represents 34 African countries and has more than 130 members. It has established a strategic linkage with the IAFI, and the IAFI's support in the development of the network will be beneficial.

4. A possible strapline for the vision could be: "Appropriate fish technologies for market access". Thereafter, a resolution – "Seychelles Resolution" – was adopted with a simple roadmap to implement the vision. The following are **the recommendations from the resolution**:

### Establishing the organization as a recognized body

5. The current members emphasized the need for this body to be legally recognized to permit or support them to play a full role as a member. It was proposed that FAO be requested to provide the support in the drafting of the legal constitution of the organization. This constitution should include the formal rules for the membership, including the rules for the election of the future board and Executive Committee and the terms of reference of such board.

### Confirming the vision, mission and plans to achieve the vision

6. The members requested clarity in the services they would receive and how this would work. This new draft vision has been simply articulated and proposes some of the key services and support needed in terms of dissemination of best practice. The committee proposed that the vision, mission and previously created

strategic plan, business plan and action plans should be reviewed by the Executive Committee and a summary document circulated to the members for comments and future adoption.

**Rapidly developing the Web site**

7. The members reported that a key step agreed in past meetings had been the creation of a Web site to showcase best practice. The Executive Committee proposed the setting up of a working group to develop a pilot Web site based on the design of the existing Web page to host the presentations and reports presented at the Seychelles meeting and, where records exist, the material from the previous meetings. It is also **recommended** that this group should then create a proposal that will specify and cost the development of a fully functional Web site.

**Creation of a marketing and communication plan**

8. The members noted that regular communication was lacking. The Executive Committee discussed this issue and proposed the creation of a communication and dissemination plan that would market the benefits of the network organization. This would provide agreed statements on the network vision, example of projects and current actions to target stakeholders. The defined benefits for countries and members of the network would be articulated by the information created by this group for use by the members. This group may lead the development of publications for journals, the Web site and standard information similar to the Fish Inspector but focused on African fish technology and market access issues.

**Review of the business plan and funding support for the next phase**

9. The members requested clarity on the operational planning and necessary budget. The Executive Committee proposed that the current business plan be updated using this roadmap and that a concept note, which could assist the group, be created for consideration by FAO. The IAFI was requested to seek support for the network from UNIDO and FAO and coordinate a meeting to review options as agreed in Washington at the World Seafood Congress in October 2011.

*Technical sessions*

10. Much effort was put into the research activities presented, but it was **recommended** again, as in 2008, that due consideration be given to the cost–benefit analysis of techniques developed in research. The Secretariat should provide guidance for future research work to be presented to the experts.

11. In order to celebrate excellence and encourage the experts in their efforts in sustaining the Expert Meeting through applied research and development initiatives, the Secretariat should explore the possibility of instituting an award (in whatever form: certificate, grant, etc.) for the best work/innovative approach, to be applicable as from the next meeting in 2014 in Ghana.

**Live and fresh fish handling and preservation**

12. The Program for Capacity Building in Good Hygiene Practices for Regional Trade within the Strengthening Fishery Products Health Conditions in ACP/OCT Countries (SFP Programme) has demonstrated positive results from community capacity-building initiatives. However, there are some challenges that need to be addressed in relation to: sustainability, infrastructure, government support, and fisheries management. The programme should continue with implementation of the planned activities, which include: development of training of trainers (ToT) manuals; support for community-level initiatives to demonstrate better practices; and additional ToT workshops; and possibly expand to other countries, e.g. Zambia. There is a need to link it with other programmes in the region for synergies and complementarities and take into consideration other ideas raised at the FAO meeting.

13. A presentation on the effect of bleeding on the keeping quality of farmed catfish (*Clarias gariepinus*) in Uganda was discussed. Noting the improvements in consumers' acceptance and extended shelf-life by bleeding of catfish, a comprehensive study should be conducted on other quality patterns (microbiological, chemical) and on other relevant species for documentation and promotion of this practice in the region.

14. A seminar was presented on refrigeration as applied to seafood – recent developments and applications. Trends indicate that the technology is growing very fast as a function of expanding international marketing of fish and aquaculture products, associated with stringent quality assurance demands.

15. The pro and cons of applying different refrigeration techniques, such as chilling-icing, super chilling, glazing technology, brine freezing, blast air freezing, plate freezing, freezing under electric field or electro-freezing, as well as the high-pressure processing technique, were thoroughly discussed.

16. Participants agreed unanimously that refrigeration technology was important for fish preservation and was critical in expanding fish trade. However, it was noted that adoption of some of these techniques could be difficult in the African context, especially due to cost implications.

17. Nevertheless, the need for developing institutional capacity to meet the recent refrigeration challenges was underscored. Thus, it was **recommended** that the practicability of the various refrigeration techniques should be explored further.

18. Similarly, the establishment of partnerships and collaboration between institutions in the regions was emphasized in order to examine the use of refrigeration technologies using solar energy, a readily available resource, in the region.

**Fish processing, value addition and nutrition**

19. Smoking of fish as an effective method to add value, reduce post-harvest losses and diversify products offered to consumers was underscored, especially in countries where this technique is being assessed for introduction.

20. Any further work in these countries should take into account existing techniques and consider the following key factors for a successful outcome:
- economics, cost–benefit analysis;
- market acceptance;
- safety aspects, especially the process-related hazard polycyclic aromatic hydrocarbons (PAHs) – in hot smoked fish;
- explore possible alternative packaging techniques (e.g. vacuum packaging) for value addition and extended shelf-life; and
- compare the consumer preferences for smoked fish products made from liquid smoke versus traditional smoking, based on the level of salt.

21 The urgent need to be prepared for the changing quality demands being imposed on cured fish products was highlighted. One such requirement is to meet emerging European Union (EU) regulations on cured fish in 2012 and 2014. This situation will require efforts to improve traditional fish processing techniques in the region.

22. Noting the trend, and under an FAO-sponsored programme, the Institute of Technological Research in Senegal has developed an improved smoking system (named the FAO-Thiaroye system) aimed at controlling the concentration of PAHs in smoked fish and, thus, improving the safety of final products. This will facilitate access to international markets, such as the EU, and also protect domestic consumers. Participants acknowledged the cost-effectiveness and suitability of the technology to small-scale fisheries and noted the positive initial laboratory test results.

23. It was **recommended** that further work be conducted to include the assessment of the additional three markers of PAHs (benzo<a> anthracene, benzo<b>fluoranthene and chrysene) in order to document properly the effectiveness of the FAO-Thiaroye smoking system to control PAHs to meet the new standards and to disseminate the results. This should also be complemented by a study of PAH-producing characteristics of the major tree species used for smoking in the region.

24. Utilization of the oil/fat extracted using this smoking system should be encouraged at the small-scale level, e.g. in the production of soap as a means of income diversification.

25. The heating unit of the FAO-Thiaroye smoker has also been shown to be effective as a mechanical drier. In order to curb the high post-harvest losses in natural drying processes during rainy seasons, the participants **recommended** that this all-weather drying technique be disseminated for adoption in the region.

26. In considering the environmental threat through deforestation by using charcoal, participants **recommended** that alternative sources of energy should be investigated.

27. The dryer has been validated for shrimps with a payback after 15 production cycles of 400 kg. Participants **recommended** that further trials should be done on other fish species.

28. Considering the nutritional composition of fish – proteins, micronutrients (an exclusive source of certain minerals), polyunsaturated fatty acids – and the scientifically established role of omega-3 fatty acids in the prevention of certain diseases (coronary heart disease) and in child brain development, the study on micronutrient enrichment of meals using simple technology provides a significant contribution to food and nutrition security, adding value to local meals and low-value fish. Within a prevailing context of malnourished/undernourished children (and especially micronutrient-deficient children), the **recommendation** was made to encourage the practice of micronutrient enrichment in school feeding, and for vulnerable children and adults in other relevant nutrition programmes.

29. The approach to nutrition should be an integral part of any development programme. For this purpose, a linkage should be established with key agencies involved in food and nutrition security, such as FAO, WHO and WFP, and various other stakeholders (e.g. local government, education and health institutions at national level) to combat malnutrition.

30. An IQ assessment should be conducted to show if there are other benefits of micronutrient enrichment of meals for children.

31. Other **recommendations** on the micronutrient enrichment study were:
   • conduct cost–benefit analysis and optimization of the techniques; and
   • transfer the technology to local producers at artisanal and, in due course, industrial levels of fish powder production.

32. It was generally acknowledged that promoting an increased consumption of fish within a static wild capture scenario requires efforts in aquaculture development, post-harvest loss reduction and promotion of the use of low-value fish species and bycatch for direct human consumption, and it was **recommended** that these issues should continue to be key areas for future research activities in Africa.

33. The cluster-based Camdeboo Satellite Aquaculture Project in South Africa uses farmed catfish to produce sterilized products in pouches for the food service sector. Meals made from pouched products had high consumer acceptance ratings. The approach demonstrated the potential for:
   • creating self-employment opportunities for women, contributing to economic growth through women empowering one another;
   • producing affordable nutritious food products; and
   • decreasing the reliance on overexploited wild stocks.

The strategic business plan of this project could be considered for use as an example for similar studies.

34. To maximize the benefits and to reduce the impact on the environment, a **recommendation** was made to explore further the utilization of by-products from processing the farmed catfish, affording further income opportunities from the farmed fish.

35. The study on press oil extraction was an interesting approach to adding value to low-value and underutilized fish and utilizing the by-products from processing. The press was developed at a low cost (EUR450). The production of oil and of flavouring cubes from the press cake followed a simple process and the products were well accepted by the consumers.

36. To market the products effectively, it was **recommended** that there would be a need for the development of adequate packaging and labelling for the products, especially the flavouring cube.

37. The challenge of maximizing utilization of bycatch, especially in inshore shrimp trawling, was discussed. Experience from the United Republic of Tanzania suggests that bycatch continues to form a high percentage of catches in the shrimp fishery despite the institution of various selective catching measures.

38. In the case of the United Republic of Tanzania, effort has been made to promote bycatch utilization by using different value-addition methods. A traditional fish smoking technique was used to preserve the bycatch in remote coastal villages where electricity is non-existent, whereas specialty products were

produced at the Fisheries Institute using modern equipment. The result from both of these initiatives suggests that it is possible to maximize utilization of bycatch through economically viable value-addition initiatives.

39. Participants **recommended** that these initiatives should be promoted, focusing on dissemination of the results to key players, including communities and owners of shrimp trawlers. In this regard, there is a need to strengthen training programmes aimed at building the capacity of community members and extension workers to participate in production of proven bycatch products.

40. As mentioned in the preceding paragraph, the industrial fish sector should be made aware of the findings so that it can be persuaded to participate in this initiative as part of rational recuperation of the resource rent it generates from the fishery.

41. Results from experiments conducted in Senegal on cottage production of marinated and semi-preserved shellfish were presented and discussed. The method applied was found to have the potential to enable smallholder women to access more-lucrative markets. Considering that shellfish collection in Africa is mostly done by women, such improvements have great potential to contribute to improved socio-economic livelihood in communities, in addition to potentially decreasing the pressure on resources.

42. However, a successful and sustainable implementation requires the development of the business management skills of the operators. It was strongly **recommended** that the study should continue in order to perfect the value-addition method and integrate marketing strategies, resource management issues, and attractive packaging methods in order to make it economically viable among target groups, especially rural women.

**Fish safety and quality**

43. A peer-reviewed new method of histamine analysis was presented and discussed. This method has been acknowledged and rewarded at several scientific fora as a method that is accurate, has low detection levels and shows improved reproducibility. It was **recommended** that FAO and WHO should explore whether this method may be considered as a Codex official method.

44. The study conducted on swordfish caught in the Western Indian Ocean (Seychelles Exclusive Economic Zone [EEZ]) has provided preliminary information relating to the levels of total mercury in swordfish of varying fish sizes. The results were comparable with previous studies in other parts of the world. There is a need to continue with the study, putting more focus on fish of 10 kg and above and having a smaller interval of different weight categories. FAO will monitor the findings of these studies and provide guidance on their implementation, which might include a possible introduction of a maximum weight limit for swordfish export.

**Socio-economic, marketing and market access issues**

45. The report on challenges to small-scale fish exports outlined the importance of fish exports to sub-Saharan Africa, even though the region is a relatively small player in the world fish export markets. This sector, however, faces serious challenges especially because of constraints faced by the exporting countries in meeting the stringent quality control and sanitary conditions of importing nations and the different requirements and approaches of various certification bodies for ecolabelling schemes. The experience by the Naturland certification process in Lake Victoria has provided a good example of how an inclusive approach, bringing in the private sector, private–public partnerships and beach management units and including the social dimensions in an ecolabelling initiative, can deliver results.

46. FAO should support adoption of these approaches in line with the Code of Conduct for Responsible Fisheries and introduction of the FAO Voluntary Guidelines for small-scale fisheries. There is also a need for better data collection to bring out the true value of intraregional trade in small-scale fisheries, which has been underestimated owing to a lack of adequate information.

47. The case study conducted in Senegal on the octopus value chain has brought out the role of different players, the revenues accrued in the system, and conflicts between the fishers and intermediaries. It has revealed that post-capture handling of octopus does not affect octopus prices and is not considered a major problem by the market based on very low reject levels.

48. There is, however, a need for more accurate baseline information, which should also bring out more clearly the role of women and how they can benefit from the octopus fishery.

49. Past interventions have concentrated on the supply side, which has been a weakness in development of the industry, and on attempts to improve the welfare of the fishing community. There is a need for interventions focused on improving market access for fishers alongside formulation of better conservation measures.

50. The study on the sun-dried Lake Victoria sardine value chain in Uganda has identified the key players in the Lake Victoria sardine value chain, margins from trade and the significant role of regional trade, feed manufacturing and supermarkets in driving the Lake Victoria sardine industry. It has also identified the main constraints, including seasonality, high input costs, inappropriate handling facilities, lack of knowledge and skills, unreliable markets, insufficient market information, inadequate extension and unfavourable credit markets, among others.

51. However, there are many opportunities for advancing the sardine trade, including low fishing labour costs, abundant fish stocks, presence of Beach Management Units (BMUs), insatiable and diverse regional markets, favourable government and regional policies.

52. For development of the industry, a number of interventions will need to be undertaken, including: formation and strengthening of fisher and trader groups; improved sanitation; improved infrastructure for fish drying, vessel holds and fish storage; product diversification; improved capacity of the key players, including extension staff, processors, fish inspectors, BMUs, local government and researchers; creation of market information centres, establishment of cooperatives, and readdressing the credit services; and enforcement of mesh-size regulations to reduce overfishing, among other measures.

53. Development of a Lake Victoria sardine quality policy for the Eastern and Central African (ECA) region should be supported.

54. In addition, there is a great need for regional cooperation among respective inspection services and border-post agencies and a detailed study across the ECA region to gain a better understanding of Lake Victoria sardine value chains and creation of market platforms for chain actors.

55. Women in Tunisia participate in various economic activities as sources of income to support their livelihoods, including the collection of clams. In spite of the fact that both men and women participate in this business, men have been responsible for marketing the products while women remain mainly as collectors of clams. Improvement in living conditions of collectors and the revenue made was a great concern to women's groups.

56. Participants noted the findings and it was **recommended** that arrangements should be made to institutionalize the clam collection groups to improve their autonomy and bargaining power for a better profit-sharing environments. The need to strengthen women's participation in small-scale fisheries should be given priority in African countries.

57. Findings from interventions made on sardine utilization in Uganda indicated that East African countries are experiencing high post-harvest losses as a result of a lack of skilled fishers, weakness in law enforcement, illegal fishing gear and practices, which partly accounted for harvest of immature fish coupled with limited capacity to land high catches.

58. It was further stipulated that harmonization of mesh size and shifting fishing activities away from the shoreline should be adopted as some of the measures to protect juvenile fish.

59. Participants noted the findings and **recommended** that experience from similar interventions in countries such as Mozambique by using solar dryers and application of a range of value-addition techniques be considered for improved sardine utilization.

60. Furthermore, participants **recommended** that regional cooperation on management and marketing of Lake Victoria sardine was a priority and that this should involve people at the grassroots level (producers, fishers, etc.), financiers and traders to ensure sustainable promotion of quality with an appropriate credit scheme.

61    Experience with regard to introduction of the Post-harvest Fish Technology Platform (PHFTP) approach in Liberia was presented. It was noted that problems in small-scale fisheries (SSFs) are mostly organizational rather than technical. Similarly, there was agreement that poor extension services are a real constraint on disseminating innovations from research institutions aimed at addressing challenges faced by small-scale fishers in Africa. Thus, addressing these issues is key to successful application of the PHFTP approach in SSF development.

62.    In view of this situation, it was strongly **recommended** that efforts should be made to establish a regional training programme aimed at providing extension training for fisheries extension workers. Potential trainees should be drawn from strategic public services, non-governmental organizations (NGOs) and other private institutions across the region. The experience gained from previous regional training programmes, such as the one conducted under the Ghana–Netherlands artisanal fish processing project, should be considered during development of the proposed programme.

63.    Mozambique presented training manuals, posters and leaflets as well as practical methods (corporal expressions, video, drawings, photos) for training of trainers, technicians and small-scale fisheries communities in order to train other members of fishing communities on hygiene, the Hazard Analysis and Critical Control Point (HACCP) system, traceability and best practice in preservation methods (icing, safety in the market, aquaculture, etc.).

64.    The need for preparing pictorial training material for on-site and distance training in small-scale fisheries was discussed.

65.    The meeting **recommended** that a concerted effort should be made to develop further the material to respond to a large demand for this type of training.

66.    A presentation provided the basic frames of reference with regard to a "seafood trade corridor approach" to improving trade performance. Case studies from Indonesia demonstrated key problems in the value chain: suppressed value due to poor handling and poor compliance; inefficient production and logistics; increasing costs; and lack of awareness and perceived risk. The trade corridor approach would mitigate these problems, problems that face many African countries.

67.    It was therefore strongly **recommended** that East African nations should consider forming trade corridors and piloting the approach including the participation of the export associations, artisanal suppliers and the government. The existence of a strong Ugandan Fish Processors and Exporters Association (UFPEA) and a regional export association, in an alliance with inspectors and government officials, would make the trade corridor approach useful.

68.    An innovative electronic market information system currently applied in Kenya's fisheries has demonstrated that the system is relevant and effective for enhancing competitiveness and marketing efficiency in small-scale fisheries. The project received support from the International Labour Organization to implement a pilot phase limited to Lake Victoria in June 2009–May 2011; and, subsequently, the EU supported an up-scaled phase covering the entire country as from March 2011. The project has achieved an increase in fish prices, reduction in post-harvest fish losses, and improvement in incomes of fishers and fish traders.

69.    The above system is relevant as it addresses the development policy objectives for fisheries in the region within the framework of the East African Community, Common Market for Eastern and Southern Africa (COMESA), and FAO, among others. Implementation of the up-scaled phase in Kenya should continue, incorporating additional information, especially export data. FAO should study the model of the project in Kenya with a view to supporting expansion of the system to other countries in Africa and adapting the model for other small-scale fisheries.

## CLOSURE OF THE WORKSHOP

13.    The next meeting on Fish Technology, Utilization and Quality Assurance in Africa will be held in November 2014 in Ghana. The meeting was officially closed on 25 November 2011 by Ms Sreekala Nair, Manager Biotechnical Services of SBS, following some remarks from Ms Diei-Ouadi, and the vote of thanks of the participants.

**ORGANISATION**

1.	La troisième réunion consacrée à la technologie, l'utilisation et l'assurance de qualité du poisson en Afrique a été organisée par le Service des produits, des échanges et de la commercialisation (FIPM) de la Division des politiques et de l'économie de la pêche et de l'aquaculture (FIP) de l'Organisation des Nations Unies pour l'alimentation et l'agriculture (FAO), en collaboration avec le Seychelles Bureau of Standards (SBS).

**OUVERTURE**

2.	L'atelier s'est déroulé à l'«International Conference Centre» de Mahé (Seychelles), du 22 au 25 novembre 2011.

3.	M^{me} Amy Quatre, Directrice générale du SBS, s'est dite ravie et honorée que son institution ait été sollicitée pour co-organiser la réunion d'experts sur la technologie, l'utilisation et l'assurance de qualité du poisson. Elle a ensuite chaleureusement salué l'ensemble des participants, avant de remercier la FAO de cette initiative, qui donnait aux pays africains la possibilité de partager leurs expériences et d'échanger des informations. Elle a précisé que le poisson et les produits de la pêche jouaient un rôle majeur dans l'économie seychelloise. La question de l'ampleur des pertes post-capture dans la pêche artisanale a été soulevée, de même que la nécessité de les réduire pour optimiser l'utilisation des ressources halieutiques, comme le prescrit le Code de conduite pour une pêche responsable élaboré par la FAO. Les activités de la FAO liées au secteur post-capture ont également été soulignées. Mme Quatre a ensuite souhaité des discussions fructueuses aux participants.

4.	M. Alejandro Anganuzzi, Secrétaire exécutif de la Commission des thons de l'océan Indien (CTOI), s'est adressé à l'atelier au nom de la FAO et a remercié les participants d'avoir pris le temps d'assister à l'atelier. Il a également remercié le Gouvernement des Seychelles, qui accueillait la manifestation et avait fourni son appui sans réserve pour en permettre l'organisation. Il a ensuite expliqué que de plus en plus d'habitants de pays en développement tiraient leurs revenus de la pêche, alors que les techniques après capture et des systèmes d'assurance-qualité sont peu fiables et les pertes importantes, autant de facteurs conjugués qui pesaient considérablement sur l'économie des ménages et des communautés de pêcheurs concernées. Il a souligné le caractère opportun de cet atelier, qui se pencherait sur des aspects socioéconomiques importants en vue d'améliorer le niveau de vie et la sécurité alimentaire des communautés de pêcheurs et, plus largement, des populations rurales, et contribuerait donc à la réalisation des Objectifs du Millénaire pour le développement en réduisant l'insécurité alimentaire et en luttant contre la pauvreté. À cet égard, il a également évoqué les initiatives passées et en cours – y compris les ateliers thématiques comme cette manifestation – dont l'Organisation des Nations Unies pour l'alimentation et l'agriculture était à l'origine depuis 30 ans.

5.	La parole a ensuite été donnée à S.E. Peter Sinon, Ministre de l'investissement, des ressources naturelles et de l'industrie, qui a déclaré que les Seychelles étaient fières d'accueillir la réunion. Il a souligné l'importance de la pêche pour le pays, qui en est le deuxième pilier économique après le tourisme et, s'agissant de la sécurité alimentaire, il a rappelé que les Seychelles étaient l'un des premiers consommateurs au monde de poisson (par habitant). Il a brièvement présenté les politiques sectorielles du Gouvernement visant à une plus grande diversification, en mettant l'accent sur l'élaboration de produits à valeur ajoutée à partir des captures accessoires, qui représentent environ 5 pour cent en volume des prises de thon industrielles. Concernant plus particulièrement les petites entreprises, S.E. l'Ambassadeur Sinon a évoqué les difficultés qu'elles rencontraient pour se conformer aux normes internationales et pour réduire leurs pertes de revenus, le gaspillage de ressources précieuses et la production de denrées de moindre qualité, d'où l'importance des délibérations qui allaient suivre.

6.	Les allocutions sont reproduites dans leur intégralité aux annexes D, E et F et le programme se trouve à l'annexe A.

**PARTICIPANTS**

7.	Ont participé à l'atelier 29 experts venus d'Afrique du Sud (1), de France (1), du Ghana (2), d'Italie (3), du Kenya (1), du Mali (1), du Maroc (1), du Mozambique (1), d'Ouganda (1), de la République-Unie de Tanzanie (2), du Royaume-Uni de Grande-Bretagne et d'Irlande du Nord (2), du Sénégal (5), des Seychelles (7), de Somalie (1) et de Tunisie (1). Douze des participants étaient des femmes. La liste des experts se trouve à l'annexe B.

**DÉSIGNATION DES MEMBRES DU BUREAU**

8.		M. Christopher Hoareau (SBS), M^me Ottah-Atikpo (Food Research Institute), M. Alain Le Bail (Institut international du froid), M. Richard Abila (KMFRI), M. Alphonse Tine (Université Cheick Anta Diop) et M. Mike Dillon (Grimsby Institute, Secrétaire de l'Association internationale des inspecteurs de poisson (AIIP)) ont animé les sessions. M. Christopher Hoareau a joué le rôle de facilitateur durant l'adoption finale des recommandations. Les noms des rapporteurs des différentes sessions figurent à l'annexe A. M^me Yvette Diei-Ouadi, M. Jogeir Toppe et M. John Ryder, de la FAO, ont assuré le secrétariat technique de la réunion.

9.		Les experts ont passé en revue les progrès accomplis en matière d'utilisation, de technologie et d'assurance de qualité du poisson après capture et formulé une série de recommandations. L'accent a été mis sur la manutention du poisson frais et vivant, la transformation du poisson, la sécurité sanitaire et la qualité, la commercialisation et les aspects socioéconomiques ainsi que sur le mécanisme de mise en réseau régional. Ces points ont été traités sur la base des éléments suivants:

- La présentation par le secrétariat d'un rapport sur les progrès et les événements ayant eu lieu depuis l'Atelier sur la technologie, l'utilisation et l'assurance de qualité du poisson en Afrique qui s'était tenu à Agadir (Maroc), en 2008. Les participants ont accepté le rapport tel que présenté.
- La présentation de 26 communications sélectionnées par le panel d'examen. Ce panel avait été établi par le secrétariat technique suite à l'appel à contributions lancé six mois avant la réunion et qui avait débouché sur 38 communications d'experts.
- Une visite de terrain à la conserverie de thon «Indian Ocean Tuna factory».

10.		Les communications sont reproduites à l'annexe C.

11.		Le secrétariat s'est appuyé sur les exposés et les débats pour rédiger un projet de document reprenant les principaux points abordés et les recommandations. Ces points et ces recommandations ont été débattus, amendés et adoptés par les experts pendant la dernière session de l'atelier.

**PRINCIPAUX POINTS ET RECOMMANDATIONS**

12. Les principaux points abordés et les recommandations faites à la FAO, à ses pays membres africains et aux instituts concernés par l'utilisation du poisson en Afrique sont les suivants:

*Réseautage régional*

1.		Les progrès concernant les recommandations formulées lors du deuxième atelier (Agadir, 2008) ont été présentés puis débattus. Note a été prise que l'African Network on Fish Technology and Safety (ANFTS) avait été constitué et qu'un logo avait été créé. Pour autant, l'ANFTS n'était pas aussi opérationnel qu'il devrait.

2.		Au terme de longues et constructives délibérations accompagnées d'exposés sur deux initiatives nationales de mise en réseau et d'un exposé sur l'avancement du réseau régional, certains membres du bureau ont été désignés pour formuler des propositions concrètes et cohérentes en vue de rendre le réseau opérationnel. Sur la base de ces discussions, un comité exécutif élu constitué de six membres [M. Christopher Hoareau (président); M. Yahya Mgawe, M. Khalifa Sylla et M^me Margaret Adzoa Ottah Atikpo (planification stratégique); M. Anass Karzazi (communication et correspondant); et M. Clifford Edmund Clifford (opérations et plan d'action)], a été confirmé dans sa mission de poursuivre l'effort, avec le soutien stratégique de la FAO.

3.		Douze autres sièges sont à pourvoir pour le conseil d'administration et l'idéal serait qu'ils soient occupés par des membres représentant les régions couvertes par le réseau. Le réseau englobe actuellement 34 pays africains et compte plus de 130 membres. Un lien stratégique a été établi avec l'AIIP et l'aide de cet acteur sera utile pour le développement du réseau.

4.		Le slogan résumant la vision envisagée pourrait être: «Des technologies du poisson appropriées pour l'accès aux marchés». Par la suite, une résolution – «Résolution des Seychelles» – a été adoptée, avec une feuille de route simple destinée à mettre la vision en œuvre. Les **recommandations** énoncées dans la résolution sont les suivantes:

**Faire de l'organisation un organisme reconnu**

5. Les membres actuels ont souligné la nécessité de la reconnaissance juridique de l'Organisation pour leur permettre de jouer un rôle à part entière en tant que membres ou les y aider. Il a été suggéré que la FAO aide à la rédaction des actes constitutifs, qui devront décrire les règles formelles relatives au statut de membre, y compris les modalités d'élection des futurs conseil d'administration et comité exécutif, ainsi que le mandat du conseil d'administration.

**Confirmer la vision, la mission et les plans de mise en œuvre de la vision**

6. Les membres ont demandé que leur soient clairement exposés les services dont ils bénéficieraient et la manière dont le réseau fonctionnerait. Le nouveau projet de vision est présenté de façon simple et suggère certains des services clés et les formes de soutien qui pourraient être nécessaires à la diffusion des bonnes pratiques. Le comité a proposé que la vision, la mission ainsi que le plan stratégique, le plan d'organisation et les plans d'action déjà élaborés soient examinés par le comité exécutif et qu'une synthèse soit diffusée aux membres pour qu'ils formulent des commentaires, avant adoption.

**Développer rapidement le site Web**

7. Les membres ont indiqué que, lors des précédentes réunions, il avait été convenu que la création d'un site Web destiné à mettre en avant les bonnes pratiques constituerait une étape clé. Le comité exécutif a proposé de constituer un groupe de travail qui, en s'appuyant sur la page Web existante, concevrait un site pilote afin de mettre en ligne les exposés et les rapports présentés à la réunion des Seychelles, ainsi que les documents utilisés lors des précédentes réunions pour peu qu'ils aient été archivés. Il est également **recommandé** que le groupe de travail dresse un plan détaillé pour la conception d'un site Web parfaitement fonctionnel, en indiquant les coûts.

**Élaborer un plan d'organisation et de communication**

8. Les membres ont remarqué la carence en communication régulière. Après discussion, le comité exécutif a proposé d'établir un plan de communication et de diffusion qui vanterait les mérites du réseau. Il contiendrait des déclarations communes sur la vision et des exemples de projets et d'actions en cours à l'intention des acteurs ciblés. Les avantages précis du réseau pour les pays et les membres seraient exposés par le biais des informations mises en ligne par le groupe de travail, à l'attention des membres. Le groupe pourrait piloter la publication d'informations destinées à des revues et au site Web et d'informations plus générales, comme celles de l'Association internationale des inspecteurs de poisson mais axées sur la technologie du poisson et l'accès aux marchés.

**Examen du plan d'organisation et soutien financier pour la phase suivante**

9. Les membres ont demandé des éclaircissements sur la planification des activités et le budget requis. Le comité exécutif a proposé de mettre à jour le plan d'organisation existant en s'appuyant sur cette feuille de route et suggéré qu'un document de réflexion soit élaboré et soumis à la FAO. Il a été demandé à l'AIIP de solliciter l'aide de l'ONUDI et de la FAO pour le réseau et de coordonner une rencontre pour examiner les options retenues, comme convenu à Washington lors du Congrès mondial des produits de la mer, en octobre 2011.

*Sessions techniques*

10. Beaucoup d'énergie a été investie dans les activités de recherche présentées mais, comme en 2008, il a été **recommandé** d'accorder l'attention requise à l'analyse de rentabilité des techniques mises au point par les chercheurs. Le secrétariat devrait fournir des lignes directrices pour les prochains travaux de recherche qui seront présentés aux experts.

11. Pour honorer l'excellence et encourager les experts dans leurs efforts d'assurer la pérennité des réunions d'experts à travers les travaux de recherche appliquée et initiatives pour le développement, le secrétariat devrait étudier la possibilité d'instituer une récompense du mérite (attestation, bourse, etc.) pour le meilleur travail ou la démarche la plus novatrice, qui serait attribuée dès la rencontre suivante, prévue au Ghana en 2014.

**Manutention et conservation du poisson vivant et du poisson frais**

12. le programme de renforcement des capacités en bonnes pratiques d'hygiène pour le commerce régional, composante du programme d'amélioration de l'état sanitaire des produits de la pêche dans les pays ACP et les PTOM (programme SFP), a montré les résultats positifs qui avaient été obtenus avec des initiatives locales de renforcement des capacités. Néanmoins, il reste un certain nombre de problèmes à résoudre en lien avec :la durabilité, les infrastructures, l'aide des pouvoirs publics et la gestion des pêcheries. Le programme devrait suivre son cours, en s'appuyant sur les activités prévues (conception de manuels de formation des formateurs, soutien aux initiatives locales visant à faire connaître les meilleures pratiques, autres ateliers de formation des formateurs, etc.) et pourrait être étendu à d'autres pays, la Zambie par exemple. Il faut établir des liens avec d'autres programmes réalisés dans la région pour exploiter les synergies et les complémentarités et tenir compte d'autres idées émises lors de la réunion de la FAO.

13. Une présentation consacrée aux effets de la saignée sur la bonne conservation du poisson-chat d'élevage (*Clarias gariepinus*) en Ouganda a été examinée. Compte tenu d'une meilleure acceptation de la part des consommateurs et de l'allongement de la durée de conservation du poisson, une étude complète devrait être réalisée sur d'autres aspects qualitatifs (microbiologie, chimie) et sur d'autres espèces pertinentes, à des fins de documentation et de promotion de cette pratique dans la région.

14. Un séminaire sur la réfrigération des produits de la mer (évolutions et applications récentes) a été présenté. Les tendances indiquent que la technologie évolue très rapidement, au rythme de l'essor du commerce international de poisson et de produits aquacoles et dans un contexte d'exigences très strictes en matière d'assurance de qualité.

15. Les avantages et les inconvénients des différentes techniques de réfrigération – refroidissement-mise en glace, surrefroidissement, technologie du glazurage, congélation en saumure, congélation par air pulsé, congélation sur plaques, congélation sous champ électrique ou électro-congélation) et technique de traitement à haute pression – ont fait l'objet d'un débat approfondi.

16. Les participants sont unanimement convenus que les techniques de réfrigération appliquées à la conservation du poisson étaient déterminantes pour le développement du commerce de poisson. Il a toutefois été observé que l'adoption de certaines d'entre elles serait peut-être difficile dans le contexte africain, en particulier pour des questions de coût.

17. Cependant la nécessité de renforcer les capacités institutionnelles pour relever les récents défis d en matière de réfrigération a été soulignée. Il a donc été **recommandé** 'd' étudier davantage la faisabilité des différentes techniques.

18. De même, l'établissement de partenariats et de collaborations entre institutions de la région afin d'étudier l'utilisation de techniques de réfrigération à base d'énergie solaire (facilement disponible dans la région) a été mis en exergue.

**Transformation du poisson, valeur ajoutée et nutrition**

19. Le fumage du poisson comme méthode efficace d'ajout de valeur, de réduction des pertes après capture et de diversification des produits proposés aux consommateurs a été souligné, surtout pour les pays où cette technique fait l'objet d'une évaluation avant introduction.

20. Toute nouvelle étude conduite dans ces pays devrait tenir compte des techniques existantes et des facteurs de réussite suivants:
- Économie, analyse de rentabilité.
- Acceptation par le marché.
- Aspects touchant à la sécurité sanitaire, notamment les risques liés à la contamination du poisson par les hydrocarbures aromatiques polycycliques (HAP) lors du fumage à chaud.
- Exploration des nouvelles techniques de conditionnement (sous vide, par exemple) susceptibles d'accroître la valeur marchande des produits et d'en allonger la durée de conservation.
- Comparaison des préférences des consommateurs (fumée liquide ou fumage traditionnel) en fonction de la teneur en sel.

21. L'urgente nécessité de se préparer aux évolutions des exigences de qualité concernant le poisson salé, séché et fumé a été soulignée. L'une de ces exigences est la conformité aux nouvelles dispositions

règlementaires de l'Union européenne (UE) qui s'appliqueront au poisson salé, séché et fumé en 2012 et 2014. Cette nouvelle situation exigera des efforts d'amélioration des techniques traditionnelles de transformation du poisson dans la région.

22. Conscient de cette tendance, le Centre national de formation des techniciens des pêches et de l'aquaculture (CNFTPA) du Sénégal a mis au point, dans le cadre d'un programme financé par la FAO, un système de fumage amélioré (système FAO-Thiaroye) qui vise à maîtriser la teneur du poisson fumé en HAP et à rendre ainsi améliorer la sécurité sanitaire du produit final.. L'accès aux marchés internationaux, notamment à l'UE, s'en trouvera facilité et les consommateurs intérieurs seront mieux protégés. Les participants ont reconnu que cette technologie était rentable et qu'elle était adaptée aux entreprises de pêche artisanale, et ils ont pris note des premiers résultats probants obtenus lors d'essais en laboratoire.

23. Il a été **recommandé** de poursuivre les travaux pour y inclure l'évaluation de trois autres marqueurs de la présence de HAP (benzo<a> anthracène, benzo<b>fluoranthène et chrysène) afin de démontrer avec certitude que le système de fumage FAO-Thiaroye permet de diminuer les HAP et de satisfaire aux nouvelles normes et afin de diffuser les résultats. Ces travaux devraient être complétés par une étude des caractéristiques d'émission de HAP des principales essences d'arbres utilisées dans la région pour le fumage du poisson.

24. L'utilisation de l'huile et de la graisse extraites lors de la mise en œuvre de cette méthode devrait être encouragée dans les productions à petite échelle, par exemple pour fabriquer du savon, ce qui permettrait une diversification des revenus.

25. L'unité de chauffage du fumoir FAO-Thiaroye a également prouvé son efficacité comme séchoir mécanique. En vue de réduire les pertes post-capture importantes qui sont enregistrées avec les méthodes de séchage naturelles pendant la saison des pluies, les participants ont **recommandé** que cette technique de séchage, utilisable en tout temps, soit diffusée pour adoption dans la région.

26. Compte tenu de la menace environnementale que constitue la déforestation liée à l'utilisation de charbon de bois, les participants ont **recommandé** de rechercher des sources d'énergie durable.

27. Le séchoir a été validé pour les crevettes, avec un amortissement au bout de 15 cycles de production de 400 kg. Les participants ont **recommandé** de procéder à des essais avec d'autres espèces.

28. Considérant la composition nutritionnelle du poisson – protéines, micronutriments (seule source de certains minéraux), acides gras polyinsaturés – et du rôle scientifiquement établi des acides gras oméga-3 dans la prévention de certaines pathologies (maladie coronarienne) et dans le développement cérébral de l'enfant, l'étude sur les repas enrichis en micronutriments grâce à une technique simple constitue une contribution notable à la sécurité alimentaire et nutritionnelle; l'enrichissement des aliments permet d'améliorer le régime alimentaire local et d'accroître la valeur marchande des poissons de moindre valeur. Dans un contexte de malnutrition/sous-alimentation infantile (et en particulier de carences en micronutriments), il a été **recommandé** d'encourager l'enrichissement des repas en micronutriments, dans les cantines scolaires mais aussi dans le cadre d'autres programmes nutritionnels qui ciblent les enfants et adultes vulnérables.

29. Tout programme de développement devrait comporter une démarche nutritionnelle. À cet effet, un lien devrait être établi avec des organismes clés s'occupant de sécurité alimentaire et nutritionnelle, comme la FAO, l'OMS et le PAM, et divers autres acteurs (collectivités locales, institutions nationales en charge de l'éducation et de la santé, etc.), afin de lutter contre la malnutrition.

30. Une évaluation de QI devrait être réalisée pour déterminer si l'enrichissement en micronutriments des repas servis aux enfants présente d'autres avantages.

31. Les autres **recommandations** concernant l'étude sur l'enrichissement en micronutriments étaient les suivantes:
- Effectuer une analyse de rentabilité et optimiser les techniques.
- Transférer la technologie de production de farine de poisson aux producteurs locaux, d'abord à l'échelle artisanale, puis industrielle le moment venu.

32.     Les participants ont généralement admis que promouvoir une consommation accrue de poisson dans un scénario de stagnation des captures sauvages impliquait de développer l'aquaculture, de réduire les pertes post-capture et de promouvoir l'utilisation des espèces de moindre valeur et des captures accessoires pour la consommation humaine directe; il a été **recommandé** que ces différents aspects soient les domaines clés des futurs travaux de recherche en Afrique.

33.     Dans le cadre du projet de fermes aquacoles de Camdeboo (Afrique du Sud) (Camdeboo Satellite Aquaculture Project), des poissons-chats d'élevage servent à l'élaboration de produits stérilisés et conditionnés en sachets destinés à l'alimentation. Les aliments à base de produits en sachets ont été très bien acceptés par les consommateurs. La démarche a prouvé que l'on pouvait:

- Créer de nouvelles possibilités d'emplois indépendants pour les femmes, ce qui contribuait à la croissance économique, les femmes se soutenant mutuellement.
- Produire des denrées alimentaires nutritives à un prix abordable.
- Réduire la dépendance vis-à-vis de stocks de poissons sauvages surexploités.

Le plan d'organisation stratégique de ce projet pourrait servir d'exemple pour des études comparables.

34.     Pour maximiser les avantages et réduire les incidences sur l'environnement, il a été **recommandé** d'étudier davantage les possibilités d'utilisation des sous-produits issus de la transformation des poissons-chats d'élevage, afin de générer de nouvelles sources de revenus.

35.     L'étude sur l'extraction d'huile à froid constituait une approche intéressante pour accroître la valeur marchande de poissons de faible valeur sous-utilisés et pour tirer parti des sous-produits de la transformation. Une presse a été mise au point pour un coût modique (450 euros). La production d'huile et de cubes aromatiques à partir du gâteau de presse résultait d'un processus simple et les produits ont été bien accueillis par les consommateurs.

36.     Pour assurer une commercialisation efficace des produits, il a été **recommandé** de mettre au point un emballage et un étiquetage adéquats, en particulier pour le cube aromatique.

37.     Le défi de la maximisation de l'exploitation des captures accessoires, surtout dans le cadre du chalutage à la crevette côtière, a été examiné. L'expérience de la République-Unie de Tanzanie semblerait indiquer que les captures accessoires représentent toujours un pourcentage important des prises dans la pêche à la crevette, et ce malgré l'instauration de diverses mesures de pêche sélective.

38.     Dans le cas de la République-Unie de Tanzanie, des efforts ont été faits pour promouvoir l'utilisation des captures accessoires en recourant à différentes méthodes d'ajout de valeur. Une technique traditionnelle de fumage du poisson a été appliquée pour la conservation des captures accessoires dans les villages côtiers reculés non alimentés en électricité, tandis que des produits spéciaux ont été élaborés à l'institut tanzanien des pêches qui dispose d'équipements modernes. Le bilan de ces deux initiatives est qu'il est possible de maximiser l'utilisation des captures accessoires par des initiatives d'ajout de valeur économiquement viables.

39.     Les participants ont **recommandé** de promouvoir ces initiatives, en mettant l'accent sur la diffusion des résultats aux acteurs clés, y compris les communautés et les propriétaires de chalutiers-crevettiers. À cet égard, il est important de renforcer les programmes de formation destinés à accroître les capacités des communautés et des vulgarisateurs à participer à l'élaboration de produits à partir de captures accessoires constatées.

40.     Comme indiqué au paragraphe précédent, il faudrait tenir le secteur de la pêche industrielle informé des résultats d'étude pour le convaincre de participer à l'initiative, dans le cadre de la récupération rationnelle de la rente tirée des ressources halieutiques.

41.     Les résultats des expériences menées au Sénégal sur la production artisanale de coquillages marinés et en semi-conserves ont été présentés et examinés. Les participants ont jugé que la méthode appliquée pouvait permettre aux petites exploitantes d'accéder à des marchés plus rémunérateurs. Étant donné qu'en Afrique le ramassage des coquillages est effectué principalement par les femmes, ces améliorations sont très susceptibles de contribuer à la hausse du niveau de vie des communautés concernées, en plus d'atténuer la pression sur les ressources.

42. Une mise en œuvre réussie et durable nécessite cependant de développer les compétences des exploitants en matière de gestion d'entreprise. Il a été vivement **recommandé** de poursuivre l'étude en vue de perfectionner la méthode d'ajout de valeur et d'intégrer les stratégies commerciales, les questions de gestion des ressources et les méthodes de conditionnement attractives afin que la méthode soit économiquement viable pour les groupes cibles, en particulier les femmes vivant en milieu rural.

**Sécurité sanitaire et qualité du poisson**

43. Une nouvelle méthode d'analyse des niveaux d'histamine ayant fait l'objet d'une évaluation par les pairs a été présentée et examinée. Dans plusieurs enceintes scientifiques, cette méthode a été reconnue comme une méthode précise, caractérisée par des seuils de détection bas et une meilleure reproductibilité et elle a été récompensée à ce titre. Il a été **recommandé** que la FAO et l'OMS se penchent sur la question pour déterminer si elle pourrait devenir une méthode Codex officielle.

44. L'étude consacrée à l'espadon pêché dans l'ouest de l'océan Indien (zone économique exclusive des Seychelles) a fourni des informations préliminaires sur les concentrations de mercure total retrouvées dans l'espadon (spécimens de différentes tailles). Les résultats étaient comparables à ceux d'études déjà menées dans d'autres régions du monde. Il faut poursuivre l'étude, en l'axant plus spécifiquement sur les poissons de 10 kg et plus et en choisissant un intervalle plus restreint entre les différentes catégories de poids. La FAO suivra les résultats de ces études et fournira des lignes directrices pour leur mise en œuvre, ce qui pourrait se traduire par l'introduction d'une limite de poids maximale pour les exportations d'espadon.

**Aspects socioéconomiques et commerciaux et accès aux marchés**

45. Le rapport sur les difficultés inhérentes aux exportations de poisson à petite échelle a souligné l'importance des exportations de poisson vers l'Afrique subsaharienne, même si la région n'est qu'un acteur relativement modeste sur le marché mondial. Mais ce secteur rencontre de sérieuses difficultés, en particulier du fait des obstacles auxquels se heurtent les pays exportateurs pour remplir les conditions drastiques imposées par les pays importateurs en matière de contrôle de la qualité et conditions sanitaires et du fait des exigences et des démarches différentes des organismes de certification dans le domaine de l'étiquetage écologique. L'expérience de la certification Naturland conduite dans la région du lac Victoria a été un exemple que pouvait produire une démarche globale combinant secteur privé, partenariats public-privé et unités de gestion des plages et intégrant les dimensions sociales dans une initiative d'étiquetage écologique.

46. La FAO devrait soutenir l'adoption de ces démarches conformes au Code de conduite pour une pêche responsable et l'introduction des Directives volontaires de la FAO pour les artisans pêcheurs. Il faut également améliorer la collecte des données pour établir la vraie valeur du commerce intrarégional lié à la pêche artisanale, qui est sous-estimé par manque d'informations pertinentes.

47. L'étude de cas sur la chaîne de valeur des poulpes qui a été menée au Sénégal a souligné le rôle des différents acteurs, les recettes générées par le système et les conflits entre pêcheurs et intermédiaires. Elle a révélé que, compte tenu du taux de rejet très bas, la manutention post-capture n'avait pas d'effet sur le prix des poulpes et n'était pas considérée comme un problème majeur par le marché.

48. Quoi qu'il en soit, il faut recueillir des informations de références plus précises, ce qui devrait aussi permettre de faire plus clairement ressortir le rôle des femmes et les avantages qu'elles peuvent retirer de la pêche céphalopodière.

49. Les interventions passées se concentraient sur l'offre, qui était un frein à l'essor du secteur, et sur les tentatives d'amélioration du bien-être de la communauté des pêcheurs. Il faut désormais aussi prévoir des interventions visant en priorité à améliorer l'accès des pêcheurs aux marchés, et édicter des mesures plus efficaces en matière de conservation des pêches.

50. Une étude a été conduite en Ouganda sur la chaîne de valeur des sardines pêchées dans le lac Victoria et conservées par séchage au soleil. Elle a identifié les acteurs clés de cette chaîne, les marges produites par l'activité commerciale et le rôle significatif des échanges régionaux, de la fabrication d'aliments pour animaux et des supermarchés dans le développement du secteur sardinier du lac Victoria. L'étude a également identifié les principales contraintes, notamment le caractère saisonnier de l'activité, le coût élevé des intrants, l'inadaptation des équipements de manutention, le manque de connaissances et de compétences, l'absence de marchés fiables, l'insuffisance des informations relatives aux marchés et du travail de vulgarisation et les conditions défavorables sur le marché du crédit.

51. Il existe néanmoins de nombreuses opportunités pour l'essor du commerce de la sardine, parmi lesquelles le faible coût de la main-d'œuvre employée à la pêche, l'abondance des stocks, la présence d'unités de gestion des plages, le caractère insatiable et la diversité des marchés régionaux ou encore l'existence de politiques nationales et régionales favorables.

52. Pour que le secteur se développe, un certain nombre d'interventions sont à programmer dans les domaines suivants: formation et renforcement de groupes de pêcheurs et de commerçants; amélioration de l'assainissement ainsi que de l'infrastructure nécessaire au séchage, aux cales à poisson et à l'entreposage du poisson; diversification des produits; renforcement des capacités des acteurs clés (vulgarisateurs, transformateurs, inspecteurs, unités de gestion des plages, collectivités locales, chercheurs, etc.); création de centres d'information sur les marchés, constitution de coopératives et amélioration des services de crédit; et respect des règles relatives à la taille des mailles de filet pour réduire la surpêche, etc.

53. Il serait souhaitable de soutenir l'élaboration d'une politique de qualité de la sardine du lac Victoria pour l'Afrique orientale et centrale.

54. En outre, une coopération à l'échelle régionale entre les services d'inspection et les autorités des postes frontières fait cruellement défaut et il faudrait mener une étude approfondie en Afrique orientale et centrale pour comprendre les chaînes de valeur de la sardine du lac Victoria et créer des plateformes de marché pour les acteurs de ces chaînes.

55. En Tunisie, les femmes participent à diverses activités économiques qui génèrent autant de revenus d'appoint pour leurs moyens d'existence; y compris le ramassage des palourdes. Bien que le ramassage soit assuré aussi bien par des hommes que par des femmes, la commercialisation des produits incombe aux hommes tandis que les femmes se limitent essentiellement au ramassage. L'amélioration des conditions de vie des ramasseuses et les revenus tirés de cette activité étaient des questions primordiales pour les groupes de femmes.

56. Les participants ont pris note des résultats de l'étude et **recommandé** que des dispositions soient prises pour officialiser l'existence des groupes de ramasseuses en vue de les rendre plus autonomes et d'accroître leur pouvoir de négociation pour parvenir à un partage plus équitable des bénéfices. Dans les pays africains, le renforcement du rôle des femmes dans le secteur de la pêche artisanale devrait être une priorité.

57. Les résultats des interventions liées à l'utilisation de la sardine en Ouganda ont montré que les pays d'Afrique orientale enregistraient des pertes après capture importantes en raison du manque de qualifications des pêcheurs, du mauvais respect de la législation et du recours à des engins et des pratiques de pêche illégaux, qui expliquaient en partie les captures de poisson trop jeune et la capacité limitée d'embarquer des captures abondantes.

58. En outre, il a été précisé qu'harmoniser les maillages et obliger les pêcheurs à s'éloigner des côtes feraient partie des mesures à prendre pour protéger les juvéniles.

59. Les participants ont pris note des résultats de l'étude et **recommandé** que les actions similaires conduites dans des pays comme le Mozambique – avec utilisation de séchoirs solaires et mise en œuvre d'une série de techniques d'ajout de valeur – soient envisagées pour améliorer l'utilisation de la sardine.

60. Qui plus est, les participants ont **recommandé** que la coopération régionale en matière de gestion et de commercialisation des sardines du lac Victoria soit une priorité, ce qui impliquerait d'associer au projet les acteurs strictement locaux (producteurs, pêcheurs, etc.), les bailleurs de fonds et les négociants afin d'assurer une promotion durable de la qualité en s'appuyant sur un mécanisme de crédit approprié.

61. L'expérience concernant l'introduction au Liberia de l'approche fondée sur les plateformes de technologie halieutique post-capture a été présentée. Il a été observé que les problèmes se posant aux entreprises de pêche artisanale étaient essentiellement organisationnels, plutôt que techniques. De même, les participants sont convenus que la médiocrité des services de vulgarisation était un réel frein à la diffusion des innovations que les instituts de recherche mettaient au point pour remédier aux difficultés rencontrées par les artisans pêcheurs africains. Par conséquent, il est capital de régler ces problèmes pour que l'approche fondée sur les plateformes technologiques contribue efficacement à l'essor de la pêche artisanale.

62. Dans ce contexte, il a été vivement **recommandé** de faire en sorte qu'un programme de formation régional soit mis sur pied pour former les vulgarisateurs de la production halieutique. Les candidats à la formation devraient être issus de services publics stratégiques, d'organisations non gouvernementales (ONG) et d'autres institutions privées régionales. L'expérience acquise lors de précédents programmes de formation régionaux, comme celui qui avait été conduit dans le cadre du projet ghanéo-néerlandais de transformation artisanale du poisson, devrait être prise en compte pendant l'élaboration du programme proposé.

63. Le Mozambique a présenté des manuels de formation, des affiches et des brochures, ainsi que des méthodes pratiques (expression corporelle, vidéos, dessins, photos); ce matériel est destiné aux formateurs, techniciens et communautés d'artisans pêcheurs qui formeront d'autres membres de communautés de pêcheurs à l'hygiène, au système HACCP (analyse des risques et maîtrise des points critiques), à la traçabilité et aux bonnes pratiques de conservation (mise en glace, sécurité sanitaire sur le marché, aquaculture, etc.).

64. La nécessité de préparer des supports de formation sous forme d'images pour la formation sur place et à distance des artisans pêcheurs a été discutée.

65. Les participants à la réunion ont **recommandé** de conjuguer les efforts pour mettre au point le matériel permettant de répondre à l'importante demande formulée pour ce type de formations.

66. Un exposé a présenté les principaux cadres de référence nécessaires à une approche de type «corridor commercial pour les produits de la mer» susceptible d'améliorer les résultats commerciaux. Des études de cas réalisées en Indonésie ont mis en lumière les principaux problèmes existants dans la chaîne de valeur: perte de valeur liée à une mauvaise manutention et un mauvais respect des règles; production et logistique inefficaces; hausse des coûts et absence de prise de conscience et de perception du risque*u*. L'ouverture de corridors commerciaux atténuerait ces problèmes, qui se posent à de nombreux pays africains.

67. Il a donc été vivement **recommandé** que les pays d'Afrique orientale envisagent de créer des corridors commerciaux et de piloter cette approche, en y faisant notamment participer les associations d'exportation, les fournisseurs artisanaux et l'État. L'existence d'une puissante association des transformateurs et exportateurs ougandais de poisson (UFPEA) et d'une association d'exportation régionale, en lien avec les inspecteurs et les représentants de l'État rendrait utile l'approche du corridor commercial.

68. Le système électronique d'informations de marchés innovant qui est actuellement utilisé par les pêcheurs kényans s'est révélé pertinent et apte à doper la compétitivité et l'efficacité commerciales de la pêche artisanale. Le projet a reçu le soutien de l'Organisation internationale du travail pour une phase pilote concernant uniquement le lac Victoria, entre juin 2009 et mai 2011; par la suite, l'UE a soutenu le passage à une plus grande échelle, avec un système couvrant tout le pays à compter de mars 2011. Le projet a permis une hausse des prix du poisson, une diminution des pertes après capture et une amélioration du revenu des pêcheurs et des vendeurs de poisson.

69. Le système susmentionné est pertinent car il répond aux objectifs de la politique de développement des pêcheries de la région, dans le cadre de la Communauté de l'Afrique de l'Est, du Marché commun de l'Afrique orientale et australe (COMESA) et de la FAO, entre autres. Le déroulement de la phase d'extension à tout le Kenya devrait se poursuivre, avec l'intégration d'informations complémentaires, en particulier de données sur l'exportation. La FAO devrait étudier le modèle du projet conduit au Kenya en vue de le transposer dans d'autres pays d'Afrique et de l'adapter à d'autres pêcheries artisanales.

## CLÔTURE DE L'ATELIER

13. La prochaine réunion sur la technologie, l'utilisation et l'assurance de qualité du poisson en Afrique se tiendra au Ghana, en novembre 2014. M^me Diei-Ouadi a formulé quelques remarques et la motion de remerciements des participants a été votée, après quoi l'atelier a été officiellement déclaré clos le 25 novembre 2011 par M^me Sreekala Nair, Directrice des services biotechniques du SBS.

**APPENDIX/ANNEXE A**

## PROGRAMME

**Tuesday 22 November 2011**

Opening address by Mrs Amy Quatre, Chief Executive Officer, Seychelles Bureau of Standards (SBS), Seychelles

Address by Mr Alejandro Anganuzzi, Executive Secretary of the Indian Ocean Tuna Commission (IOTC), on behalf of FAO

Welcoming address by H.E. Peter Sinon, Minister for Investment, Natural Resources and Industry, Seychelles

Election of Chairman and meeting officers

FAO report on progress made since the first FAO Workshop on Fish Technology, Utilization and Quality Assurance in Africa
Presented by Yvette Diei-Ouadi, FAO, Rome, Italy

The effect of bleeding on the keeping quality of farmed catfish (*Clarias gariepinus)* in Uganda
Presented by Margaret Masette, FBRC, Uganda

Refrigeration applied to seafood: recent developments and applications
Presented by Alain Le Bail, Oniris, France

Promoting value addition to Indian mackerel (*Rastrelliger kanagurta*) in the Seychelles by hot smoking and assessment of its microbiogical and sensory qualities
Presented by Christopher Hoareau, SBS, Seychelles

Smoking healthy and eating healthy fish: Performance of the FAO-Thiaroye system, an improved design of kiln with particular focus on the control of polycyclic aromatic hydrocarbons (PAH)
Presented by Oumoukhairy Ndiaye, CNFTPA, Senegal

Maximizing by-catch utilization in prawn fishery: case of inshore fishery in Tanzania
Presented by Ambakisye Mwanjala Simtoe, Ministry of Livestock Development and Fisheries, Tanzania

Micronutrient enrichment of meals fed to pupils using highly nutritious and low-cost underutilized fish under the school feeding programme in Ghana
Presented by Margaret Ottah Atikpo, CSIR, Ghana

Marinades of shellfish and seafood of the Saloum Islands and Fadiouth in Senegal
Presented by Momar Yacinthe Diop, Institut de Technologie Alimentaire, Senegal

**Rapporteurs:**
am: Mr Ndiaye/Mrs Arthur
pm: Mr Mgawe/Mrs Ndiaye

**Tuesday 23 November 2011**

Camdeboo Satellite Aquaculture Project: Canning catfish for the food service industry
Presented by Liesl de la Harpe, Blue Karoo Trust, South Africa

Improved *Brycinus leuciscus* oil extraction technology and utilization of the cake obtained after extraction
Presented by Oumou Cissé Traore, IER, Mali

Eating fish: Enjoying the benefits and avoiding the risks
Presented by Jogeir Toppe, FAO, Italy

Improved process for the reduction of post-harvest losses related to natural drying during the rainy season
Presented by Oumoulkhairy Ndiaye, CNFTPA, Senegal

Proximate composition and levels of polycyclic aromatic hydrocarbons (PAHs) in catfish (*Clarias gariepinus*) using different smoking systems
Presented by Yvette Diei-Ouadi, FAO, Italy

New methods for measuring the level of histamine in fisheries products
Presented by Alphonse Tine, Faculté des Sciences et Techniques, Université Cheikh Anta Diop, Senegal

The level of total mercury in swordfish (*Xiphias gladius*) caught in the western Indian Ocean
Presented by Christopher Hoareau, SBS, Seychelles

Challenges to Sub-Saharan fish exports
Presented by John Ryder, FAO, Italy

Octopus value chain and implementation of an upgrading strategy: Key findings in the case study on Nianing and Pointe Sarene (Mbour Senegal)
Presented by Papa Gora Ndiaye, REPAO, Senegal

Sun-dried Mukene (*Rastrinebola argentea*)
value-chain analysis in Uganda
Presented by Margaret Masette, FBRC, Uganda

**Rapporteurs:**
am: Mr Diop/Mrs Masette
pm: Mr Frimpong/Ms Traore

**Thursday 24 November 2011**

Improvement of women's earnings in the clam fishing
sector: Value chain approach and clam farming in the
Gulf of Gabes.
Presented by Amine Ibn Chbili, Agence de Formation
et de Vulgarisation Agricole, Tunisia

A holistic approach to resource sustainability: The
interventions in lake sardine fisheries in the Uganda
part of Lake Victoria
Presented by Yvette Diei-Ouadi, FAO, Italy

Potential socio-economic benefits of Post-Harvest Fish
Technology Platforms (PHFTP) approach in
small-scale fisheries development
Presented by Yahya Mgawe, Ministry of Livestock
& Fisheries Development, Tanzania

Alternative communication techniques for training in
fishing communities
Presented by Luisa Arthur, Ministry of Fisheries,
Mozambique

A seafood trade corridor approach to driving economic
performance
Presented by Mike Dillon, IAFI, United Kingdom

Enhancing fish marketing through ICT: Experiences
of EFMIS project in Lake Victoria
Presented by Richard Abila, Kenya Marine and
Fisheries Research Institute, Kenya

Networking and exchange of information in fisheries:
Moroccan experiences
Presented by Anass Karzazi, Maritime Fishery
Department, Ministry of Agriculture and Maritime
Fishery, Morocco

**Rapporteurs:**
am: Mr Karkazi/Mr Simtoe
pm: Mr Sylla/Mrs de la Harpe

**Friday 25 November 2011**

Field trip (Indian Ocean Tuna factory)

Resolution, discussion and adoption of the
recommendations of the Workshop

Closing

**Technical Secretariat**

Mrs Y. Diei-Ouadi, FAO, Rome
Mr J. Ryder, FAO, Rome
Mr J. Toppe, FAO, Rome

## LIST OF PARTICIPANTS/LISTE DES PARTICIPANTS

| Surname, name and title | Institution | Address | E-mail | Telephone | Fax and Skype ID |
|---|---|---|---|---|---|
| ABILA, Mr Richard Oginga | Kenya Marine and Fisheries Research Institute | PO Box 1881 Nkrumah Road Kisumu 40100 Kenya | abilarichard@yahoo.com | +254-733922643 Mob: +254-733922643 | abilarichard |
| AGRICOLE, Ms Amanda Quality Control Supervisor | Oceana Fisheries | PO Box 71 Victoria Mahe Seychelles | amanda050481@hotmail.com | +248-4224712 Mob: +248-2521614 | |
| ARRISOL, Mrs Rona | Seychelles Fishing Authority | Fishing Port PO Box 449 Victoria Mahe Seychelles | ralbert@sfa.sc | +248-4670300 | |
| ARSHE, Ms Amina Farah | Somali Fishery Network (Somfin) | Bada Cas Hargesra Somali Land | aminafarah22041@yahoo.com | +25224428044 Mob: +25224428044 | |
| ARTHUR, Mrs Luisa Advisor | Ministry of Fisheries | Maputo Mozambique | luisa.arthur@gmail.com | +258-824583310 | |
| BEN CHEBILI, Mr Amine | Agence de Formation et de Vulgarisation Agricole | CFPP Sfax Tunisia | chbili_amine@yahoo.fr | +216-23246060 | sahiliano73 |
| CONFIANCE, Mrs Sindy | Fish Inspection Quality Control unit Seychelles Bureau of Standards | PO Box 953 Victoria Mahe Seychelles | agnetteduval@hotmail.com | +248-4324768 Mob: +248 2725810 | |
| DIEI-OUADI, Mrs Yvette Fishery Industry Officer | FAO Products Trade and Marketing Service (FIPM) | Viale delle Terme di Caracalla 00153 Rome Italy | yvette.dieiouadi@fao.org | +39-0657053251 Mob: +39-3408583771 | mummyyvie |
| DE LA HARPE, Mrs Liesl Project Coordination | Blue Karoo Trust | PO Box 534 Wolwas Road 6280 Graaff-Reinet South Africa | liesl@blue-karoo.co.za | +27-49-8930960 Mob: +27767054731 | liesl.de.la.harpe |

| Surname, name and title | Institution | Address | E-mail | Telephone | Fax and Skype ID |
|---|---|---|---|---|---|
| DILLON, Mr Michael IAFI Secretary | The International Association of Fish Inspectors (IAFI) | 10 Scartho Road Grimsby DN33 2AD United Kingdom | mikedillon2010@hotmail.co.uk | +447734193127 | mikedillon2010 |
| DIOP, Mr Momar Yacinthe Chef Atelier Poisson | Institut de Technologie Alimentaire | BP2765 Route Des Pères Maristes Hann Dakar Senegal | monaryacinthe@yahoo.fr mdiop@ita.sn | +221-338590754 Mob : +221-775367942 | +221-8328295 |
| FRIMPONG, Mr Clifford Edmund | Food and Agriculture Laboratory Ghana Standards Authority | PO Box MB245 Off Tetteh Quarshie-Legon Road Okponglo near Gulf House Accra Ghana | frimcliff@yahoo.co.uk | +233-302501494 Mob: +233-244548076 | +233-302500231 |
| GOVINDEN, Mr Graham | Quality Control Sea Harvest Ltd | Fishing Port Victoria, Mahé Seychelles | quality-control@seaharvest.sc | +248-4224880 Mob: +248-2777215 | |
| HOAREAU, Mr Christopher Chief Fish Inspector | Fish Inspection and Quality Control Unit Seychelles Bureau of Standards | PO Box 953 Victoria Mahe Seychelles | vetfiqcu@seychelles.net | +248-4324768/15 Mob: +248-2530535 | +248 4373826 christopher.hoareau1 |
| KARKAZI, Mr Anass Chief Service and Marketing Communication | Maritime Fishery Department Ministry of Agriculture and Maritime Fishery | PO Box 476 Haut-Agdal Rabat Morocco | karzazi@mpm.gov.ma | +212 -5378688274 Mob:+212-661756941 | +212 -537688394 |
| LAURETTE, Mrs Ladis | Fish Inspection and Quality Control Unit Seychelles Bureau of Standards | PO Box 953 Victoria Mahe Seychelles | Ladis_laurette@hotmail.com | +248-4324768 Mob: +248 -2529710 | |
| LE BAIL, Mr Alain | Oniris | BP 82225 Rue de la Geraudière F-44322 Nantes France | alain.lebail@oniris-nantes.fr | +33-251785454 Mob : +33-663086237 | alain.lebail |

| Surname, name and title | Institution | Address | E-mail | Telephone | Fax and Skype ID |
|---|---|---|---|---|---|
| MASETTE, Ms Margaret | Food Biosciences Research Centre (FBRC) National Agricultural Research Laboratories (NARL) | PO Box 7852 17 Bombo Road Kampala Uganda | mmasette@gmail.com | +256 414 566844 Mob: +256 772 394298 | +256 414 566849 margaret.masette |
| MGAWE, Mr Yahya Ibrahim | Ministry of Livestock & Fisheries Development (Fisheries Education & Training Agency) | PO Box 83 Bagamoyo United Republic of Tanzania | ymgawe@yahoo.com | Mob: +255 755 492988 | |
| NDIAYE, Ms Oumoulkhairy | CNFTPA | BP 2241 10 km de Rufisque Dakar Senegal | oumoulinda@yahoo.fr | +221-775786326 +221-338340546 | oumour.ndiaye1 |
| NDIAYE, Mr Papa Gora | Réseau sur les Politiques de Pêche en Afrique de l'Ouest (REPAO) | PO Box 47076 Liberte 4, No 5000 Dakar Senegal | gndiaye@gmail.com | +221-338252787 Mob: +221-776443473 | +221-338252799 |
| OTTAH ATIKPO, Mrs Margaret Adzoa | Council for Scientific and Industrial Research (CSIR) Food Research Institute | PO Box M20 Accra Ghana | magatik@yahoo.co.uk | +233-20-8161431 Mob: +233-20-8161431 or +233-20-5368505 | maggieadzoa |
| RYDER, Mr John | FAO Products Trade and Marketing Service (FIPM) | Viale delle Terme di Caracalla Rome 00153 Italy | john.ryder@fao.org | +39-0657052143 | |
| SERAPHINE, Mr Karl Senior Fish Inspector | Fish Inspection and Quality Control Unit Seychelles Bureau of Standards | PO Box 953 Victoria Mahe Seychelles | lkaseraphine@yahoo.com | +248-4324768 | |
| SIMTOE, Mr Ambakisye Mwanjala | Fisheries Division Ministry of Livestock Development and Fisheries (Fisheries Education & Training Agency (FETA)) | PO Box 83 Bagamoyo Tanzania | ambakisyes@yahoo.com | +255-784443566 Mob: +255-713443566 | ambakisyes |

| Surname, name and title | Institution | Address | E-mail | Telephone | Fax and Skype ID |
|---|---|---|---|---|---|
| SYLLA, Mr Khalifa Serigne Babacar | Laboratoire HIDAOA Ecole Vétérinaire de Dakar (EISMV) | BP 5077 Fann Dakar Senegal | khsylla@refer.sn | +221-338343354 Mob : +221-775318466 | +221-338254283 |
| TINE, Mr Alphonse Professor | Faculté des Sciences et Techniques Université Cheikh Anta Diop | BP 16404 Dakar-Fann Senegal | alphtine@yahoo.fr | +221-7754190253 +221-703088249 | +221-338246318 |
| TOPPE, Mr Jogeir | FAO Products Trade and Marketing Service (FIPM) | Viale delle Terme di Caracalla 00153 Rome Italy | jogeir.toppe@fao.org | +390657056490 Mob : +393466798762 | jtoppe |
| TRAORE, Ms Cissé Oumou Chercheur | Institut Economie Rurale (IEK) | BP 258 Rue Mohamed V Bamako Mali | oumouni2006@yahoo.fr | +223 20231905 Mob: +223 66750271 | +223 20223775 |
| WARD, Mr Ansen | Fisheries Development Specialist | 1 Coes Cottages Beaneys Lane Shottenden CT4 8JA Kent United Kingdom | ansenward@hotmail.com | +44(0) 1227730127 Mob: +44752982746 | ansenward |

**APPENDIX/ANNEXE C**

**PRESENTED PAPERS/DOCUMENTS PRÉSENTÉS**

# THE EFFECT OF BLEEDING ON THE KEEPING QUALITY OF FARMED CATFISH (*CLARIAS GARIEPINUS*) IN UGANDA

## *[L'EFFET DE SAIGNEMENT SUR LE MAINTIEN DE LA QUALITÉ DU POISSON-CHAT D'ÉLEVAGE (CLARIAS GARIEPINUS) EN OUGANDA]*

by/par

Margaret Masette[1] and Pius Olowo-Ochieng

### Abstract

Bleeding especially of marine fish has been extensively applied to improve product appearance and probable shelf life extension for decades. However, it has not been practiced in Uganda despite its commendable success elsewhere in the world. Probably this lack of practice has contributed to the slow pace of farmed catfish *(Clarias gariepinus)* commercialization in Uganda. As a result some farmers have benefited less from their aquaculture enterprises due to low consumer acceptance and limited local market access because of the dark appearance of processed products. Despite the preservative effect of smoke, catfish ostensibly has a limited shelf life, which has undermined its availability as a source of animal protein in certain parts of the country. Considering the declining per capita fish consumption in Uganda and the need to popularize aquaculture-based fishery products, it was deemed necessary to undertake the present study. About 60 farmed catfish were divided into four equal batches. Two batches, one bled and the other unbled were kept on ice while the other two batches, similarly treated, were held at ambient (28 °C) conditions. The results indicated that fish that was bled had a longer shelf life than fish that was not bled, regardless of the holding temperature. Under chilled conditions the shelf life of bled fish was extended by 11 days while fish kept under ambient temperatures remained acceptable for an extra 5 hours compared to fish that were not bled. It was therefore concluded that bleeding and chilling intervention measures improved and extended shelf life of catfish. Application of these strategies in aquaculture systems would undoubtedly enhance local as well as international catfish trade for whole or processed products, improved food security and the fight against malnutrition in Uganda by increasing consumer acceptance.

*Key words: C. gariepinus, Bleeding, Keeping quality*

### Résumé

Le saignement notamment des poissons marins a été largement appliqué pendant des décennies pour améliorer l'apparence du produit et l'extension probable de la durée de vie. Cependant, il n'a pas été pratiqué en Ouganda en dépit de son remarquable succès ailleurs dans le monde. Sans doute ce manque de pratique a contribué à la lenteur de la commercialisation du poisson-chat *(Clarias gariepinus)* d'élevage en Ouganda. En conséquence, certains éleveurs ont bénéficié moins de leurs entreprises d'aquaculture du fait de la faible acceptation du consommateur et l'accès au marché local limité en raison de l'aspect sombre des produits transformés. Malgré l'effet conservateur de la fumée, le poisson-chat a apparemment une durée de vie limitée ce qui a compromis sa disponibilité en tant que source de protéine animales dans certaines parties du pays. Compte tenu de la baisse de la consommation de poisson par habitant en Ouganda et de la nécessité de vulgariser les produits d'aquaculture, il a été jugé nécessaire d'entreprendre la présente étude. Environ 60 poisson-chats d'élevage ont été divisés en quatre lots identiques. Deux lots, un saigné et l'autre non saignéont été conservés dans de la glace tandis que les deux autres lots, traités de la même façon, sont tenus à températures ambiantes (28 °C). Les résultats ont indiqué que le poisson qui a été saigné avait une plus longue durée de conservation que le poisson qui ne l'a pas été, quelque soit la température de conservation. Sous des conditions froides la durée de vie du poisson saigné a été prolongée de 11 jours alors que le poisson gardé à température ambiante est resté acceptable pendant 5 heures supplémentaires par rapport au poisson qui n'a pas été saigné. Il a été conclu que les mesures d'intervention de saignée et de refroidissement ont amélioré et augmenté la durée de vie du poisson-chat.

L'application de ces stratégies dans les systèmes aquacoles améliorera sans doute le commerce local de même qu'international de poisson-chat pour des produits entiers ou transformés, la sécurité alimentaire et la lutte contre la malnutrition en Ouganda en augmentant l'acceptation par le consommateur.

*Mots clés: C. gariepinus, Saignement, Maintien de qualité*

[1] Food Biosciences Research Centre (FBRC), National Agricultural Research Laboratories (NARL), PO Box 7852, 17 Bombo Road, Kampala, Uganda. mmasette@yahoo.com

# 1. INTRODUCTION

Bleeding is rarely done in freshwater fisheries although it is a common practice for marine fish and in the animal industry. In fisheries, struggling fish are sometimes stunned by a blow on the head or by electric shock (Trestven and Patten, 1981) or immersed in chilled water to facilitate the killing procedure. Adequate research has not been done to show the effect of bleeding and stunning on fish flesh quality although the literature indicates advantages from these practices. The removal of blood by making cuts or by evisceration has been noted as a means of accelerating death (Howgate, 2003) and retarding spoilage (Fellows, 1988) and preserving the quality (Braker, 1992). The incision may be made at the caudal peduncle or nape depending on fish species. A cut at the isthmus severs the ventral aorta; at the nape, it severs the dorsal aorta and the spinal cord as well, thereby immobilizing the fish; and at the caudal peduncle, it excises the tail and severs the dorsal aorta. In *Salmo gairdneri*, the greatest amount of bleeding occurred with the caudal peduncle cut (Trestven and Patten, 1981)

Catfish (*Clarias gariepinus*), is the leading farmed fish in Uganda (FAO, 2000, 2003, 2008) followed by tilapia and mirror carp in order of commercial importance. Owing to the fish farming pattern in Uganda, catfish is usually marketed live or in processed (particularly smoked) form. The recent Department of Fisheries Resources (DFR) report indicates that there were 20,000 fish farmers with an average pond size of 500 $m^2$ per farmer and mean production yield of 1.5 $kg/m^2$ for catfish and less than 1.0 $kg/m^2$ for tilapia (Wadunde-Owori, *pers. Com.*) which accounts for its number one status as source of income for fish farmers. Besides, catfish grows faster than tilapia species (Haylor, 1992). As such, many entrepreneurial farmers have massively invested in aquaculture to take advantage of the high production yield and the market niche provided by the African diaspora in Europe. According to Onega *et al.* (2005) DFR has embarked on a promotional campaign to popularize aquaculture in Uganda as a strategy to augment the declining wild fisheries stocks (Nuwagira, 2007). Despite the absence of standards for catfish, Uganda is one of the five countries allowed to export products from farmed fish to the European Union (EU). Indeed according to DFR (2011) significant quantities of smoked products have been exported to Europe since 2005 to cater for the African ethnic minorities.

Quality is a subjective parameter which varies with type of product and end-user. In the fisheries sector, it may be defined as the degree of excellence or spoilage undergone by fish (Afolabi, 1984; Huss, 1995). It is influenced by a number of major factors that include; high temperatures, handling practices and initial microbial load. In aquaculture systems and particularly in catfish, additional influential quality factors include fish feed, bleeding and stress. Bleeding soon after capture or harvest contributes substantially to consumer acceptance, based on appearance (colour). Although the national production of catfish was estimated at 33,000 tonnes, local consumption is fairly low due to its red discolouration of fillets and varying geographical preferences within the country. The unappealing discolouration of the flesh has been attributed to post-mortem retention of blood. The retained blood provides sufficient nutrition for the evading post-mortem spoilage bacteria (Frenly, 2007) which may ultimately shorten the keeping quality of catfish (Howgate, 2003; Huss, 1995). According to Martin (1996), the iron component in the haemoglobin molecule catalyses oxidation and formation of yellow spots in salted fish. Although significant quantities of smoked catfish are exported as well as locally traded, they have limited shelf life, despite the preservative effect of smoke (Masette, 1990). Consequently, the limited shelf life compromises the marketing time for both processed and unprocessed catfish which in turn affects its availability as an important source of animal protein to reduce malnutrition (James, 1986) in certain parts of the country. The current status quo negatively impacts on the financial benefits of some farmers and traders of processed fish and as a result undermines the campaign to promote aquaculture enterprises in Uganda. On the contrary, if the blood was removed by bleeding the resultant product would be acceptable to majority of consumers and it would invigorate the campaign. According to Outdoor (2007) bled fish results in a white, acceptable fillet and extends shelf life, which is attributed to the reduced quantities of haemoglobin that reduces susceptibility to oxidative rancidity (Martin, 1996). It has been observed that fish blood remains in fluid form for about 30 minutes after capture (Howgate, 2003) and therefore it can easily be removed by the jerking action of a dying fish. In some marine fish species, bleeding is a recommended procedure for colour enhancement and shelf life extension (Ahimbisibwe *et al.,* 2009). Bleeding of fish is required in some tuna-like species to preserve fish quality (Braker, 1992). Despite the benefits of bleeding, this significant step is hardly practiced in Uganda, probably due to lack of knowledge or appropriate facilities

The generation of data on the effect of bleeding on keeping quality of catfish will contribute substantially to the international export trade and will also enhance marketability, and acceptability of catfish fillets among local consumers. If technical information to backstop the Uganda National Bureau of Standards (UNBS) in the formulation of local standards is not undertaken as a matter of urgency, it will jeopardize putting Ugandan catfish on the international market in the near future. Besides, the per capita fish consumption, which had declined from 13 kg in 1980 to the present 6 kg would be substantially increased by consumption of catfish. In addition, use of ice during storage of post-mortem farmed fish will give impetus to promotional campaigns for aquaculture that are being conducted incessantly in Uganda.

### Overall objective of the study
To assess the effect of bleeding on the keeping quality of catfish held at designated temperatures.

### Specific objectives
  a)  To determine iron content in bled and unbled catfish at ambient and chilled temperatures.
  b)  To establish the shelf life of catfish fillets using organoleptic evaluation

## 2. MATERIALS AND METHODS

### Materials
Relevant materials included knife, ruler, insulated containers, disposable gloves, microwave, plates, chopping board, timing device (clock) and fish samples

### Methods
#### Sample collection
Catfish samples (64) with an average age of 4 months, weighing 300 g and harvested using a non-stress method (scoop net) were collected from Umoja fish farm located 76km Northwest of Kampala in Busunju Sub-county, Wakiso District in Uganda. A total of (23) fish samples were bled and a similar number was left unbled. From the bled batch, 13 fish were kept on ice in an insulated box while the remaining 10 fish were held at ambient temperatures (28 °C). From the unbled batch 13 were similarly kept in another ice box and the remaining 10 were left at ambient temperature.

### Bleeding
Each catfish was firmly held down on a chopping board then, with a sharp pointed knife, an incision was made behind the operculum to cut through the artery to allow blood to flow out. Blood flow was accelerated by hanging the fish upside down and the jerking action of the fish. After 30 minutes, the blood stopped dripping and the fish was gutted prior to storage either on ice or in ambient conditions depending on treatment. The unbled batch of catfish were humanely killed by immersion in chilled water for 30 minutes and then also gutted prior to treatment.

### Transportation of samples
After all preliminary treatments at the farm, all samples were transported to the Makerere Department of Zoology laboratory for subsequent tests.

### Sampling plan
One piece of fish from each batch of samples kept at room temperature was sampled every two hours and subjected to sensory assessment by a panel of 12 trained individuals. While one fish from each batch held at chilled temperatures was sampled once after 7 and then 14 days. Subsequently, after the 14th day, sampling was every other day.

### Sensory assessment
The two fish sampled (1 bled and 1 unbled) were held at ambient temperature and sampled every two hours. They were cut into 12 steaks weighing approximately 5g each, incised from the middle part of the fish and steamed in a closed beaker for 10 minutes before serving to a taste panel of trained individuals. Each panelist was requested to score the general acceptability of the sample on a hedonic scale of 1–5.

### Determination of iron content
About 0.5 g of the remaining portion of the fish after sampling for sensory evaluation was ashed at 500°C for two hours. The ash was dissolved in aqua regia at 3: 1 ratio (Hydrochloric acid: Nitric acid). The solution was aspirated by atomic absorption spectrometer (AAS) to quantify the iron content.

## 3. RESULTS AND DISCUSSION

The practice of bleeding in some fisheries has been known to extend shelf life (Ahimbisibwe *et al.*, 2009) and preserve quality (Braker, 1992). Essentially, the procedure removes significant amounts of blood that would have provided sufficient nutrition for the invading post-mortem spoilage bacteria (Frenly, 2007). Besides, the iron from the haemoglobin molecule in the blood is one of the precursor ionic free radicals that normally catalyze lipid oxidation in fish (Huss, 1995) and therefore its removal in significant quantities should prolong the onset of oxidative rancidity. In catfish, the bleeding procedure reduced the iron content by a factor of 1 in fish held at ambient temperatures and by almost 1.5 in fish kept on ice (Table 1). Indeed the t-Test: Two-Sample Assuming Equal Variances indicated that the difference between bled and unbled catfish samples was significant (p= 0.525). The difference may be attributed to the rinsing effect of melting ice which ultimately contributed to bleaching of the flesh. However, there was no significant difference attributed to holding conditions. Implying that temperature does not affect the levels of iron content (Martin, 1996)

**Table 1**. Variation of iron content in muscle of catfish held at different temperatures

| Storage conditions | Mean concentration of iron (ppm) n= 6 | |
|---|---|---|
| | **Bled Catfish** | **Unbled catfish** |
| Ambient (28 °C) | 2.827 [0.121] | 3.835 [0.178] |
| Chilled (0+/-1 °C) | 2.475 [0.117] | 3.973 [0.159] |

The keeping quality of fish is usually objectively assessed by chemical and microbiological methods (Huss, 1995) but since the consumer is the ultimate decider on whether the fish is acceptable for consumption or not, sensory evaluation methods, albeit subjective, are always applied to supplement the objective methods (Connell and Shewan, 1980). However, in the present study the focus was on sensory evaluation due to limited funding to purchase chemicals for chemical and microbiological methods. In the study, point of rejection at mean score 2 was 14 hours and 19.5 hours post-mortem for unbled and bled samples respectively for catfish held at ambient conditions (Figure 1). Ice may not be easily available in the vicinity of remote fish farms and even if it was available, the cost may be prohibitive for the majority of fish farmers. The difference of 5.5 hours would be ample time for a trader or a farmer to sell his/her farmed uniced fish, without fear of rejection by intending customers. Consistent transaction of high quality bled catfish at least-cost holding temperatures would be another avenue of promoting aquaculture in Uganda.

On the contrary, catfish held at chill conditions in ice was rejected at score 2 on the 17th day for unbled and the 29th day for bled when both batches of catfish were chilled (Figure 2). The difference of approximately 12 days offers ample time for sales on local as well as international markets without incurring losses attributed to quality deterioration. Implying that bleeding together with chilling as compared to chilling alone has a significant effect on the shelf life of catfish.

**Figure 1. Organoleptic assessment of Clarias held at ambient temperatures (27 °C)**

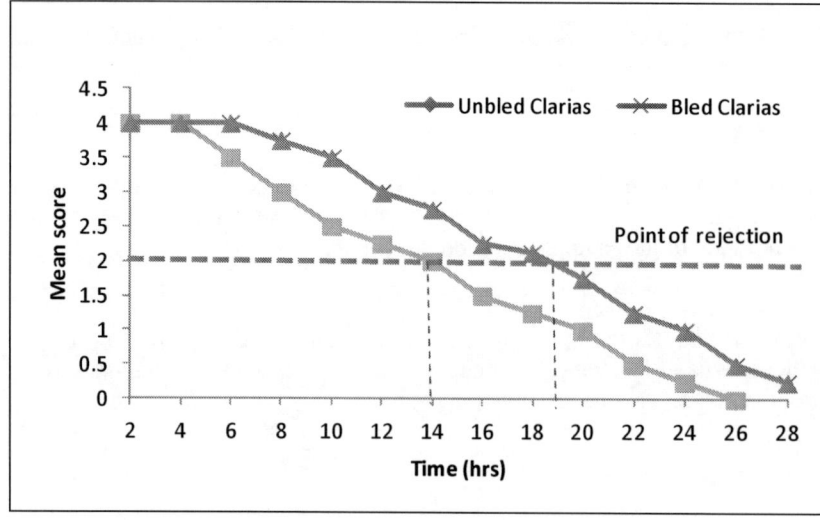

**Figure 2. Organoleptic assessment of *Clarias* held at chill temperatures (0+/-1 °C)**

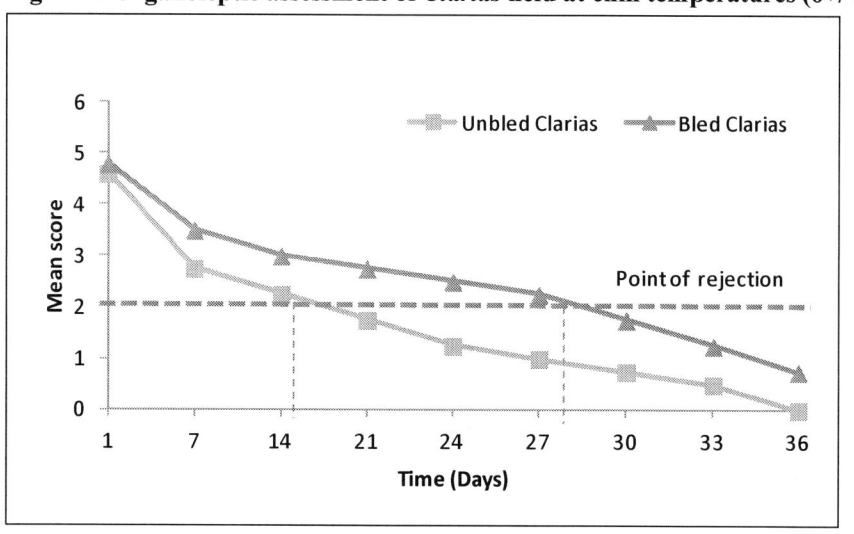

Several reasons can be advanced to explain the effect of bleeding on the keeping quality of catfish kept at different temperatures. According to Martin (1996), removal of blood delays collagen fibril degradation and muscle softening of pelagic fish (Ando *et al.*, 1999) which explains the extended shelf life. Several authors ascribe the extended shelf life to reduction of nutrients for the invading spoilage bacteria (Frenly, 2007; Venugopal, 1990) while Huss (1995) highlights the reduction of the iron content. In iced fish samples, shelf life may also be extended by the washing effect of ice (Chen and Chai, 1982). Since the three factors appeared to happen concurrently, it is plausible to suggest that synergistic tendencies between the three factors have a combined effect as observed in both batches of catfish that were bled but kept at different temperatures.

## 4. CONCLUSIONS

The practice of bleeding farmed catfish immediately after capture has a twofold advantage: enhanced customer acceptance and extended shelf life of the resulting product, regardless of the holding temperature. However, the synergistic effect of blood removal; reduction of nutrients for invading spoilage microbes and reduction of haemoglobin iron, coupled with the effect of chilling increases the keeping quality of catfish by almost two weeks compared to similar batches that were not bled and kept at room temperature.

## 5. RECOMMEDATIONS

- The study should be repeated to include objective chemical as well as microbiological methods for quality assessment. Effort should also be made to incorporate the effect of feed regimes practiced by different farmers on the quality of farmed fish products;
- Farmed catfish should be bled to enhance its local marketability;
- Bleeding procedures should be inserted in the Quality Assurance rules of 2008 which have been incorporated in the 1970 Fish Act as a mandatory practice for farmed catfish; and
- The qualities of bled catfish should be used as a strategy to promote fish farming in Uganda.

## 6. REFERENCES

**Afolabi, O.A.** 1984. Quality change of traditionally processed fresh water fish species. Nutritive and organoleptic changes. *Journal of Food Technology*, 19(3): 333–340.

**Ahimbisibwe J. B., Inoue T, Shibata T and T. Aoki** (2010). Effect of bleeding on the quality of amberjack *Seriola dumerili* and red sea bream *Pagrus major* muscle tissues during iced storage. *Fisheries Science*. 76: 389–394.

**Ando, M., Nishiyabu, A., Tsukamasa, Y. & Makinodan, Y.** 1999. Post-mortem softening of fish muscle during chilled storage as affected by bleeding. *Journal of Food Science,* 64(3): 423–428.

**Braker, M.** 1992. Handling Sport Caught Fish, Oregon State University, Washington County, USA. Extension Sea Grant Program. EC 1414.

**Chen, H.C. & Chai, T.** 1982. Micro flora of drainage from ice in fishing holds. *Applied Environmental Microbiology*, 43: 1360–1365.

**Connell, J.J. & Shewan, J.M**. 1980 Sensory and non-sensory methods of Assessment of fish freshness. *In* J.J. Connell, ed. *Advances in fish science and technology.* pp. 56–65. Farnham, UK, Fishing News Books Ltd.

**FAO.** 2000. *The state of world fisheries and aquaculture,* Publication related to aquaculture of Uganda, Rome. FAO yearbook. Fishery statistics. Vol. 86/2. Rome. 169p.

**FAO.** 2003. Aquaculture production, Yearbook of Fishery Statistics 96(2), Rome, Italy.

**FAO.** 2008 *National Aquaculture Sector Overview*, Fisheries and Aquaculture Department, Uganda. FAO yearbook. Fishery and Aquaculture Statistics. 2006. Rome. 57p.

**Fellows, P.** 1988. Principles of food preservation (http://www.fst.uq.edu.au/staff/rmason). Food Processing Technology.

**Frenly, G.** 2007. Explaining the techniques of bleeding fish. (http://www.helium.com/items).

**Haylor, G.S.** 1992. African Catfish Hatchery Manual, Central Region and Northern Region Fish Farming Malawi, Stirling Aquaculture Scotland, pp.86.

**Howgate, P.** 2003. Review of the public heath safety of products from aquaculture. Article first published online: 4 FEB 2003. http://onlinelibrary.wiley.com/doi/10.1046/j.1365-2621.1998.3320099.x/abstract

**Huss, H.H. 1988.** *Quality changes in fresh fish.* Manual prepared for FAO/DANIDA Training Project.*FAO Fisheries Series* No.29, FAO, Rome.

**James, D.G.** 1986. The prospects of fish for the undernourished. *FAO Food and Nutrition,*12: 20–27.

**Jasen, E.G.** 1997. Rich Fisheries, Poor fish folk, Socio-economics of L.Victoria fisheries. Publication by IUCN East African Program.

**Martin, A.M.** (Ed). 1996. Fisheries Processing: Biotechnological application http://www.fmcfoodtech.com

**Masette, M.** 1990. Liquid smoke treatment as an alternative method to traditional smoking in fish processing. MSc. Thesis. University of Lincoln (UK).

**Nuwagira, S.** 2007. Fisheries Development and aquaculture. *East African Business Week.* Kampala (Uganda). Issue 22.

**Onega, D.N., Mwanja, W.W. & Mushi, V**. 2005. Meeting the increasing demand for fish in the Lake Victoria Basin through development of aquaculture. *Lake Victoria Fisheries Organisation Conference*, Entebbe, Uganda.

**Tretsven, W. & Patten, B.G.** 1981. Effect of Arterial Incisions on the Amount of Bleeding and Flesh Quality of Rainbow Trout. *Marine Fisheries Review,* 43(4).

**Venugopal, V.** 1990. Extra cellular proteases of contaminant bacteria in fish spoilage: A review. *J. Food Prot.,* 53: 341–350.

# REFRIGERATION APPLIED TO SEAFOOD: RECENT DEVELOPMENTS AND APPLICATIONS

## [RÉFRÉGÉRATION APPLIQUÉE AUX PRODUITS DE LA MER: RÉCENTS DÉVELOPPEMENTS ET APPLICATIONS]

by/par

Alain Le Bail[1], M. Orlowska, M. Havet, A. Adedeji, J. Abadie, A. Beaufort, G. Bourdin and M. Cardinal

**Abstract**

The challenge of preserving seafood after capture is most of the time associated with refrigeration and the cold chain. Depending on the place of capture, different technologies can be used. This paper presents an overview of selected technologies, mostly used on land. Freezing addresses different issues, particularly in terms of energy demand; consideration of the recommendations of the international institute of refrigeration, can result in substantial energy savings. Focus is then placed on recently developed technologies: superchilling, high pressure processing and freezing under static electric field (electro-freezing). Superchilling consists of preserving products at temperatures close to the initial freezing temperature; such a strategy permits reduction of water activity and potentially leads to extension of shelf life (gelation). High pressure processing can be used to freeze products and also to process seafood. High quality frozen foods can be obtained with pressure shift freezing process with refined ice crystallization. More recently, freezing under an electric field has been developed and applied to seafood; recent results demonstrate the advantages of electro-freezing in terms of microstructure.

*Key words: Freezing, Seafood, Superchilling, Electric field, High pressure, Microstructure*

**Résumé**

Le defi de la conservation des produits de la mer est le plus souvent étroitement lié à la réfrigération et à la chaîne de froid. Différentes technologies peuvent être mises en œuvre selon le lieu de capture. Cet article présente une vue d'ensemble de quelques technologies destinées à un usage à terre. La congélation est associée à des enjeux de consommation d'énergie ; la mise en œuvre des recommandations de l'Institut International du froid est à même de conduire à des économies d'énergie substantielles. Le focus des technologies présentées porte sur des technologies récentes ; le surefroidissement, le procédé haute pression et la congélation sous champ électrique (électrocoagulation. Le procédé de surefroidissement consiste à préserver des produits à une température proche de la température de congélation ; cette stratégie permet de réduire l'activité de l'eau et conduit à un accroissement de la durée de conservation. Le procédé de traitement à haute pression peut être mis en œuvre pour congeler des produits ou pour les traiter (gélification, texturation). Des produits congelés de haute qualité peuvent être obtenus par congélation par détente haute pression avec une microstructure de cristaux de glace accrue (cristaux fins). Plus récemment, la congélation sous champ électrique a été développée et appliquée aux produits de la mer; des résultats récents ont mis en évidence des avantages en termes d'affinement de la microstructure des cristaux de glace.

*Mots clés: Congélation, Produits de la mer, Sur-refroidissement, Champ électrique, Haute pression, Microstructure*

# 1. INTRODUCTION

Refrigeration technology and refrigeration equipments have been developed primarily for preserving and for transporting foods, in particular from distant countries, for example from Australia, New Zealand, South America etc. to Europe. One of the examples of the first refrigerated transportation by Ship is "Le Frigorifique" in 1876 (see Figure 1), which transported meat from Buenos Aires in Argentina to Rouen in France. Other examples regarding food preservation are quoted in Table 1.

---

[1] Representing the IIR. LUNAM University, Oniris, UMR 6144 GEPEA, BP 82225, 44322 Nantes, Cedex 3/CNRS, Nantes, F-44307, France. alain.lebail@oniris-nantes.fr

**Figure 1. Refrigerated food transport by the ship "Le Frigorifique", 1876**
**Extrait du livre Hommage à Tellier - © Association Française du Froid**

**Table 1. A little bit of history on preservation of seafood and on preservation techniques**

| Year(s) | Bullet Points |
|---|---|
| 1812 | Appertisation - Nicolas APPERT - Canning technology |
| 1822–1895 | Pasteur |
| 1866 | On board freezing in USA - First tests |
| 1877 | French trawler equipped with freezers |
| 1914–1918 | Canada and USA provide frozen fish to Europe during the 1st World-war |
| 1919 | Whale meat freezing in Norway |
| 1930 | Cod fish liver is frozen before oil extraction (USA) |
| 1930 | Recommended storage temperature: -10°C |
| 1938 | Recommended storage temperature: -20°C |

Due to the growing importance of refrigeration in the modern world, a pioneering organization specializing in refrigeration was established in Paris in 1908 during the first world congress on refrigeration: the "International Association of Refrigeration". A few years later, in 1920, an intergovernmental association was created from the International Association of Refrigeration; the "International Institute of Refrigeration" (IIR) or in French the "Institut International du Froid" (IIF). (IIR-IIF, 177 Bd Malesherbes 75017 Paris, France. www.iifiir.org).

The IIR is an Intergovernmental organization with more than 60 member countries representing over 60 percent of the total world population. The International Institute of Refrigeration (IIR) is the only independent intergovernmental organization which promotes knowledge of refrigeration and associated technologies that are necessary for life in a science-based, cost-effective and environmentally sustainable manner including:

- Food quality and safety from farm to consumer
- Comfort in homes and commercial buildings
- Health products and services
- Low temperatures and liquefied gas technologies
- Energy efficiency
- Use of non-ozone depleting and low global warming refrigerants in a safe manner

The IIR is organized around 10 commissions as presented in Figure 2. The journal of the IIR is the International Journal of Refrigeration (Elsevier), which is ranked 19th out of 115 Journals in mechanical engineering and 13th out of 49 Journals in thermodynamics. The IIR website provides a unique data base gathering all the technical data needed and also many facilities such as Newsletters.

**Figure 2. Organisation of the 10 technical commissions of the International Institute of Refrigeration**

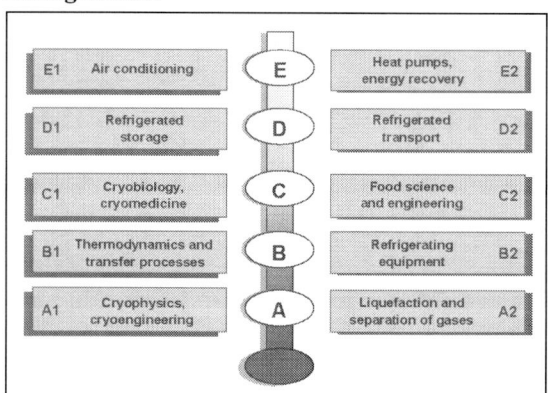

**Figure 3. Scheme of different technologies used to extend the shelf life of seafood**

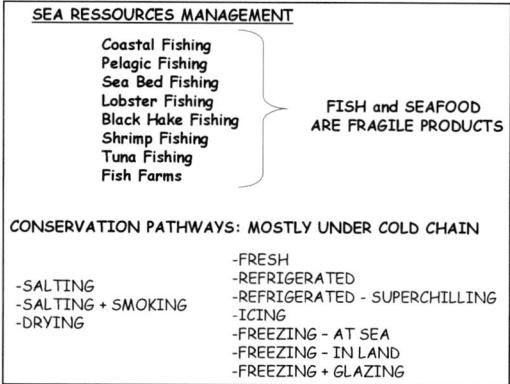

Refrigeration and in particular freezing are broadly used to extend the shelf life of seafood, which is a fragile product. Freezing is one of the preferred methods, especially for long-term preservation. In particular when seafood is used by the food manufacturing industry, frozen products offer convenience even though freezing may lead to substantial changes in the organoleptic quality. Figure 3 presents an overview of different seafood sources and the different technologies that can be used for preserving them. Apart from salting and drying, and also in some cases smoking, refrigeration is often required to extend the shelf life of seafood.

In this article, we provide an insight into selected technologies associated with refrigeration. There are different issues related to freezing, in particular its energy demand. Based on the consideration of the IIR, a recommendation can be proposed resulting in substantial energy savings. Such examples of recently developed technologies that may result in energy saving are superchilling, high pressure processing and freezing under a static electric field. Superchilling consists of preserving products at temperatures close to the initial freezing temperature; such a strategy permits reduction of water activity and potentially leads to extension of product shelf life. High pressure processing can be used to freeze products and also to process seafood for better quality products. High quality frozen foods can be obtained with the pressure shift freezing process with refined ice crystallization. More recently, freezing under an electric field has been developed and applied to seafood; recent results demonstrate the potential advantages of this technique terms of microstructure and reduction in drip loss.

## 2. ENERGY DEMAND IN FREEZING

Refrigeration systems use up to 15 percent of the total energy consumed worldwide. Therefore, there is a stake in optimizing the energy demand for the refrigeration process applied to any food. The freezing process usually starts with a non-frozen product at chill temperature. Freezing is sometimes done cryogenically, which permits very high freezing rates. However, the availability of cryogenic fluids (liquid nitrogen, solid carbon dioxide) on production sites is not always possible and cryogenic freezing is restricted to selected cases. Compression units are mainly used for the freezing of seafood. Freezing can be carried out with unwrapped products using air or brine as a refrigeration media. In some cases, such as onboard freezing, seafood is usually frozen using plate freezers. The fish (mostly filleted) are placed in plastic pouches that are put in a carton. The carton is closed when full and is frozen directly. The impact of the set point temperature of the freezing equipment has a direct impact on the energy efficiency of the compression unit (COP = coefficient of performance = ratio between net refrigeration energy and electrical energy). Even though a lower temperature yields a reduction of the COP, in some cases, it allows global energy savings due to a reduced freezing time (see for example Le Bail *et al.*, 2008, in the case of bread freezing). Indeed, the global refrigeration energy needed to freeze a food encompasses the energy taken by the food and also different "collateral" energy consumption such as losses to the environment, ventilators (blast air freezing), circulating pumps (brine freezing) or others. The final temperature of the product at the end of the freezing process has also a big impact on the global energy demand. Indeed, the set point temperature of the ambient temperature during freezing is always much lower than the set point temperature in frozen storage. Therefore, the freezing equipment has a lower COP in comparison to the frozen storage. According to the criterion proposed by the IIR, a product can be considered as frozen if, either: 80 percent of the freezable water is effectively frozen or the temperature of the product is -12 °C Le-Bail *et al.*, 2010 for fresh seafood, the criteria of -12 °C (to be reached at the end of the freezing process) can be considered as sufficient. Frozen food can then be transferred to frozen storage. In the case of smoked salmon, with an initial freezing temperature of -5 °C, the criteria to consider is rather "80 percent of freezable water frozen", which corresponds to a final temperature at the end of freezing of around -15 °C (Le-Bail *et al.,* 2008). This way of managing the final freezing temperature may result in substantial energy savings without altering the final quality of the frozen products. Indeed, most of the ice crystallization takes place between the initial freezing temperature and a few degrees Celsius below this temperature. Freezing and the completion of the water to ice transition can be achieved in frozen storage with a lower energy demand than for freezing equipment.

## 3. SUPERCHILLING

Superchilling was described in 1920 by Le-Danois, (Le-Danios, 1920). It consists of keeping a product to 1 °C or 2 °C below or just above the initial freezing temperature. An overview of the superchilling process was proposed recently by Kaale *et al.,* 2011, showing a strong interest of the food industry in this process. The objective of superchilling is an extension of the shelf life because of the lowering effect on water activity of the products. Accordingly, a reduction of the growth rate of microorganisms is expected, due to a synergistic effect of the reduction of water activity and temperature. Superchilling can be considered as an intermediate process in between chilling and freezing. The implementation of superchilling technology is on the rise. In particular it has been adopted by several smoked salmon producers. Figure 4 shows the depression of water activity of ice as a function of temperature. Considering a frozen product, the laws of thermodynamics control the equilibrium between each phase. In particular, in the case of freezing, the partial vapour pressure of the aqueous solution present in the product must be equal to the water activity of ice. For example in the case of smoked salmon with an initial freezing temperature of -5 °C, one can expect a water activity of 0.953.

**Figure 4. Water activity of ice as a function of temperature (°C)**

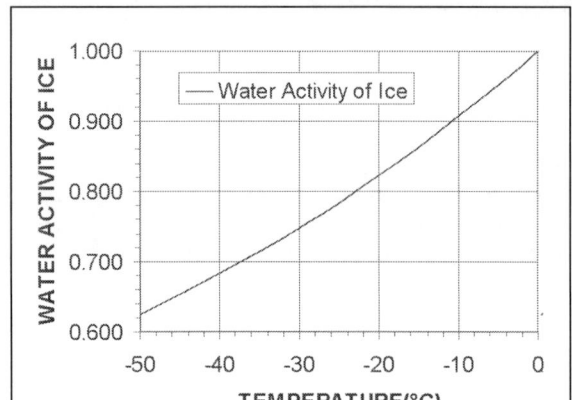

Super chilling addresses different issues in terms of organoleptic properties, and is often presented as a technique better able to preserve the initial quality of the product. In the case where freezing is expected to lead to substantial damage to the flesh of seafood products, superchilling is expected to result in improved quality. There is still a controversy between authors regarding the real benefits and drawbacks of superchilling. For example Beaufort *et al.,* 2009 studied the super chilling of smoked salmon by applying temperatures just above the initial freezing temperature. They found that the storage of cold-smoked salmon at −2 °C during 14 days didn't show any serious consequences on the quality in comparison to a control. After 28 days of storage, a positive effect on the development of *Listeria monocytogenes* was observed (lower growth than control). Regarding the sensory evaluation, all sensory parameters were not affected. They concluded that further studies related to superchilling are needed to better understand the changes observed in the period ranging between day 14 and day 28.

Hauland *et al.,* 2005, worked on roasted pork meat. They concluded that superchilling allows a doubling in shelf life (chemical, microbiological and sensory). These authors also concluded that there is a large need for help with implementation of superchilling in the industrial process. The major challenge of superchilling lies in the precise control of the temperature; indeed, slight departure from the set point may result in HACCP risks (Hazard Analysis Critical Control Points). Since temperature control is very difficult and often inadequate, a new principle based on the control of the ice fraction in the food would be necessary. Moreover superchilling requires the establishment of a new channel among the existing conventional channels of the cold chain. However, at the scale of a company storing its own production, superchilling can be relatively easily set in operation to extend the shelf life of products. This is critical for example in the smoked salmon industry, which faces peak demands during certain periods of the year (Christmas, Easter, etc).

## 4. HIGH PRESSURE PROCESSING

Several researchers pointed out the effect of high hydrostatic pressure on biological material in the early 20th century. Bridgman, (1911), for example showed that a pressure of 700 MPa was able to coagulate albumin, and also studied phase changes under pressure. After almost a century of latency, the use of high hydrostatic pressure has been reinvestigated since the end of the 80's by scientists all over the world. The High Pressure Process (HPP) consists mainly of placing a sample in a high-pressure (HP) vessel. After closing the vessel, a pressure intensifier is used to increase the pressure up to the required level. Due to the compressibility of water: the major component of most biological systems, the temperature of the sample increases during pressurisation and decreases during depressurisation. The order of magnitude of these changes is an increase in temperature of 4K for an increase in pressure of 100 MPa. There are basically three main areas of application of HPP to food systems: sanitation/stabilisation (by modification of the enzymatic activity or inactivation of microorganisms), texturization (by protein denaturation) and phase change. Some technological barriers limit the pressure level. Nowadays, industrial applications are currently working at pressures of 400-500 MPa, whereas scientists and researchers might investigate up to 900 MPa. Higher pressures can be obtained (several orders of magnitude of kilobars) with special systems but with miniature volumes.

HPP of seafood has been investigated by several authors. Among the most promising applications already implemented in the industry is the shucking of oysters by some companies in Australia and in USA. This process permits the extension of shelf life of oysters in refrigerated conditions after HPP. The shell is usually circled with an elastic ring. This process permits the opening of large oysters very easily under pressure and also inactivates specific microorganisms (*Vibrio*). Another interesting application has been developed in Canada to process lobsters. Using this process the lobster meat can be removed very easily in one piece. Other fields of application have been developed for meats, ready to eat meals, rice, fruits, sauces, etc.

Several reviews on the interest of HPP in refrigeration are available such as Le-Bail *et al.*, 2002 and Cheftel *et al.*, 2000. Regarding freezing, high pressure technology permits very high freezing rates, due to the pressure shift freezing process. The phase change temperature of water is depressed with increasing pressure until 200 MPa. Above this pressure, the opposite phenomenon is observed (see Figure 5). The pressure shift freezing process is based on the very specific property of the depression of the freezing point. The food is first pressurized in a high pressure vessel at negative temperature. Then the pressure is released as shown in Figure 6.

**Figure 5. Phase diagram of water under pressure. Ice I = hexagonal structure Ice III = cubic structure (Bridgman, 2012)**

**Figure 6. Scheme of the pressure shift freezing process 1–2:pressurisation/2–3: cooling under pressure without phase change/3–4–5–6: pressure release (< 10 seconds) SC : supercooling**

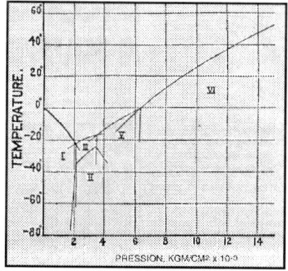

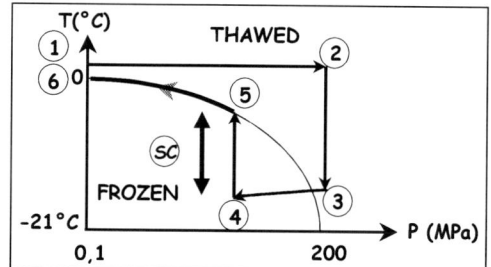

During pressure release, numerous, fine ice crystals are formed, resulting in a better preserved microstructure. However, the HP at 200 MPa exposes the proteins contained in the food to partial denaturation, resulting in increased toughness (similar to semi cooked products). Even though there is evidence showing a reduction of ice crystals size (see Figure 7 and Figure 8 – from Alizadeh *et al.*, 2007) and of drip loss in the case of high pressure freezing (extended to high pressure thawing), there is no evidence at the moment that this process will be adopted by the industry. Indeed, the conventional freezing techniques are relatively satisfactory and the benefit brought by HPP is not sufficient to overcome the cost of the equipment. Only in the case of few products such as oysters and lobsters has HPP been effectively implemented by the industry. HPP raises some regulatory issues as not all countries necessarily accept this technology. In particular in Europe, there are regulatory constraints which limit the use of the technology, whereas in the USA, HPP has been accepted for selected products.

**Figure 7. Blast air frozen salmon fish. White spots are ghosts of ice crystals [10]**

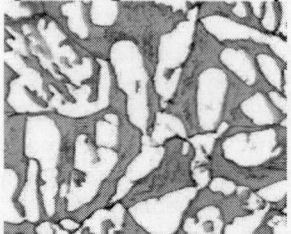

**Figure 8. Pressure shift frozen salmon fish. White spots are ghosts of ice crystals [10]**

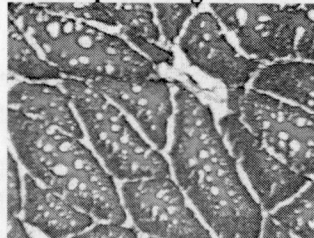

**100 μm**

## 5. FREEZING UNDER A STATIC ELECTRIC FIELD

There are connections between the freezing of water (and in general with most phase changes) and electrical phenomena. Several authors have related the presence of currents during formation of crystals. Pruppacher *et al.*, 1968, showed that a voltage of several volts develops in the vicinity of the freezing front between ice crystals and the solution (currents of $\sim 10^{-7}$ A). Indeed, during freezing, the surface of an ice crystal is covered by a thin 40 Å semi-liquid layer (Fletcher, 1970 and Fletcher, 1968). The voltage is explained by the embedding of ions in the semi crystalline structure, resulting in a diffusion of electrical charges and consequently in the presence of a current and of a potential difference.

The application of an external electrostatic field is thus likely to affect the phase change phenomena. This has been demonstrated theoretically. For example Goldman *et al.*, 2005, present an analysis of the stability of water clusters. Due to the re-orientation of water molecules, hydrogen bonds are stronger along the field than along orthogonal directions (Wei *et al.*, 2008). Some authors reported that with increasing field strength, a water system has a more perfect structure that is similar to an ice-like structure. The possibility of obtaining liquid water with more structural regularity by applying an electric field was also suggested by Jung *et al.*, 1999. The application of electric disturbance during freezing in aqueous systems may thus be of interest in terms of control of the crystallization phenomenon. This has been investigated by some authors. The impact of external electrostatic fields on crystallization has been seen experimentally for several systems, such as synthetic polymers (Marand & Stein, 1989). In the case of aqueous systems, several references are available as well. However, most existing experimental work is based on the application of a moderate current flowing through an aqueous solution during freezing. In a recent study (Wei *et al.*, 2008) the impact of an external electric field on water to ice solidification was evaluated. The studies of molecular dynamics simulation showed that at a critical electrostatic field strength of E = $1.5 \times 10^7$ V/m the re-orientation of the dipolar water molecules along an external electrostatic field starts. Wei *et al.*, 2008, observed that a field of around $1.0 \times 10^5$ V/m was in fact sufficient.

Recent application of freezing under high electric and magnetic field has been adopted by the industry, using the ABI's, Cells Alive System (CAS), also called "useful freezer", manufactured by a Japanese company. The design of the equipment is based on the patent of Owada *et al.*, 2001. Recent studies done in our laboratory with selected fish products frozen in this equipment showed that a slight reduction of the size of the ice crystals was effectively obtained. Histological sections of cod fish fixed by a Carnoy solution (in a frozen state – see Alizadeh *et al.*, 2007 for protocols) are shown in Figure 9 (control) and Figure 10 (frozen in high electric field freezer).

The industrial application of this technology is on the increase, especially for high value products that demand a high freezing quality. It is likely that similar advantages in terms of microstructure can be achieved using cryogenic freezing. The economic and technological comparison has to be done to conclude on this specific technique. This technology also introduced concerns on some safety issues, such as high voltages (several kV) which are in operation in a humid environment. However, high voltage may lead to some degree of microbial inactivation thanks to the presence of ozone under selected operating conditions.

**Figure 9. Cod fish frozen in a blast air freezer. White spots are ghosts of ice crystals. Unpublished from Prof. Le-Bail in collaboration with Grimsby University, UK**

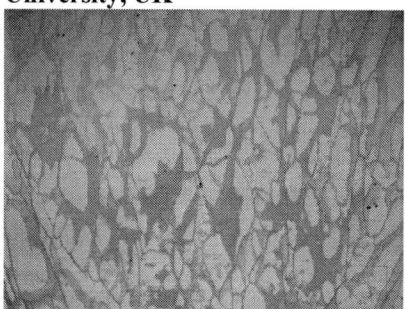

**Figure 10. Cod fish frozen under electric field. White spots are ghosts of ice crystals. Unpublished from Prof. Le-Bail in collaboration with Grimsby University, UK**

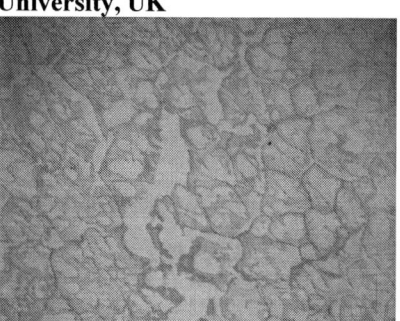

## 6. CONCLUSIONS

Refrigeration is a key technology for seafood products that experience rapid deterioration after harvest. Chilling is extensively used for short term preservation. However, long term preservation is needed to meet the expectations of the industry and of consumers. Freezing addresses different issues in terms of quality, of cost and also of environmental impact. Even though freezing has a larger environmental impact than chilling technology, it is the only technology that can limit the food loss to a minimum.

Several refrigeration innovations are still being developed to overcome the damage caused by freezing. Among the most promising technologies are high pressure, superchilling and freezing under high electric field, which are expected to provide new cost saving avenues and quality retention in sea foods to the industry and to the global seafood market. However, it is likely that these technologies will remain as niche applications for some time, and that freezing using conventional compression units will remain in the long term the most preferred freezing technology, mainly for the economic reasons.

## 7. ACKNOWLEDGEMENTS

The authors acknowledge the financial support of the CNRS (Centre National de la Recherche Scientifique, Delegation Bretagne et Pays de Loire, France), of OFIMER, of the European Union (European project SAFE ICE) and of the Ministry of Agriculture for partly funding this research. Special thanks are addressed to Luc Guihard, Christophe Couëdel, Delphine Queveau and Sylvie Chevallier (ONIRIS) for technical support. The samples presented in the section on freezing under electric field have been prepared using ABI's, Cells Alive System (CAS); they were supplied by Dr Simon Derrick at the Grimsby Institute in the UK.

## 8. REFERENCES

**Alizadeh, E., Chapleau, N., De-Lamballerie-Anton, M. & Le-Bail, A**. 2007. Effect of different freezing processes on the microstructure of Atlantic salmon (Salmo salar) fillets. *Innovative Food Science and Emerging Technologies*, 8: 493–499.

**Beaufort, A., Cardinal, M., Le-Bail, A. & Midelet-Bourdin, G**. 2009. The effects of superchilled storage at - 2 °C on the mircobiological and organoleptic properties of cold-smoked salmon before retail display. *International Journal of Refrigeration*. 32(7): 1850–1857.

**Bridgman, P.W**. 1912. Water in the liquid and five solid forms under pressure. *Proceedings of the American Academy of Arts and Sciences*, 47: 411–558.

**Cheftel, J.C., Lévy, J. & Dumay, E**. 2000.Pressure-assisted freezing and thawing : principles and potential applications. *Food Review International,* 16(4): 453–483.

**Fletcher, N.H**. 1968. Surface structure of water and ice. II. A revised model. *Phil. Mag.,* 18: pp. 1287–1300.

**Fletcher, N.H**. 1970. Liquid water and freezing, in *The chemical physics of ice*. Cambridge University Press: London. pp. 73–103.

**Goldman, N., Leforestier, C. & Saykally, R.J**. 2005. A 'first principles' potential energy surface for liquid water from VRT spectroscopy of water clusters. *Philosophical Transactions of the Royal Society A*, 363: 493–508.

**Haugland, A., Aune E.J. & Hemmingsen, A.K.T**. 2005. Superchilling - innovative processing of fresh foods. EUROFREEZE - Individual Quick Freezing of Foods, Sofia (Bulgaria) - *Proceedings of EU Workshop (Project QLK1-CT-2002-30544)*, pp. 1–8.

**Jung, D.H., Yang, J.H. & Jhon, M.S**. 1999. The effect of an external electric field on the structure of liquid water using molecular dynamics simulations. *Chemical Physics*, 244(2–3): 331–337.

**Kaale, L.D., Eikevik, T.M., Rustad, T. & Kolsaker, K**. 2011. Superchilling of food: A review. *Journal of Food Engineering*, 107(2): 141–146.

**Le-Bail, A., Chevalier, D., Mussa, D. & Ghoul, M**. 2002. High pressure freezing and thawing of foods; a review. *International Journal of Refrigeration*, 25: 504–513.

**Le-Bail, A., Chapleau, N., De-Lamballerie, M. & Vignolle, M**. 2008. Evaluation of the mean ice ration as a function of temperature in an heterogenous food; application to the determination of the target temperature at the end of freezing. *International Journal of Refrigeration*. 31(5): 816–821.

**Le-Bail, A., Dessev, T., Jury, V., Zuniga, R., Park, T. & Pitroff, M**. 2010. Energy demand for selected bread making processes: Conventional versus Part Baked frozen technologie*s. Journal of Food Engineering*, 96: 510–519.

**Le-Danois, E**. 1920. Nouvelle méthode de frigorification du poisson. French Patent No. 506.296.

**Marand, H. & Stein, R.S**. 1989. Isothermal crystallization of poly(vinylidene fluoride) in the presence of high static electric fields. II. Effect of crystallization temperature and electric field strength on the crystal phase content and morphology *Journal of polymer science: Part B: Polmer physics*, 27: 1089–1106.

**Owada, N. & Kurita, S**. 2001. *US PATENT US - 6-250-087-B1*. 26th June 2001.

**Pruppacher, H.R., Steinberger, E.H. & Wang, T.L**. 1968. On the electrical effects that accompany the spontaneous growth of ice in supercooled aqueous solutions. *Journal of Geophysic Research*, 73: p. 751.

**Wei, S., Xiaobin, X., Hong, Z. & Chuanxiang, X**. 2008. Effects of dipole polarization of water molecules on ice formation under an electrostatic field. *Cryobiology*, 56(1): 93–99.

# PROMOTING VALUE ADDITION TO INDIAN MACKEREL *(RASTRELLIGER KANAGURTA)* IN THE SEYCHELLES BY HOT SMOKING AND ASSESSMENT OF ITS MICROBIOLOGICAL AND SENSORY QUALITIES

## *[PROMOUVOIR LA VALEUR AJOUTEE DE MAQUEREAU (RASTRELLIGER KANAGURTA) DANS LES SEYCHELLES PAR LE FUMAGE À CHAUD ET L'EVALUATION DE SES QUALITÉS MICROBIOLOGIQUES ET SENSORIELLES]*

by/par

Christopher Hoareau[1] and Luckson Accouche

**Abstract**

Fish is the main source of protein in the Seychellois diet and the annual per capita consumption is 64.3 kg (FAO, 2008). The Indian mackerel, Rastrelliger kanagurta is a popular fish in Seychelles, much liked by the consumers when still in the fresh state. An excess of this species is harvested during the good season from November to April. During this period, the local market quickly becomes saturated and a lot of the fish goes very cheaply for use as fishing bait. At times the fishers would also avoid harvesting due to lack of an outlet for the catch.

This project looks at adding value to the Indian mackerel by hot smoking with a view to providing local consumers with a different way of appreciating the fish. It is expected to subsequently create a market for this product, thus increasing the earnings of the small fishers and the manufacturers as well. It was necessary to obtain the opinion of the consumers with regards to quality and acceptance, taking into consideration that this is a relatively new product in the context of Seychelles. A small sample of the population, who are well versed with fish products, was requested to assess the sensory quality after smoking, the result of which will give an indication if the general public will appreciate hot smoked mackerel. The product shelf life that is strongly influenced by the initial microbiological quality was assessed after storage at chill and ambient temperatures. The development of histamine was investigated over the length of the storage period.

The responses received on the sensory quality were very positive and encouraging. All the scores for the parameters assessed were above 90 percent with the exception of the degree of dryness. This results show that there are markets for hot smoked mackerel in the Seychelles, however a cost benefit analysis should be properly undertaken to ensure that a commercial venture will be viable. Microbial analysis showed that early spoilage with yeasts and moulds can occur when stored in chill condition, if strict hygiene is not observed during post process handling. There was no development of histamine during the storage period both at ambient temperature and under chill condition.

*Key words: Seychelles, Mackerel, Value-added, Indian Ocean*

**Résumé**

Le poisson est la principale source de protéines dans le régime alimentaire des seychellois et la consommation annuelle par habitant est de 64.3 kg (FAO, 2008). Le maquereau indien, Rastrelliger kanagurta, est un poisson populaire aux Seychelles, très apprécié par les consommateurs lorsqu'ils sont encore à l'état frais. Un excès de cette espèce est peché pendant la bonne saison de novembre à avril. Au cours de cette période, le marché local devient rapidement saturé et beaucoup de poissons est très bon marché pour servir comme appâts de pêche. Parfois les pêcheurs d'éviteraient de pêcher en raison du manque de débouché pour la capture.

---

[1] Fish Inspection and Quality Control Unit, Seychelles Bureau of Standards, PO Box 953, Victoria, Mahe, Seychelles. vetfiqcu@seychelles.net

Ce projet vise à ajouter de la valeur au maquereau indien par le fumage à chaud en vue de fournir aux consommateurs locaux une manière différente d'apprécier le poisson. Il est attendu de créer par la suite un marché pour ce produit, augmentant ainsi les revenus des petits pêcheurs et aussi des transformateurs. Il est nécessaire d'obtenir l'opinion des consommateurs en ce qui concerne la qualité et l'acceptation, prenant en considération qu'il s'agit d'un relativement nouveau produit dans le contexte des Seychelles. Un petit échantillon de la population, qui a une grande experience des produits de la pêche, a été prié d'évaluer la qualité sensorielle après le fumage, ce qui donnera une indication si en général le public pourra apprécier le maquereau fumé. La durée de conservation du produit qui est fortement influencée par la qualité microbiologique initiale a été évaluée après le stockage à température ambiante et refroidie. Le développement de l'histamine à été étudié au cours de la durée de la période du stockage. Les réponses reçues sur la qualité sensorielle ont été très positives et encourageantes. Tous les scores pour les paramètres évalués ont été au-dessus des 90 pour cent à l'exception du degré de sécheresse. Ces résultats montrent qu'il existe des marchés pour les maquereaux fumés aux Seychelles, cependant une analyse coûts-bénéfices devrait correctement être entreprise afin d'assurer la viabilité d'une entreprise commerciale.. Des analyses microbiologiques ont montré que la détérioration précoce de levures et de moisissures peuvent survenir lors du stockage sous condition de refroidissement, si l'hygiène stricte n'est pas examinée au cours de la manutention post traitement. Il n'y a aucun développement d'histamine pendant la période du stockage à température ambiante ainsi que sous condition de refroidissement.

*Mots clés: Seychelles, Maquereau, Valeur ajoutée, Océan Indien*

## 1. INTRODUCTION

The Indian mackerel, *Rastrelliger kanagurta*, locally known as "makro dou" is a pelagic shoaling scombroid fish very popular in the Seychelles. It is found in the coastal waters of mainly the granitic islands predominantly during the north east monsoon. Harvesting is done by small fishing vessels operated with outboard engines using small purse seines nets. The average catch is 250–300 tonnes annually, representing between 6-8 percent of total artisanal marine landings (Seychelles Fishing Authority, 2011). More than 90 percent of the catch is landed during the north east monsoon between November and April, a period characterized by calm seas. So far there has been limited interest in trying to export this species either fresh or frozen and according to reliable sources, its perishable nature and low economic value makes it unattractive for the export market.

This fish is a major source of protein in the diet of the ordinary Seychellois families, since it is very cheap during the good season and well appreciated among the locals. Currently, the local market is unable to absorb the catch landed during the good seasons. In addition, the habits of local consumers are such that the mackerel is not appreciated after freezing. Despite its low cost, consumers would purchase only what can be consumed fresh immediately. The average weight of the round fish is around 350 g and it retails in packets of 6 to 8 fish at a cost of SR15.00 to SR50.00 (SR 13.25 = US$ 1.00) depending on availability.

So far little has been done to add value to this product, as a means of increasing its commercialization and improving the earnings of the fishers. Hot smoking is one of the methods of preservation that would add value to the product and present consumers with an alternative option to appreciate the fish. Hot smoking means curing the fish by smoking at a temperature of 60-70 °C at some stage in the process in order to cook the flesh; hot smoked fish products do not require further cooking before consumption (Torry Research Station, 2010). The Codex International Code of Practice for Smoked Fish defines hot smoking as "smoking fish at temperatures and for sufficient period of time to obtain heat coagulation of the protein throughout." However in hot smoking, issues such as the microbial quality, development of histamine and formation of polycyclic aromatic hydrocarbons (PAH) namely benzo-a-pyrene need to be well controlled to prevent the products becoming a hazard to the consumers. The microbial quality will strongly influence the shelf life and also the safety of the product. Scombroid fish such as mackerel, under conditions of temperature abuse is known to produce a significant quantity of histamine due to the presence of an appreciable quantity of histidine in its tissues. PAH can contaminate food during smoking, heating, and drying processes that allow combustion products to come into direct contact with food (Regulations 1881/2006/EC). PAH molecules are produced during pyrolysis of organic material that is used for generation of the smoke, mainly wood. It has been established that the content of carcinogenic PAHs in smoked fish mostly depends on smoking temperature and smoking time.

## 2. OBJECTIVES

The main objectives of this project are:

- To offer local consumers a different way of appreciating mackerel thus increasing its consumption and the intake of protein in the local diet.
- To produce a hot smoked mackerel with an acceptable keeping quality and having a flavour, taste and salt level appreciated by the local consumers.
- To produce hot smoked mackerel with a level of histamine that would not cause adverse health effects to the consumers.
- To provide the mackerel fishers with more opportunities to market their catch during the good season, consequently increasing their earnings and in the longer term improving their livelihoods.

## 3. MATERIALS AND METHODS

The study involved the production of two batches of hot smoked mackerel with the same product formulation. The first batch was used for determining the acceptance of the products by the consumers through the assessment of sensory qualities. The second batch was used for testing the product shelf life extended over a period of thirty two days from production. This was assessed through the determination of the overall microbiological quality and the histamine level after storage under ambient temperature and chill condition.

### Materials

The following materials were required:

- Freshly caught Indian mackerel
- Common salt and molasses for brine preparation
- Plastic tank for brining
- Smoking kiln with technology capable of regulating the temperature.
- Weighing scale to measure the weight of ingredient
- Standard container for volume of water

### Methods

#### Preparation of smoked mackerel

##### Raw material (fish)

Fresh fish were purchased in the morning from the fishing vessel upon landing at Victoria Fishing Port. The fish were caught off-coast of Mahe Island by the small fibreglass boats using small purse seine nets. Upon transportation to the processing plant, cooling in ice slush was immediately done, dropping the temperature from ambient to 2 °C–0.5 °C.

##### Preparation of fillets

The fish were scaled, washed and filleted. As much as practicable, care was taken to avoid damaging the gut, thus preventing contamination of fillets with gut content. Fillets were placed on a clean plastic sheet on the table underneath which sufficient ice was used to maintain the cold chain.

##### Fillets yield

The fish available for the exercise were rather small weighing 250 g on the average. Indian mackerel found along the coast of the Seychelles islands ranges in size from 200 g to 400 g. One of the fillets weighed on average 55g thus giving an average yield of 44 percent.

- Average weight of fish = 250 g
- Average weight of fillets = (2 x 55g) =110 g
- Yield per fish = $\frac{110 \text{ g} \times 100}{250 \text{ g}}$ = 44%

*Brine preparation*

The brine was prepared by dissolving 2 kg of salt in 20 litres of water. This produced a brine concentration of roughly 100 g of salt/litre of water. Using the table below, the brine-ometer (salinometer) reading would be close to **40 degrees**. This concentration has been chosen mainly based on analytical data available on the percentage of salt in several batches of smoked fish, mainly marlin, produced and sold on the local market. The salt level of this product is perceived as generally satisfactory by the local consumers. Strong brine above 80 brine-ometer degrees would reduce immersion time but has the disadvantages that, after the fish are dried, salt can crystallize on the surface of the skin in unattractive white patches (Torry Research Station, 2010). To improve flavour and impart an attractive colour to the finished product, molasses was added to the brine.

**Table 1. Brine strength**

| Brine-ometer degrees | Weight of salt g/litre brine | Salt (lb/gallon) | % salt by weight |
|---|---|---|---|
| 10 | 26.4 | 0.22 | 2.64 |
| 20 | 53.8 | 0.46 | 5.28 |
| 30 | 79.2 | 0.71 | 7.91 |
| **40** | **105.6** | **0.98** | **10.55** |
| 50 | 132.0 | 1.26 | 13.19 |
| 60 | 158.4 | 1.56 | 15.83 |
| 70 | 184.8 | 1.88 | 18.47 |
| 80 | 211.2 | 2.23 | 21.11 |
| 90 | 237.6 | 2.59 | 23.75 |
| 100 | 264.0 | 2.98 | 26.39 |

*Source:* Adam Mariansk, 2011.

*Brining*

The fillets were carefully immersed in the brine tank and left for a period of 15 minutes. Based on the thickness of the mackerel fillet, about 1 cm on average, and the experience of the operator, 15 minutes was deemed sufficient for the brining process. Salt penetrates fish easier in places that are open or cut than through the skin, hence the short brining time used. A stop watch was used to time the process.

*Smoking*

The liquid smoke, product name, "SmoKez Supreme C" is produced by Red Arrow International LLC, a United States- based company specialized in production of condensed natural smoke. The production of liquid smoke involves the drying of hardwood sawdust to optimal moisture level, burning the sawdust to generate the smoke, condensing the smoke with water, phase separation, filtration and containerization. The healthy and flavourful parts of the smoke are water soluble whereas the unhealthy carcinogenic chemical compounds known as polycyclic aromatic hydrocarbons (PAH) and other impurities, are not. These unhealthy components are filtered and removed.

The fillets were laid with the skin facing downward on the stainless steel trays on the rack and placed in the smoking kiln. The smoking kiln, "BASTRAMAT - System liquid smoke technology" has a Microprocessor control MC 500 (Computerized smoking program) consisting of different cycles. This allows the choice of an appropriate cycle based on the type of fish and the sizes of fillets or loins to be smoked and the fat content. In this case the smoking process selected involved the following steps:

- Drying for 30 minutes at 55 °C.
- Smoking for 16 minutes at 60 °C.
- Drying for 15 minutes at 60 °C.
- Smoking 16 minutes at 55 °C.
- Drying 10 minutes at 55 °C.

The drying period is used to toughen the skin and prevent subsequent breakage (Torry Research Station, 2010). It also significantly reduces the moisture content of the fish to improve its keeping quality. The drip liquid - brine and fish juice - is collected in a tray below to prevent it dirtying the floor of the kiln.

*Cooling*

Fish was allowed to cool to room temperature inside the kiln within a period not exceeding one hour. The racks were then placed in a chill room at a temperature of about 0 °C before packaging.

*Packaging*

The smoked fillets were packaged in polystyrene trays, five fillets per tray and wrapped with cling film. The products were packed in carton boxes and stored for one day in the cold room at -20 °C.

**Figure1. Filleting of mackerels**

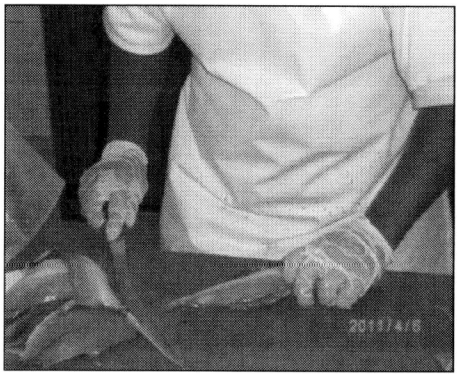

**Figure 2. Mackerel fillets immersed in brine**

**Figure 3. Racking of fillets**

**Figure 4. Smoked fillets packaged in polystyrene trays**

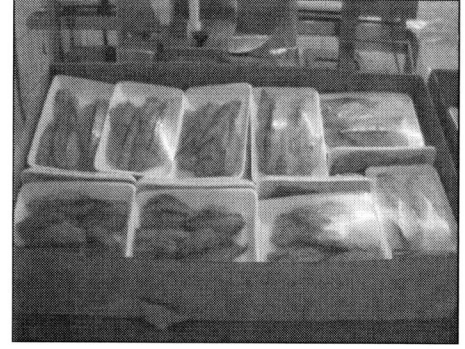

*Sensory assessment*

On the next day the smoked products were distributed to different employees selected from the processing plant and also to all the Fish Inspectors working with the Competent Authority. Each assessor was given a form (Appendix 1) which was designed to rate the sensory quality of the products based on the levels of saltiness, their appreciation of the flavour, the appearance as defined by the colour and their rating of the juiciness or dryness.

***Shelf life assessment based on microbial quality and histamine development.***

A second batch of products was produced using the same formulation and packed in the same manner. This time the products were used for the shelf life test at ambient temperature and under chill condition. The product qualities were assessed based on the microbial count as an indication of spoilage and testing of histamine level.

## 4. RESULTS AND DISCUSSIONS

*Sensory assessment*

Each person selected for the evaluation was briefed on how to proceed with the exercise so that the best result could be obtained. They were asked to taste the fish without the addition of any condiments, sauces or any other food substances to ensure that the true taste of the fish was reflected. A total of nineteen questionnaires were returned following the tasting and evaluation.

As mentioned earlier the aim of the sensory evaluation was to assess if consumers in general would find this product acceptable, therefore giving an indication of success if placed on the market. Below are the results of the sensory assessment extracted from the nineteen respondents to the questionnaires.

**Table 2. Result on assessment for level of salt**

| Criteria | Scores | Percentage |
|---|---|---|
| Acceptable level of salt | 17 | 90% |
| Slightly salty but acceptable | 2 | 10% |
| Too salty, unacceptable | 0 | 0 |

The salt level was found very acceptable by 90 percent of the assessors. Only 10 percent found it slightly salty whereas nobody found it unacceptable due to too much salt. The brine concentration was estimated at around 40 degrees on the brine-ometer (100 g salt per litre) and analysis of the smoked fish revealed a concentration of 1.5 percent salt. The use of salt as an ingredient in smoked fish is mainly for imparting taste and flavour. Although salt has an inhibitory effect on the growth of micro organisms, at a level of 1.5 percent it is obviously inadequate to have an effect on controlling bacterial growth. Preservation of the product depends largely on the synergistic effect of the antibacterial properties of chemicals found in the smoke and the chill storage temperature.

**Table 3. Result of assessment for flavour**

| Criteria | Scores | Percentage |
|---|---|---|
| Pleasant odour and flavour, typical of smoked fish | 19 | 100% |
| Slight odour and flavour of smoked fish | 0 | 0 |
| No odour and flavour of smoked fish | 0 | 0 |

In terms of flavour, 100 percent of the assessors rated the product as having a pleasant flavour, typically that of hot smoked fish. The flavour and odour of a product are highly influential factors for its acceptance by consumers. The flavour is imparted by the presence of compounds found in the smoke.

**Table 4. Result of assessment of colour**

| Criteria | Scores | Percentage |
|---|---|---|
| Light brown, attractive | 18 | 95% |
| Slightly dull | 1 | 5% |
| Dark brown, over smoked, unattractive | 0 | 0 |

The opinion of the assessors was sought where the colour of the product is concerned. An attractive colour has a strong influence on the consumer's perception of a quality product. It will attract the consumer to at least taste the product, whereas a dull and unattractive colour will usually have the opposite effect. As can be seen from the table, the responses from the assessors were very positive with 18 of the 19 (95 percent) rating the product attractive.

**Table 5. Result on assessment of dryness**

| Criteria | Scores | Percentage |
|---|---|---|
| Very juicy | 10 | 53% |
| Medium | 8 | 42% |
| Too dry | 1 | 5% |

The juiciness of the product is related to its softness and wetness. This characteristic will vary from consumer to consumer depending on what extent they appreciate the softness of smoked fish. The drying phase of the smoking cycle is aimed at reducing the amount of moisture in the product and the smoking method used had a good control over the drying period. In general the response from the assessors was good as only one assessor (5 percent) rated the product too dry.

*Result of microbial analysis*

The smoked fish were tested for *APC*, *Yeast and Mould* and *E. coli*. The organisms mentioned are used mainly as indicators of product quality rather than for safety. In effect, microbial quality indicators are spoilage organisms whose increasing numbers results in loss of product quality. (Jays 1992). The products were tested after storage at ambient temperature in the range of 24 °C–28 °C and in the refrigerator where temperature was recorded between 4 °C–6 °C.

**Table 6. Reference Methods for Microbiological Analysis**

| Microbiological Test | Reference Methods |
|---|---|
| *Aerobic Plate Count* | ISO 4833:2003 |
| *E. coli* | ISO 1649-2:2001 (E) |
| *Yeast and Mould* | ISO 7954:1987 (E) |

*Storage at ambient temperature*

Despite the fact that hot smoked fish is not a shelf stable food and should not be ideally stored at ambient temperature, it was found useful to have an idea how well this product will keep at ambient temperature. The storage period was stretched to at least 96 hours (4 days) with samples tested approximately every 24 hours.

As shown in Figure 4, there was slow growth for both the APC and *Yeast and Mould*. Although the temperature was ideal for growth of mesophilic bacteria, the bactericidal property of the chemicals in the smoke, such as the organic acids and aldehydes, and the reduced moisture content in the fish from the drying effect during the smoking process have had an inhibitory effect on bacterial multiplication. The APC increased by almost 2 log counts between 24 hours to 96 hours. The rate of multiplication increased more significantly between 72 and 96 hours where an increase of more than one log cycle was observed. It was noted that the product was not yet decomposed at the end of this period.

The growth of yeast and mould increased tenfold (1 log) from $1.5 \times 10^3$ to $1.5 \times 10^4$ from 24 hours to 48 hours and remained constant until 96 hours. The growth pattern suggests that the logarithmic growth phase was between 24 to 48 hours after which the organisms entered a stationary phase.

**Figure 5. APC and Yeast and Mould growth in hot smoked mackerel at ambient temperature (24 °C–28 °C)**

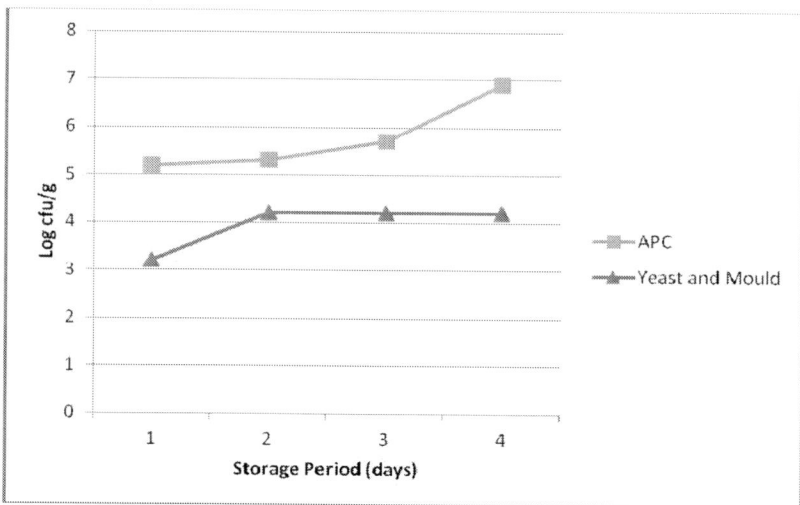

*Storage under chill condition*

The main objective of the chill temperature storage was to assess the keeping quality in terms of the approximate length of time it takes for the smoked mackerel to become spoiled or otherwise considered unfit for consumption under that particular condition of storage. Apart from having an acceptable organoleptic quality, it is important that the products possess a reasonably good keeping quality when kept under chill conditions.

**Figure 6. APC and yeast and mould growth in hot smoked mackerel under chill temperature (4 °C–6 °C)**

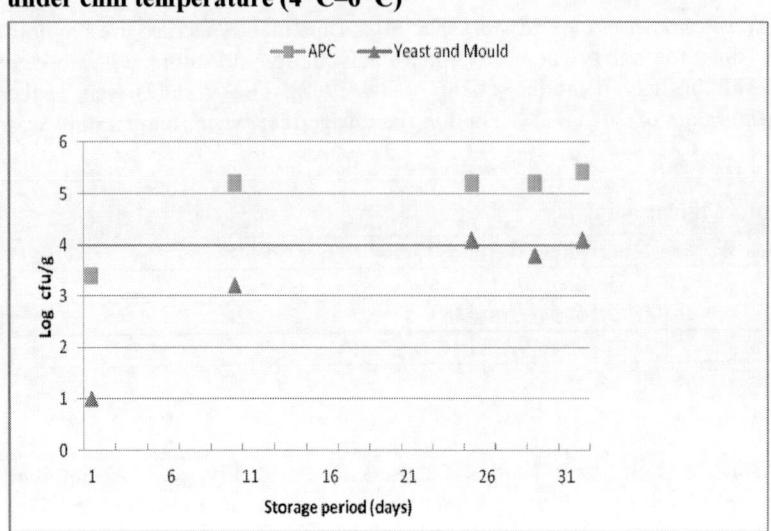

In the case of APC, growth of bacteria was significant on the samples tested on day 1 to day 10 showing an increase of 2 logs or 20 fold. The bacterial count remained uniform for samples tested on the 25th and 29th day and finally an increase in the sample tested on the 32nd day. Although produced under identical condition, the level of contamination with micro organisms of the individual fish fillets could have been different. This was affected by manipulation of the fillet during preparation and also during packaging of the finished products.

The count of yeasts and moulds increased exponentially, noting that on day 1 it was almost undetected and reached $1.5 \times 10^3$ cfu/g on the 10[th] day and $1.5 \times 10^4$ cfu/g on the 25[th] day. At that time there was visible growth of mould on the products and the growth tended to be more intense at the point where there was direct contact between the plastic film covering the trays and the smoked fillets.

The absence of *E. coli* in all samples was a good indication that there was negligible post process contamination. As indicated above, the normal micro flora would be dominated by mesophilic organisms that would have been significantly reduced by the smoking temperature at 60 °C.

### Result of histamine analysis

Mackerel is one of the species belonging to the scombridae family, which is well known to be histamine formers due to the large quantity of histidine found in their muscle tissue. Histamine is produced by the decarboxylation of the histidine in the muscle of this group under conditions of temperature abuse. Histidine decarboxilase is produced by bacteria such as *Morganella morganii, Proteus spp, and Klebsiella spp.* Sufficient levels of histamine may be produced without the product being organoleptically unacceptable, with the result that scombroid poisoning may be contracted from both fresh and organoleptically spoiled fish. (Jay, 1992).

Samples of the products were tested for histamine using the HISTAMARINE, Enzyme Immunoassay Kit Ref. IM2369 method. Tests were done on the first day of production and eventually after one month storage at chill temperature. Histamine level was negligible ranging from <1.00 ppm to a maximum of 1.55 ppm after thirty two days storage at chill temperature. The low level of histamine is not only attributed to the chill storage temperature but also to the effect of the temperature of the smoking process which is expected to have significantly reduced the bacterial count including those producing histidine decarboxylase, most of which have an optimal growth range between 15 °C and 30 °C. The lowest temperature for production of significant levels of histamine was found to be 30 °C for *Hafnia alvei*, and *E. coli*; 15 °C for two strains of *Morganella morganii*; and 7 °C for *Klebsiella pneumonia* (Jay, 1992).

## 5. CONCLUSIONS AND RECOMMENDATIONS

Although a limited number of assessors responded to the questionnaire on the sensory quality of the hot smoked mackerel, the responses received were very positive and encouraging. The scores for level of salt, flavour and colour were all 90 percent and above, whereas only the degree of dryness was rated lower. Several assessors verbally expressed their satisfaction with regards to the quality of the product, especially the taste. This result shows that there are potential markets for hot smoked mackerel in the Seychelles, and this needs to be exploited. Better information can be obtained with regards to consumer preference by doing more trials based on varying the level of salt and the length of smoking process. However one of the most important factors to consider would be the cost benefit of such a venture. Although this had not been one of the objectives of this project, any commercial venture will have to firstly assess the cost of production in relation to the benefits to be expected to make the project commercially viable. The seasonal nature of Indian mackerel in the Seychelles is an important factor that will influence a potential processor to concentrate on the smoking of other types of fish.

The shelf life test performed shows that the products should not be left for more than two days at ambient temperature, unless they are to be consumed within that period, as there was substantial increase in both the APC and the yeast and mould count. These organisms are more of spoilage types having a direct effect on the shelf life but not necessarily having public health significance.

At ordinary refrigerator temperature (4 °C–6 °C), as would be seen in many homes, yeast and mould could be a real problem as after ten days the count had reached $1.5 \times 10^3$ cfu/g. Spoilage with mould growth, however, was visible after 10 days storage despite the fact that the APC count ($1.5 \times 10^5$ cfu/g) was still within an acceptable range. The absence of *E. coli* is indicative that there was no direct or indirect post process contamination of faecal origin. The chill storage temperature should be maintained around 0 °C to 1 °C to have a better control of the growth of yeast and mould.

It is strongly recommended that a further trial is done to include the testing of the final product for levels of benzo(a)pyrene, one of the carcinogenic PAHs. The maximum level for this contaminant is set at 5.0µ/kg, wet weight basis (Regulation 1881/2006/EC). Assessing the sensory attributes can also be extended at market level to obtain the opinion of the general public considering that the rating given by the small group of assessors, especially from the processing plant, who took part in this exercise, may be perceived as bias and also not a representative sample. The use of vacuum packaging can also be tried in an effort to increase the shelf life by reducing spoilage with aerobic microorganisms.

## 6. REFERENCES

**Commission Regulations.** (EC) 1881/2006. Setting maximum levels for certain contaminants in foodstuff.
**FAO.** 1995. **Quality and** Quality Changes in Fresh Fish, *FAO Fisheries Technical Paper* 348. FAO, Rome.
**FAO.** 2005. FAO yearbook of fishery statistics. Commodities. Vol. 97.Rome: FAO. 235 p.
**FAO.** 2008. http://www.fao.org/docrep/013/i1890t/i1890t.pdf
**Jay, J. M.** 1992. Modern Food Microbiology. pp 660–661.
**Marianski, A.** The Amazing Mullet, 2011, Chapter 5, pp 56. Bookmagic LLC. www.bookmagic.com. 122p.
**Seychelles Fishing Authority.** 2011. Annual Reports, 2005–2009.

<u>**Questionnaire used for the sensory evaluation of the smoked mackerel**</u>

Form Number:
Date:
Time:

As a consumer, please assess the product based on the following criteria: Taste, saltiness, colour, juiciness. Provide your rating by inserting a tick next to the characteristic you select for each criterion.

| Criteria Assessed | Characteristic of each criteria being assessed | | |
|---|---|---|---|
| Saltiness | Acceptable level of salt | Slightly salty but acceptable | Too salty, unacceptable |
| Flavour | Pleasant odour and flavour, typical of smoked fish | Slight odour and flavour of smoked fish | No odour and flavour of smoked fish |
| Colour | Light brown, attractive | Slightly dull | Dark brown, over smoked, unattractive |
| Juiciness | Very juicy | Medium | Too dry |

Thank you for your honest opinion.

**Result of microbiological analysis of hot smoked mackerel stored at ambient temperature**

| Manufacturing date | Storage period (hours) | Storage temperature | APC (cfu/g) | Yeast and Mould (cfu/g) | Escherichia coli (cfu/g) |
|---|---|---|---|---|---|
| 02/06/2011 | 24 | 24–28 °C | $1.5 \times 10^5$ | $1.5 \times 10^3$ | <10 |
|  | 48 | 24–28 °C | $1.8 \times 10^5$ | $1.5 \times 10^4$ | <10 |
|  | 72 | 24–28 °C | $5.7 \times 10^5$ | $1.5 \times 10^4$ | <10 |
|  | 96 | 24–28 °C | $9.9 \times 10^6$ | $1.5 \times 10^4$ | <10 |

**Result of microbiological analysis of hot smoked mackerel stored at chill temperature**

| Manufacturing date | Storage period (days) | Storage temperature | APC (cfu/g) | Yeast and Mould (cfu/g) | Escherichia coli (cfu/g) |
|---|---|---|---|---|---|
| 02/06/2011 | 1 | 5–8 °C | $1.5 \times 10^3$ | <10 | <10 |
|  | 10 |  | $1.5 \times 10^5$ | $1.5 \times 10^3$ | <10 |
|  | 25 |  | $1.5 \times 10^5$ | $1.5 \times 10^4$ | <10 |
|  | 29 |  | $1.5 \times 10^5$ | $7.1 \times 10^3$ | <10 |
|  | 32 |  | $2.6 \times 10^5$ | $1.5 \times 10^4$ | <10 |

# FUMER SAINEMENT ET MANGER DU POISSON SAIN: PERFORMANCE DU SYSTEME FAO-THIAROYE, UN MODÈLE AMÉLIORÉ DE FUMOIR, AXÉE SUR LA MAÎTRISE DES HYDROCARBURES AROMATIQUES POLYCYCLIQUES (HAP)

## [SMOKING HEALTHY AND EATING HEALTHY FISH: PERFORMANCE OF THE FAO-THIAROYE SYSTEM, AN IMPROVED DESIGN OF KILN WITH PARTICULAR FOCUS ON THE CONTROL OF POLYCYCLIC AROMATIC HYDROCARBONS (PAH)]

by/par

Oumoulkhairy Ndiaye[1] andYvette Diei Ouadi

**Résumé**

Un cadre de coopération entre la FAO et le Centre National de Formation des Techniciens en Pêche et Aquaculture (CNFTPA) du Sénégal a été établi depuis la réunion d'experts organisé à Agadir (Maroc) en 2008, pour promouvoir la sécurité sanitaire tout en sécurisant les revenus des entreprises à petite et moyenne échelle. Ceci a conduit à un premier prototype de fumoir se focalisant sur la réduction du dépôt d'hydrocarbures aromatiques polycycliques (HAP), qui fut testé avec succès dans un programme financé par l'UE. Un accord fut ensuite signé entre la FAO et CNFTPA pour une amélioration plus poussée, testant différents combustibles (bois, charbon, bourres de coco, gaz) et artifices-un générateur de fumée et un collecteur de graisse- pour réduire davantage le facteur de formation d'HAP.

Cet effort collaboratif a été concrétisé par le développement et l'essai du système FAO-Thiaroye de fumage (SFTF). Les données montrent que les produits finis sont de meilleure qualité en termes sensoriel (texture, apparence, saveur) et chimique. Dans le cas particulier des composés HAP, le niveau de benzo(a)pyrène dans les méthodes améliorées de fumage à chaud les plus utilisées en pêche artisanale dépasse en général la limite réglementaire de 5 µg/kg (5ppm), alors qu'il oscille entre 0,15 et 1,40 ppm dans le mode de fumage à combustion complète du charbon lors de la cuisson avec un système de fumage indirect en utilisant les artifices. Le type de bois influence significativement la teneur en benzo(a)pyrène.

Les autres avantages de ce système sont la réduction d'inhalation de fumée et d'exposition à la chaleur par les transformateurs, de la pénibilité du travail, l'amélioration des revenus avec la baisse de consommation en combustible, mais aussi celle des couts liés à la non-conformité et non qualité.

*Mots clés: Fumage, Benzopyrène, Santé publique, FAO-Thiaroye, Senegal*

**Abstract**

A collaborative framework between FAO and the Centre National de Formation des Techniciens en Pêche et Aquaculture (CNFTPA) in Senegal was established since the 2008 Expert meeting organized in Agadir (Morocco), to ensure food safety while securing incomes from small and medium scale enterprises. This has led to the development of a preliminary sketch of a smoking kiln tested with success within an EU-funded programme. An agreement was then signed with FAO and CNFTPA for further improvement, testing different fuels (wood, charcoal, coconut, gas) and devices- a smoke generator and a fat collector- to lessen the PAH formation factor.

This cooperative effort was materialized in the development and testing of the FAO-Thiaroye fish smoking system. The data reveal that the end-products are of better quality in terms of sensory (texture, appearance, taste) and chemical attributes. In the particular case of PAH compounds, the benzo(a)pyrene level in existing improved kilns for hot smoking is generally well beyond the regulatory limits of 5 µg/kg (5ppm) while it ranged between 0.15-1.40 ppm in the new prototype, with different devices used. The type of firewood also significantly influences the level of benzo(a)pyrene.

---

[1] CNFTPA, BP 2241, Dakar, Sénégal. oumoulinda@yahoo.fr

Additional advantages of this system are a reduced inhalation of smoke and reduced exposure to heat during smoking by fish processors. To this occupational health advantage should be added the improved incomes due to reduction in fuelwood used and in non-compliance/non quality costs.

*Key words: Smoking, Benzopyrene, Public health, FAO-Thiaroye, Senegal*

# 1. INTRODUCTION

Le fumage à chaud du poisson est une activité importante dans les petites et moyennes unités post-capture du poisson des pays en développement. En Afrique en particulier une diversité de fours de fumage, du traditionnel aux modèles plus améliorés de fours, sont ainsi utilisés.

Si ce procédé a été largement utilisé pendant des décennies sans aucun souci majeur pour la sécurité sanitaire des produits générés, il est depuis les 5 dernières années sous le feu de critiques suite au progrès enregistré dans les méthodes analytiques, et l'évolution de la science alimentaire. Cette tendance est plus perceptible au niveau du marché de l'Union européenne, un créneau particulier pour ces produits fumés à chaud dont la plupart sont destinés à la diaspora africaine. En effet des dispositions sanitaires spécifiques des composés cancérigènes appartenant au groupe des hydrocarbures aromatiques polycycliques (HAP), principalement le benzo(a)pyrène ont été prises depuis 2006. Il convient de rappeler que dans le règlement de la Commission CE 1881/2006, l'UE a établi un niveau maximum acceptable de 5 ppm pour le benzo (a) pyrène dans la chair de poisson fumé, 2 ppm pour le poisson non fumé, 5 ppm pour les crustacés et céphalopodes non fumés, et 10ppm pour les bivalves vivants .

Comme conséquence immédiate de ces mesures, l'exportation des produits de pêche et d'aquaculture fumés vers le marché européen a été prise dans un cycle préoccupant de rétentions et rejets de produits ou des interdictions d'exportation d'unités de transformation vers ce débouché commercial. A titre d'illustration, au Sénégal, au Togo, en Côte d'Ivoire, de nombreuses entreprises exportatrices ont été suspendues, entrainant une perte de revenus potentiels au détriment de milliers d'opérateurs dépendant du poisson comme principal moyen de subsistance. De plus, les informations en notre possession lors de la finalisation de ce document convergent vers de nouvelles dispositions plus contraignantes qui devraient s'appliquer dès le 1$^{er}$ septembre 2012. Un projet de règlement modifiant le règlement (CE) n ° 1881/2006 portant sur les teneurs maximales admissibles de HAP dans les denrées alimentaires est en cours d'élaboration. Non seulement à partir du 1er Septembre 2012 il serait tenu compte de 3 autres HAP comme marqueurs, le benzo(a)anthracène, benzo(b)fluoranthène et chrysène, la limite réglementaire en benzo(a)pyrène serait ramenée dès 2014 à 2 ppm au lieu de 5 ppm autorisée présentement.

Que ce soit une dynamique guidée par le consommateur ou émanant de la volonté des pouvoirs publics pour protéger leur population par le biais de réglementations sanitaires et normes de plus en plus strictes, le besoin de produits sains et nutritifs appelle à un changement drastique dans la façon dont le fumage à chaud des produits de la pêche et d'aquaculture est conduite, pour répondre aux préoccupations de sécurité sanitaire, environnementales et de santé des transformateurs.

Des contributions portant sur la réduction de la teneur en HAP dans les produits fumés ont été discutées lors des réunions des experts en technologie, utilisation et assurance qualité du poisson organisés par la FAO, à Bagamoyo, Tanzanie, du 14 au18 novembre 2005 et à Agadir du 24 au 28 Novembre 2008. L'expérimentation du fumoir Bidule mis au point par l'Institut de Recherches Technologiques (IRT) du CENAREST du Gabon appuyée par les analyses effectuées a donné des résultats encourageants quant à l'amélioration de la qualité, le goût et la réduction du benzo (a) pyrène dans les produits fumés. Pour une meilleure adoption de ce fumoir de faible capacité et de maniement pas tout à fait aisé en milieu traditionnel, il a été recommandé que des études plus approfondies soient effectuées sur des modèles plus appropriés en fumage traditionnel en Afrique, sur les performances techniques, socio-économiques et une réduction significative de contaminants nocifs notamment les HAP des produits fumés.

C'est dans ce cadre qu'un protocole d'accord entre la FAO et le département de technologie du CNFTPA du Sénégal a été établi en vue de poursuivre les activités de recherche déjà entamées dans d'autres instituts.

Cet effort collaboratif a abouti à la conception du système FAO-Thiaroye de fumage basé sur le procédé indirect de fumage avec une plaque à graisse, et un accent sur l'utilisation de la braise. La présente contribution présente les résultats des essais effectués (n'incluent pas pour l'instant l'expérimentation), et donne une orientation sur l'optimisation de l'utilisation de cette technique améliorée.

## 2. MATÉRIEL ET MÉTHODE

### Matériel

Il est constitué par le fumoir, de claie et de parties annexes. Un système de filtrage de la fumée a été confectionné et incorporé dans le tuyau permettant de retenir les particules nocives de la fumée.

**Photo 1 et Figure 1. Vue générale du fumoir indirect (four parpaing du CNFTPA réfectionné connecté au générateur de fumée)**

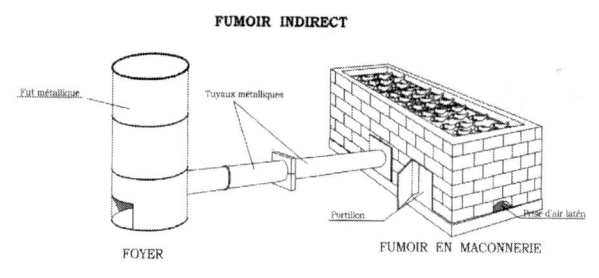

**Photo 2. Claies de fumage en matériaux inoxydables**

### La réfection du four parpaing du CNFTPA

Elle a pris 5 jours avec une main d'œuvre locale de 2 personnes.

### La confection du générateur de fumée

Le générateur de fumée, appareil servant à la production de fumée est composé d'un fût métallique relié au casier à filtre à l'aide d'un tuyau métallique de 26 cm de diamètre et de longueur 146 cm. Le système de fumage est indirect.

**Figure 2 et Photo 3. Vue générale du générateur de fumée**

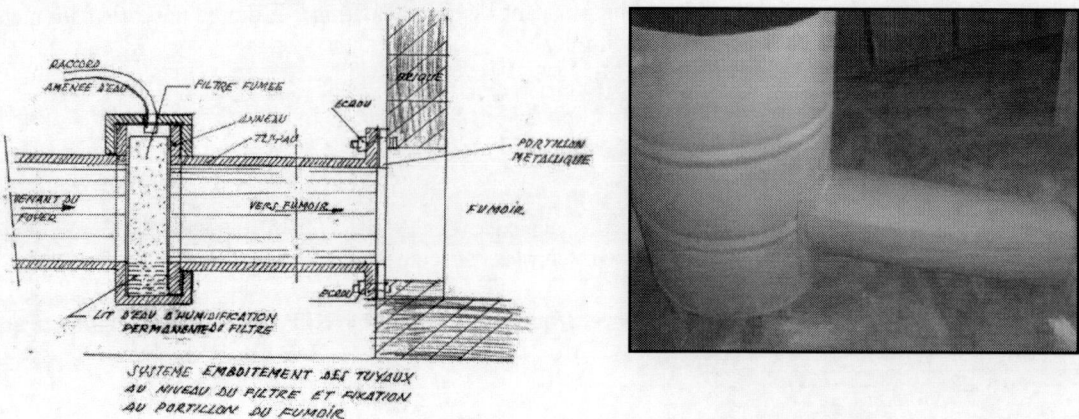

**Main d'œuvre:** 2 personnes
**Durée de confection:** 3 jours

Le coût moyen de confection du générateur de fumage indirect est de 200$ et sa durée de vie est au moins de 5 ans.

Les figures ci-dessous présentent les détails de la confection du prototype y compris les principales étapes de la réalisation:

**Figure 3. Le générateur de fumée raccordé au four**

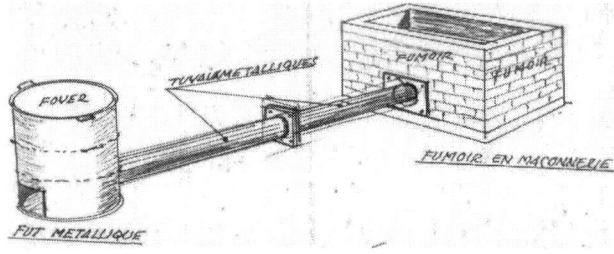

*Le casier métallique* sert à insérer le filtre de manière à faciliter son humidification et ainsi retenir l'exsudat.

**Figures 4 et 5. Les plans détaillés du casier à filtre et du système de filtrage**

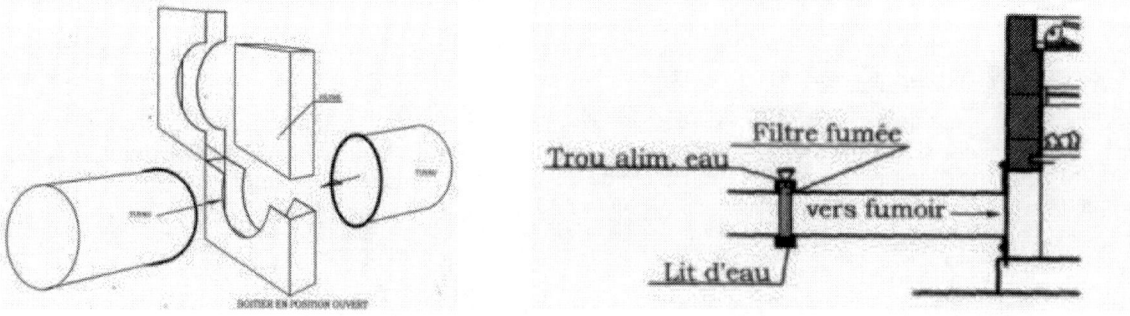

**Photo 4. Casier métallique à filtre**

**Photos 5, 6 et 7. Réceptacle de graisse/plaque à graisse**

 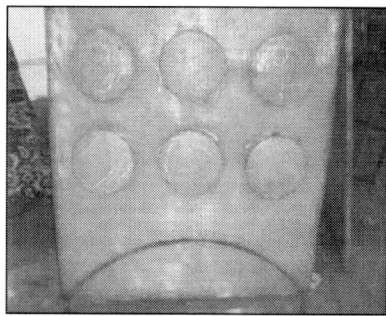 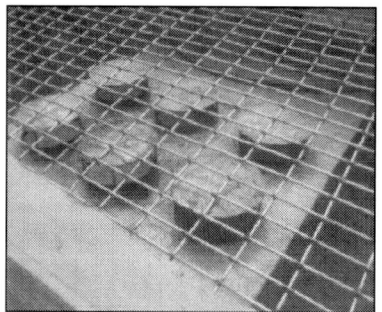

La confection *d'un réceptacle/plaque métallique **à graisse*** de 1 m de longueur et de 1 m de largeur avec des trous de 10 cm de diamètre couverts. La plaque conçue pour que la graisse ne tombe pas dans le feu à travers les trous est placée légèrement inclinée à 5 cm de la claie chargée en poisson. Pour éviter toute combustion de la graisse pouvant retomber sur le feu conduisant au dépôt du contaminant benzopyrène, la graisse et l'exsudat provenant du poisson durant la cuisson à la braise rouge sont recueillis sur cette plaque et évacués vers l'extérieur à l'aide deux tuyaux. Ces matières grasses pourront être purifiées en huile pour consommation humaine directe ou friture ou transformées en savon.

Le coût moyen de la plaque à graisse est de 100 $ et la durée d'amortissement est en moyenne de 5 ans.

*Des claies confectionnées en matériaux inoxydables.* Les couvercles ont été enduits de peinture alimentaire antirouille. Le fourneau à braise sert de source de chaleur lors du fumage mais aussi pour le séchage mécanique.

**Photos 8, 9 et 10. Fourneau pour la braise**

Le coût moyen du fourneau est de 140$.

Pour mener à bien les essais de fumage, le petit matériel (balance, thermomètre à sonde, caisse en polystyrène) a été utilisé.

*Combustibles utilises*

Quatre différents combustibles parmi les plus utilisés au Sénégal (charbon, bourres de coco, filao et *Cordyla pinnata*) ont été utilisés dans les essais en vue de les comparer, étant que la génération de HAP dans les produits fumés à chaud varie selon l'essence d'arbre.

**Photos 11 et 12.** *Cordyla pinnata* **(dimb)**

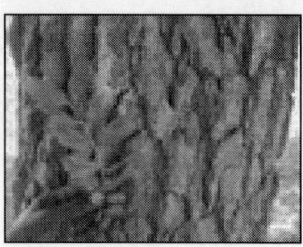 

**Nom scientifique:** *Cordyla pinnata* ; **Synonyme:** *Calycandra pinnata* ; **Nom vernaculaire:** Dimb/poirier du Cayor ; **Famille:** Césalpinacées

**Informations supplémentaires:** Arbre moyen, bois jaune, dur, lourd qui se prête au fumage L'arbre *Cordyla Pinata* est très menacé du fait de la fabrication de tam-tam traditionnels (« Djembés »)

**Photos 13 et 14. Filao et feuilles mortes du filao utilisées pour la cuisson**

 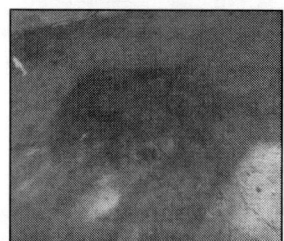

**Nom scientifique:** *Casuarina equisetifoLia*; **Nom vernaculaire:** Filao; **Famille:** Casuarinacées

**Informations supplémentaires:** Ayant l'aspect des tamaris et de certains conifères, bon fixateur de dunes, utilisé pour la reforestation grâce à sa croissance rapide, « arbre de fer » pour la dureté de son bois. Arbre apparemment résineux, l'écorce fournit un tannin (substances naturelles phénoliques), graine toxique

**Photos 15 et 16. Bourres de coco et Cocos nucifera**

**Nom scientifique :** *Cocos nucifera*; **Fruit :** Noix de coco

**Bourre de coco**: déchet de la noix de coco

**Informations supplémentaires :** Palmier présent dans toute la zone intertropicale humide. Surtout cultivé le long des côtes, mais n'y reste pas confiné. La longévité de la plante dépasse un siècle. Les bourres de coco sont utilisées comme combustible familial et les feuilles mortes comme paillage dans les plantations. Aucun aspect de phytotoxicité n'a été relevé dans la bibliographie.

*Méthode*

Vingt quatre (24) séries successives d'essais de cuisson du poisson avec quatre combustibles différents ont eu lieu au CNFTPA durant l'année 2011. Le tableau 1 présente les différents types d'essais et les échantillons prélevés pour analyse.

**Tableau 1. Plan des essais de fumage**

| Essais/Combustibles | ESSAIS | Nombres d'essais | Espèce de poisson | Nombres d'échantillons |
|---|---|---|---|---|
| **Dimb** (*Cordyla pinnata*) | Cuisson- fumage direct sans plaque | 3 | Machoiron | 1 |
| | Cuisson- fumage direct avec plaque | 3 | Machoiron | 1 |
| | Cuisson avec plaque et fumage indirect | 3 | Machoiron | 2 |
| **Bourres de coco** | Cuisson fumage direct sans plaque | 3 | Machoiron | 1 |
| | Cuisson avec plaque et fumage indirect | 3 | Machoiron | 1 |
| **Braise** | Cuisson avec plaque et fumage direct | 1 Machoiron 2 Machoiron 3 Sardinelle | Machoiron Sardinelle | 2 (1 échantillon machoiron et 1 échantillon sardinelle) |
| | Cuisson avec plaque et fumage indirect | 1 Machoiron 2 Machoiron 3 Sardinelle | Machoiron Sardinelle | 2 (1 échantillon machoiron et 1 échantillon sardinelle) |
| **Filao** (*Casuarina equisetifoLia*) | Cuisson- fumage direct sans plaque | 3 | Machoiron | 2 |
| | Cuisson- fumage direct avec plaque | 3 | Machoiron | 2 |
| **Marché** (combustible et méthode de fumage inconnus) | | | Machoiron | 1 |
| **TOTAL** | | **24** | | **15 échantillons** |

**Figure 6. Diagramme de flux des produits de pêche fumés à chaud**

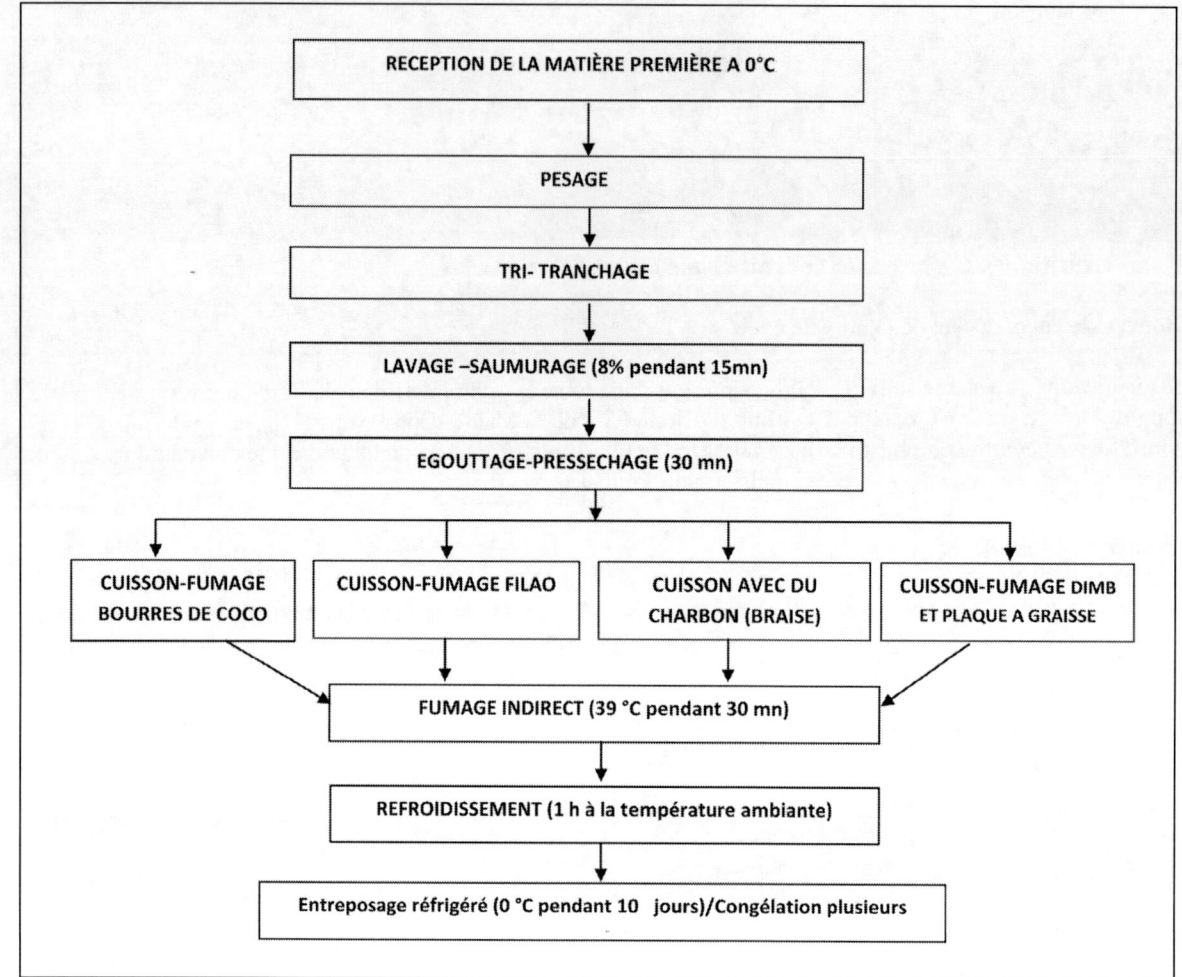

*Cuisson avec les différents combustibles*

Deux cent (200) kg de machoiron et 20 kg de sardinelles de bonne qualité ont été utilisés durant les essais.

La cuisson s'est effectuée séparément avec le charbon et les trois différents bois à combustion incomplète.

Aucun dégagement de flamme et de fumée n'a été décelé durant la cuisson du poisson à la braise. Le concept « zéro flamme et zéro fumée » a été respecté. Par contre, une fumée dense de couleur noirâtre est observée avec le dimb, et beige/incolore avec le fumage à l'aide du filao.

Afin d'évaluer l'effet de la plaque à graisse utilisée comme barrière entre les poissons et la source de chaleur ou des flammes, des essais ont été réalisés avec le dimb, le filao et les bourres de coco.

En début de cuisson, les poissons sont remués régulièrement afin d'éviter qu'ils ne se collent sur les claies et obtenir ainsi un produit homogène.

Après la cuisson, les poissons sont récupérés et transférés dans une autre claie destinée uniquement pour la fumaison et séchage.

*Fumage indirect*

Le même prototype et le même système de fumage précédemment illustré sont utilisés pour fumer les poissons cuits à combustion incomplète avec le filao, les bourres de coco et le cordylla pinnata pendant 30 mn. La durée de fumage a été réduite étant donné que les produits étaient déjà fumés lors de la cuisson.

**Photos 17 et 18. Le système de filtrage avec un exsudat après une session de fumage**

 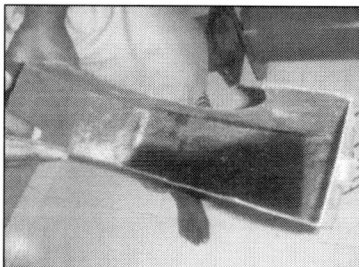

*Maîtrise des opérations de production*

Les essais réalisés ont permis de paramétrer les opérations de production (température, durée, quantité de combustible, teneur en sel).
Le filtre à la texture d'une éponge placé au niveau du casier métallique retient les résidus toxiques et acides goudronneux qui communiquent au poisson fumé une couleur noirâtre et une saveur de produit chargé particulièrement en HAP.

*Evaluation sensorielle des produits*

Un panel constitué de professionnels halieutes du CNFTPA et d'opératrices/transformatrices de poisson a conduit les tests de dégustation des différents produits. Les vingt cinq panélistes (25) ont reçu chacun deux fiches où sont énumérés les paramètres á évaluer (peau, couleur, odeur, goût, texture et acceptabilité) et la notation sur une échelle de 10 à 0 (de très bon à mauvais).

*Recherche du benzo(a)pyrène*

Des échantillons (Plan d'échantillonnage, Directive 2005/10/CEE)[2] de 300 grammes chacun prélevés dans les essais effectués, ont été emballés individuellement dans du papier aluminium et hermétiquement fermés dans des sachets en plastique puis congelés pour être acheminés au laboratoire central pour l'hygiène alimentaire et Agro-Industrie (LCHAI) du Laboratoire national d'appui au développement agricole (LANADA) à Abidjan en Côte d'Ivoire. Ce dernier a conduit la phase de lyophilisation en collaboration avec le laboratoire du Centre Suisse de Recherche Scientifique de Côte d'Ivoire puis procédé à la recherche du benzo (a) pyrène, marqueur de HAP.

*Le plan d'échantillonnage pour le poisson fumé*

Le plan d'échantillonnage est celui de la Directive 2005/10/CEE (Tableau 2).

**Tableau 2. Nombre minimal d'échantillons élémentaires à prélever sur le lot**

| Poids du lot (en kg) | Nombre minimal d'échantillons élémentaires à prélever |
|---|---|
| < 50 | 3 |
| 50 à 500 | 5 |
| > 500 | 10 |

---

[2] Directive 2005/10/CE de la commission du 4 février 2005 portant fixation des modes de prélèvement d'échantillons et des méthodes d'analyse pour le contrôle officiel des teneurs en benzo(a)pyrène des denrées alimentaires.

Les échantillons élémentaires ont un poids semblable, chacun pesant au moins 100 g et formant un échantillon global d'au moins 300 g, prélevé de manière à assurer la représentativité du lot à contrôler.

La méthode d'analyse s'est conformée à la norme ISO/DIS 15753 avec l'utilisation de HPLC avec détection UV – Vis.

## 3. RESULTATS ET DISCUSSIONS

### *Performances techniques et économiques*

Sur le plan de la gestion de l'environnement, la cuisson avec le charbon et les bourres de coco comparée à la méthode traditionnelle (dimb et filao), réduit la pollution par la fumée et prédispose moins les transformateurs à certaines maladies pulmonaires. La consommation en combustible par kg de poisson est réduite (0,6/1 pour le dimb, 0,8/1 pour le filao et 1/1 pour les bourres de coco) avec l'utilisation du four amélioré. Les feuilles mortes de filao s'enflamment plus vite que le dimb et dégagent beaucoup de fumée (claire) mais emmagasinent par la suite de la chaleur. Les bourres de coco produisent du feu à une hauteur ne dépassant pas 50 cm et sont une source de rétention de chaleur mais la fumée dégagée était supportable.

**Tableau 3. Amortissement du matériel**

| Matériel | Valeur en $ | Durée de vie | Amortisse-ment annuel en $ | Amortisse-ment mensuel en $ | Nombre moyen de sessions mensuel | Amortisse-ment par session en $ |
|---|---|---|---|---|---|---|
| Four parpaing de capacité de 400 kg | 600 | 10 ans | 60 | 5 | 32 | 0,16 |
| Fourneau | 140 | 5 ans | 28 | 2,33 | 32 | 0,07 |
| Plaque à graisse | 100 | 5ans | 20 | 1,7 | 32 | 0,05 |
| Générateur de fumée | 200 | 5 ans | 40 | 3,33 | 32 | 0,1 |
| **Total** | | | | | | **0,38** |

Les paramètres identifiés permettent de comparer les procédés de production de poisson fumé. La variation des modes de cuisson et l'utilisation de différents types de combustibles prouvent que la cuisson à la braise est plus aisée que celle avec le bois (Tableau 4).

La durée de cuisson est plus longue avec les bourres de coco comparée au dimb qui est de 3h30. La réduction en eau du poisson avec les bourres de coco est lente durant les deux premières heures de cuisson.

Le ramassage des filaos et des bourres a contribué à réduire les coûts de production et à rendre l'environnement plus propre.

Les tableaux 5 à 8 présentent les éventuels comptes d'exploitation selon le type de combustible utilisé.

**Tableau 4. Comparaison des procédés de production de poisson fumé**

| Paramètres | Dimb sans plaque à graisse | Dimb avec plaque à graisse | Filao | Bourres de coco | Braise avec plaque à graisse | Braise sans plaque à graisse |
|---|---|---|---|---|---|---|
| Système de cuisson | Combustion incomplète | Combustion incomplète | Combustion incomplète | Combustion incomplète | Combustion complète | Combustion complète |
| Distribution de la chaleur | Hétérogène | Hétérogène | Hétérogène | Plus ou moins Hétérogène | Homogène | Homogène |
| Température de cuisson | 90 °C | 85 °C | 85 °C | 80 °C | 85 °C | 85 °C |
| Durée de cuisson | 3h30 | 4h30 | 4h30 | 5h | 4h45 | 4h30 |
| Durée de production | 5h05 | 6h05 | 6h05 | 6h35 | 7h35 | 7h20 |
| Contrôle des opérations de production | Difficile à cause de la fumée et les flammes | Difficile à cause de la fumée et les flammes | Difficile à cause de la fumée plus dense et les flammes | Moins difficile, flammes moins hautes | Facile | Facile |
| Qualité finale du poisson fumé | Peu apprécié | Très peu appréciée | Peu apprécié | Appréciée | Très appréciée | Appréciée |
| % en perte d'eau finale | 51% | 46% | 45% | 44% | 49% | 55% |
| Rendement | 49% | 54% | 55% | 56% | 51% | 45% |
| Degrés de sécheresse | Sec | Légèrement mou | Légèrement mou | Légèrement mou | Légèrement mou | Trop sec |
| Consommation en combustible (ration combustible/ | 0,6/1 | 0,8/1 | 0,8/1 | 1/1 | 0,6/1 | 0,4/1 |
| Durée de conservation | Longue | Courte | Courte | Courte | Courte | Longue |

**Tableau 5. Compte d'exploitation de production de machoiron cuit à la braise ensuite fumé indirectement**

| Charges pour une session de fumage | | |
|---|---|---|
| **Rubriques** | **Quantité** | **Coût total $** |
| Achat Poisson frais | 50 kg | 50 |
| Charbon | 15 kg | 6 |
| Sel | 1 sachet | 0,2 |
| Bourres de coco | 1/2 sac | 0,5 |
| Main d'oeuvre | 1 jour X 2 personnes | 4 |
| Transport des intrants | Forfait | 5 |
| Produits de nettoyage | 1 bouteille | 1 |
| Amortissement matériel | 1 session | 0,33 |
| Entretien matériel | | 1 |
| Coût total des charges | | 68,03 |
| Total prix de vente | 50 kg (rendement 50%) | 150 |
| Bénéfice net | | **81,97** |
| Bénéfice par kg | | **1,64** |

*1$ est échangé à environ 500 FCFA.*

*NB: Le four a la capacité de produire 200 kg en une session de durée de 6 heures. En une journée de production, au moins 3 sessions pourraient être réalisées avec 600 kg de poisson frais fumés.*

Quoique l'investissement initial soit relativement important pour le four et les accessoires qui est d'environ 600$, les frais d'amortissement sont faibles et même le système pourrait être complètement amorti en 15 sessions de production avec un peu d'effort.

**Tableau 6. Compte d'exploitation avec les bourres de coco**

| Charges pour une session de fumage | | |
|---|---|---|
| **Rubriques** | **Quantité** | **Coût total en $** |
| Achat Poisson frais | 50 kg | 50 |
| Sel | 1 sachet | 1 |
| Bourres de coco | 2 sacs | 2 |
| Main d'œuvre | 1 jour x 2 personnes | 4 |
| Transport des intrants | FF | 5 |
| Produits de nettoyage | 1 bouteille | 1 |
| Amortissement | | 0,26 |
| Entretien matériel | | 1 |
| Coût total des charges | | 64,26 |
| Total prix de vente | 30 kg | **120** |
| Bénéfice net | | **55,74** |
| Bénéfice par kg | | **1,4** |

**Tableau 7. Compte d'exploitation avec le dimb**

| Charges pour une session de fumage | | |
|---|---|---|
| **Rubriques** | **Quantité** | **Coût total en $** |
| Achat Poisson frais | 50 kg | 50 |
| Sel | 1 sachet | 1 |
| Dimb | 1 tas | 5 |
| Main d'oeuvre | 1 jour X 2 personnes | 4 |
| Transport des intrants | Forfait | 5 |
| Produits de nettoyage | 1 bouteille | 1 |
| Amortissement | | 0,26 |
| Entretien matériel | | 1 |
| Coût total des charges | | 67,26 |
| Total prix de vente | 30 kg | 120 |
| Bénéfice net | | **52,74** |
| Bénéfice par kg | | **1,08** |

**Tableau 8. Compte d'exploitation avec le filao**

| Charges pour une session de fumage | | |
|---|---|---|
| **Rubriques** | **Quantité** | **Coût total en $** |
| Achat Poisson frais | 50 kg | 50 |
| Sel | 1 sachet | 1 |
| Filao | Ramassage | - |
| Main d'oeuvre | 1 jour x 2 personnes | 4 |
| Transport des intrants | Forfait | 5 |
| Produits de nettoyage | 1 bouteille | 1 |
| Entretien matériel | | 1 |
| Amortissement | | 0,26 |
| Coût total des charges | | 62,26 |
| Total prix de vente | 30 kg | 120 |
| Bénéfice net | | **57,74** |
| Bénéfice par kg | | **1,8** |

Les bénéfices pour le poisson cuit à la braise est plus élevé (1,7 $/kg) du fait de sa bonne qualité, de sa couleur et de son odeur; pour le dimb, il est de 1,8 $ et les bourres de coco 1,4 $. La durée de fumage offre des possibilités de réaliser deux à trois sessions de fumage par jour. Il importe aussi de relever des contraintes lies à l'utilisation de bois comme combustible:

- La surveillance en permanence du feu dégageant une fumée très dense pour le dimb et surtout le filao en vue d'éviter la carbonisation des produits;
- Beaucoup de dégagement de fumée et de flammes avec des risques de santé publique, de pollution de l'environnement et d'incendies surtout avec le dimb et le filao.

### *Appréciation organoleptique*

Lorsqu'on compare le temps de fumage, la teneur en eau du produit fini, la qualité du poisson fumé on voit que les méthodes de cuisson sans dégagement de flamme et de fumée (braise) donnent des résultats satisfaisants. Les poissons fumés cuits à la braise sont secs et se conservent facilement pendant un temps relativement long, tandis que les produits légèrement mous sont aussi appréciés, ont un rendement plus élevé (dimb avec plaque à graisse, filao, bourres de coco) mais se conservent difficilement.

Selon les résultats de l'évaluation organoleptique, les poissons cuits avec le dimb ont une couleur plus sombre et un goût apparemment amer. Les machoirons braisés ou cuits avec les bourres de coco et les filao sont très appréciés.

**Photos 19 et 20. Les panelistes en séance d'évaluation organoleptique de produits finis**

L'analyse des résultats montre une corrélation significative entre les modes de fumage et la qualité des produits.

Les machoirons et les sardinelles cuits à la braise et fumés avec le prototype de four par fumage indirect ont une apparence lustre appréciable, une texture ferme, une odeur caractéristique bonne. La peau est bien adhérente, de couleur dorée et un goût agréable.

Les machoirons fumés cuits avec les filaos et le dimb ont une couleur apparemment sombre, en corrélation avec le mode de fumage (combustion incomplète du bois, de dépôt de fumée chargée en goudrons et particules nocives), le goût légèrement amer, la texture dure. Ils sont moyennement appréciés.
Avec le bois, des pics intempestifs de températures sont constatés, les produits sont hétérogènes et le plus souvent calcinés.

Les machoirons cuits et fumés avec les bourres de coco sont plus humides et la texture est molle. La couleur est bien dorée et le goût est apprécié par les panélistes.

### *Teneur en benzo(a)pyrène*

Les teneurs en benzo(a)pyrène obtenues en µg/kg:

- la braise: 0,15 à 0,29 pour le machoiron et 0,26 à 1,38 pour la sardinelle;
- le Dimb (*Cordyla pinnata*): 0,22 à 107,71. Le taux le plus élevé est surtout en lien direct avec l'apport de la combustion incomplète du bois et en absence de plaque à graisse favorisant ainsi avec les dépôts de la combustion de la graisse du poisson directement sur le feu;
- le filao (*Casuarina equisetifoLia*): 3,47 à 13,22;
- les bourres de coco: 6,12 à 31, 01.

**Tableau 9. Teneur en benzo(a) pyrène et l'humidité selon le mode de cuisson et de fumage à chaud**

| | Taux d'humidité (%) | Benzo(a)pyrène (µg/kg) Poids frais | Benzo(a)pyrène (µg/kg) Poids sec |
|---|---|---|---|
| **Machoiron braise N°1 :** Cuisson avec plaque et fumage direct | | | |
| **Machoiron braise N° 2 :** Cuisson avec plaque et fumage indirect | 57,94 | 0,29 | 0,29 |
| **Machoiron braise N° 3 :** Cuisson sans plaque et fumage direct | 47,98 | 0,21 | 0,4 |
| **Sardinelle Braise N°4:** Cuisson avec plaque et fumage direct | 42,51 | 1,38 | 2,41 |
| **Sardinelle Braise N°5:** Cuisson avec plaque et fumage indirect | 51,96 | 0,26 | 0,54 |
| **Machoiron Dimb N°6:** Cuisson fumage direct avec plaque | 54,91 | 0,22 | 0,48 |
| **Machoiron Dimb N°7:** Cuisson fumage direct sans plaque | 51,3 | 107,71 | 221,18 |
| **Machoiron Dimb N°8:** Cuisson avec plaque fumage indirect, | 45,79 | 2,89 | 5,33 |
| **Machoiron Dimb N°9:** Cuisson avec plaque fumage indirect | 51,93 | 28,51 | 59,31 |
| **Machoiron Coco N°10:** Cuisson fumage direct sans plaque | 49,31 | 13,22 | 31,01 |
| **Machoiron Coco N°11:** Cuisson avec plaque et fumage indirect | 57,36 | 6,12 | 12,09 |
| **Machoiron Filao N°12:** Cuisson fumage direct sans plaque | 52,43 | 10,21 | 21,46 |
| **Machoiron Filao N°13:** Cuisson fumage direct avec plaque | 57,81 | 7,87 | 18,65 |
| **Machoiron Filao N°14:** Cuisson fumage direct avec plaque | 55,84 | 3,47 | 7,78 |
| **Machoiron Marché N°15: produit acheté au marché** | 55,84 | 1,67 | 3,77 |

Le taux de benzo(a)pyrène est très faible pour la braise. Elle tend vers zéro, donc difficile à matérialiser dans les graphiques. La cuisson à la braise, à combustion complète avec un mode de fumage indirect avec plaque donne des teneurs variant entre 0,15 et 0,29 µg/kg nettement inférieures à la norme (5 µg/kg).

La combustion incomplète du bois (dimb, filao, bourres de coco) contribue au dépôt de goudrons sur les produits engendrant ainsi des taux HAPs élevés (ex : Dimb sans plaque 107,71 µg/kg).

Le mode de cuisson, de fumage et le type de combustible ont des effets directs sur le taux de benzo(a)pyrène.

Quoique toute information relative au mode de fumage du machoiron du marché soit inconnue, le consommateur devrait être soulagé de noter que cet échantillon présente une teneur en benzo(a)pyrène très satisfaisante.

Les résultats de la teneur en benzo(a)pyrène font ressortir les paramètres suivants:

*Effet du mode de fumage*

**Tableau 10 et Figure 7. Influence du mode de fumage sur la teneur en HAP des produits fumés**

|  | **Fumage direct** | **Fumage indirect** |
|---|---|---|
| Machoiron fumé avec la braise | 0,21 | 0,29 |
| Sardinelle fumé avec la braise | 1,38 | 0,26 |
| Machoiron fumé au Dim | 107,71 | 28,51 |
| Machoiron fumé bourre de coco | 13,22 | 6,12 |

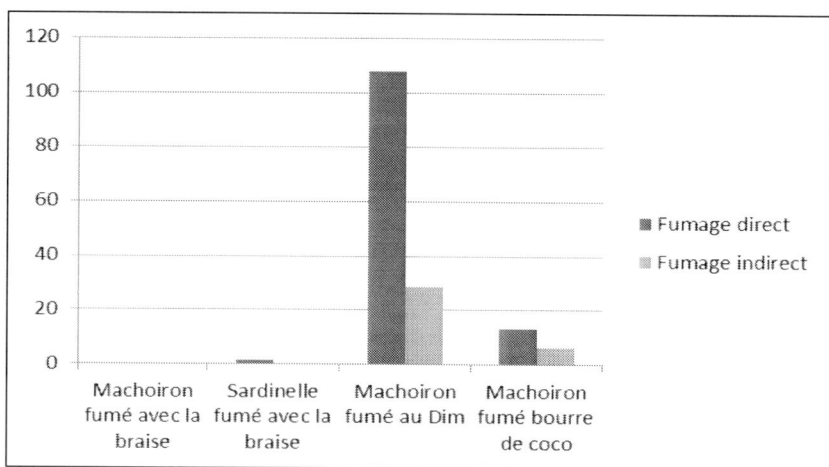

La plus faible teneur en HAP a été obtenue avec le fumage indirect. Les goudrons acides qui communiquent au poisson fumé des taux élevés de Benzo(a)pyrène sont en partie retenus par le filtre intercalé entre le générateur de fumée ou de chaleur et les produits sur la claie, ce qui est reflété par la couleur de l'exsudat et de l'éponge végétale.

Outre le fumage avec la braise au charbon où la différence entre les deux modes (direct et indirect) n'est pas significative, les résultats montrent que le fumage indirect réduit au moins de moitié, sinon plus, la teneur en benzo(a)pyrène du produit fini.

*Le type de combustible*

**Tableau 10 et Figure 8. Teneur en benzo(a)pyrène des produits fumés en fonction du type de combustible**

|  | Teneur moyenne en HAP |
|---|---|
| Machoiron fumé avec la braise | 0,22 |
| Sardinelle fumé avec la braise | 0,82 |
| Machoiron fumé au Dim | 34,83 |
| Machoiron fumé au filao | 7,18 |
| Machoiron fumé bourre de coco | 9,67 |

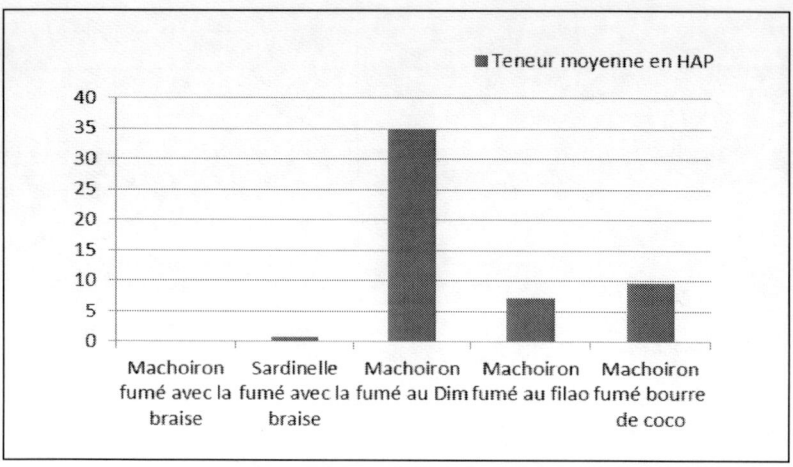

Les teneurs obtenues avec **le charbon** à combustion complète (braise rouge) ont donné des résultats très satisfaisants par exemple 0,15 µg/kg, 33 fois inférieur à la norme et aussi loin des 2 µg/kg qui sera exigée en 2014 par le marché communautaire.

**Le dimb**, bois massif, à combustion complète donne de la braise et avec l'effet de la plaque, les dépôts de particules nocives et de goudrons sur les produits sont réduits durant la cuisson. La teneur relevée est de 0,22µg/kg. Par contre le dimb à combustion incomplète, dégage énormément de fumée goudronneuse et sans plaque à graisse, la teneur en benzo(a)pyrène atteint une teneur record de l'ordre de 107,71 µg/kg ;

**Le filao, bois résineux,** sec à combustion complète et avec la plaque à graisse a donné des taux moyens, mais dont un échantillon est hors de la limite réglementaire actuelle (Tableau 8).

**Les bourres de coco,** dégagent de la fumée non goudronneuse, mais sont restés durant tous les essais à combustion incomplète compte tenu de la texture fibreuse de ce combustible. Les teneurs dépassent la norme admissible.

Alors que dans le fumage à l'aide de la braise les produits semblent homogènes en matière de teneur en benzo(a)pyrene, l'on observe une grande variation entre des échantillons fumés de la même manière, dans les cas de combustibles comme le bois ou coco.

*L'effet de la plaque à graisse*

La plaque maintenue propre sans reste de croûte de graisse a un effet sur le dépôt de HAP dans les produits fumés particulièrement pour les poissons gras.

**Tableau 11 et Figure 9. Effet plaque sur la teneur en HAP des produits fumés avec un mode de fumage identique dans la même espèce de poisson**

|  | Sans plaque | Avec plaque |
|---|---|---|
| Machoiron fumé avec la braise | 0,21 | 0,15 |
| Machoiron fumé au Dim | 107,71 | 0,22 |
| Machoiron fumé au filao | 10,21 | 7,87 |
| Machoiron fumé bourre de coco | 13,22 | 6,12 |

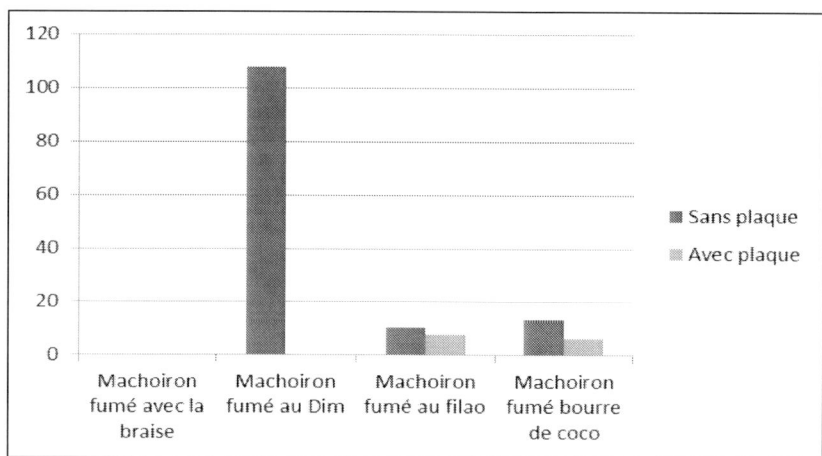

**La cuisson sans plaque** effectuée avec le dimb, le filao et les bourres de coco a donné des teneurs en benzo(a) pyrène élevées, qui dépassent largement la norme. La valeur obtenue avec de la braise à combustion complète est faible.

L'utilisation de la plaque a permis de réduire significativement la teneur en benzo (a)pyrène comparée aux produits cuits sans plaque, notamment lorsque le bois et les bourres de coco sont impliqués.

Les produits cuits directement sans plaque sont apparemment plus secs mais plus riches en benzo(a)pyrène.
Il convient toutefois de nettoyer la plaque entre les sessions de fumage, pour éviter qu'elle soit trop chargée en matières grasses, et donc favoriser l'augmentation progressive de la teneur en benzo(a)pyrène.

La suite du programme collaboratif porte sur la restitution des informations aux opérateurs en les associant à un atelier de validation. Ce partage d'expériences marquera le départ de la promotion des « nouveaux produits de pêche fumés ». Un programme d'informations (foire, publicité, plaquettes, rencontres avec les consommateurs, séances de dégustation) en collaboration avec les producteurs, permettra aux consommateurs de comprendre l'intérêt de ces technologies innovantes.

## 4. RECOMMANDATIONS ET CONCLUSIONS

Les essais ont établi la réduction significative de la teneur en benzo(a)pyrène des produits avec le système FAO-Thiaroye de fumage en mettant en évidence l'influence du type de combustible utilisé, l'importance du charbon de bois, et surtout avec les autres combustibles la nécessité du fumage indirect et l'utilisation de la plaque de récupération de la graisse, permettant d'obtenir des produits répondant aux exigences sanitaires tout en respectant les préférences organoleptiques des consommateurs.

Les teneurs en Benzo(a)pyrène sont nettement inférieures à la norme réglementaire pour les poissons cuits à la braise. Dans l'état actuel des dispositions réglementaires européennes, cette technique mérite d'être documentée et disséminée en pêche artisanale pour exploiter ce créneau.

Toutefois des investigations plus poussées sont requises en vue de :

- faire une analyse incluant les 3 autres HAPs qui seront en vigueur en 2012 au niveau du marché communautaire européen, en mettant l'accent sur la maitrise des opérations dans l'utilisation du système FAO-Thiaroye de manière à avoir des produits homogènes et avec des teneurs en dessous des limites;
- déterminer le nombre de sessions de fumage pour un nettoyage de la plaque à graisse sans risque majeur d'élever la teneur en HAPs dans le produit fini.
- tenir un répertoire de tous les bois combustibles utilisés et ces derniers testés pour leur innocuité.

De ce qui suit, des lignes directrices ou codes d'usage devront être développés et des formations organisées pour la vulgarisation de la transformation du poisson avec la technique de fumage à chaud

## BIBLIOGRAPHIE

**Amending Regulation (EC).** No 1881/2006 as regards maximum levels for polycyclic aromatic hydrocarbons in foodstuffs.

**Codex Alimentarius.** 1979. Code d'Usages International Recommandé pour le Poisson Fumé - CAC/RCP 25–1979.

**Directive 2005/10/CE** de la commission du 4 février 2005 portant fixation des modes de prélèvement d'échantillons et des méthodes d'analyse pour le contrôle officiel des teneurs en benzo(a)pyrène des denrées alimentaires.

**Doornaert, B., Pichard, A. & Gillet, C.** 2003. Hydrocarbures aromatiques polycycliques (HAPs). Evaluation de la relation dose-réponse pour des effets cancérigènes : Approche substance par substance (facteurs d'équivalence toxique – FET) et approche par mélanges. Evaluation de la relation dose-réponse pour des effets non cancérigènes : Valeurs Toxicologiques de Référence (VTR). INERIS-DRC-03-47026-ETSC-BDo-N°03DR177.

**EFSA.** 2008. "Polycyclic Aromatic Hydrocarbons in food Scientific Opinion of the Panel on Contaminants in the Food Chain" (Question N. EFSA-Q-2007-136).*The EFSA Journal* (2008) 724 : 1–114.

**Ekomy, A. S.** 2008. Vulgarisation d'un nouveau concept amélioré de séchage et de fumage artisanal des aliments : application en milieu pêche artisanale au Gabon. In report and papers presented at the second workshop on fish technology, utilization and quality assurance in Africa. Agadir, Morocco, 24–28 November 2008. FAO Fisheries and Aquaculture Report No. 904. ISSN 20706987.

**Règlement (CE).** 1881/2006 fixant les niveaux maximum de certains contaminants.

**Sérot, T., Baron, R., Cardinal, M., Cataneo, C., Knockaert C., Le Bizec B., Prost, C., Monteau, F. & Varlet, V.** 2009. Assessment of the effects of the smoke generation processes and of smoking parameters on the organoleptic perception, the levels of the most odorant compounds and PAH content of smoked salmon fillets. *In* Report and papers presented at the second workshop on fish technology, utilization and quality assurance in Africa. Agadir, Morocco, 24–8 November 2008. FAO Fisheries and Aquaculture Report No. 904. ISSN 20706987.

**Šimko, P.** 2002. Determination of polycyclic aromatic hydrocarbons in smoked meat products and smoke flavouring food additives. *Journal of Chromatography* B 770: 3-18.

# MAXIMIZING BY-CATCH UTILIZATION IN PRAWN FISHERY: CASE OF INSHORE FISHERY IN TANZANIA

## [MAXIMISER L'UTILISATION DES CAPTURES ACCESSOIRES DANS LA PÊCHE CREVETTIERE: CAS DE LA PÊCHE CÔTIÈRE EN TANZANIE]

by/par

Ambakisye Mwanjala Simtoe[1]

**Abstract**

Sub-Saharan countries face persistent poverty and food insecurity. With such a scenario there has been a need to maximize by-catch utilization both as sources of income and protein. Prawn fishing is associated with high by-catch; most industrial fishing operators would prefer to discard at sea to reduce fishing costs. The prawn fishery in Tanzania is conducted by both artisanal and commercial fishers. Having such a high market value inside and outside Tanzania, the prawn fishery has attracted a number of industrial fishing vessels, which in turn produce a significant amount of by-catch. Results indicate that exploitation of prawn resources is associated with fishing of small, medium and large non-target fish that are considered as by-catch. Results obtained from 30 fishing trips showed that the average percentage of by-catch was above 75 percent with a ratio of 1:4 to that of the target species (prawns). The fish were mostly small, with an average weight of 156 g. Major species of big fish commonly encountered in the catch were sharks, rays, emperors, groupers and mackerels, while the prawns were dominated by white shrimp (*Fenneropenaeus indicus*): over 67 percent of shrimp composition. Other species included giant black shrimp (*Penaeus monodon*) and brown shrimp (*Metapenaeus monoceros*). The study generally recommends production of speciality fish products as a means to add value to low priced by-catch and hence increase utilization alternatives.

Main challenges for by-catch utilization were found to be price alteration within fishing cycles in respective months (new and full moon). It was generally revealed that there were possible ways of utilizing this high percentage of prawn by-catch so as to reduce the fishing cost and improve the socio-economic and livelihood of many coastal communities.

*Key words: Bycatch, Shrimps, Value addition, Income, Market, Livelihoods, Tanzania*

**Résumé**

Les pays de l'Afrique sub-saharienne sont confrontés à la pauvreté persistante et à l'insécurité alimentaire. Avec un tel scénario il y a eu une nécessité d'augmenter l'utilisation des captures accessoires aussi bien en tant que source de revenu que de protéine. La pêche aux crevettes est associée à des captures accessoires élevées; la majorité des opérateurs industriels de pêche préfèrent jeter en mer pour réduire les coûts de la pêche. La pêche aux crevettes en Tanzanie est menée par les artisans et les pêcheurs commerciaux. Ayant une grande valeur commerciale sur le marché intérieur et extérieur de la Tanzanie, la pêche crevettière a attiré un certain nombre de navires de pêche industrielle, qui à leur tour produisent une quantité importante de captures accessoires. Les résultats indiquent que l'exploitation de ressources crevettières, est associée avec la pêche de petite, moyenne et grande taille de poissons non-ciblés qui sont considérés comme des captures accessoires. Les résultats obtenus des 30 voyages de pêche ont montré que le pourcentage moyen de la capture accessoire était supérieur à 75 pour cent avec un ratio de 1:4 à celui de l'espèce cible (crevettes). Le poisson était pour la plupart petit, avec une moyenne de poids de 156 g. Les principales espèces de gros poissons couramment rencontrés dans la capture accessoire étaient des requins, raies, empereurs, mérous et les maquereaux, pendant que les crevettes étaient dominées par la crevette blanche (Penaeus monodon) à plus de 67 pour cent de composition de crevette. D'autres espèces incluant la crevette géante tigrée (Penaeus monodon) et la crevette grise (Metapenaeus monoceros). L'étude recommande généralement la production de produits de poisson de spécialité comme un moyen d'ajouter de la valeur au bas prix de la capture et donc d'augmenter les alternatives d'utilisation.

---

[1] Mbegani Fisheries Development Centre, Ministry of Livestock, Development and Fisheries, PO Box 83, Bagamoyo, Republic of Tanzania. bakisyes@yahoo.com ; mbeganifdc83@yahoo.com

Les principaux défis pour l'utilisation des captures accessoires étaient la dégradation du prix pendant les cycles de pêche dans les mois respectifs (nouvelle et pleine lune). Il a été généralement révélé qu'il existe des moyens possibles d'utiliser ce pourcentage élevé de captures accessoires de crevettes afin de réduire le coût de la pêche et d'améliorer la situation socio-économique et les moyens de subsistances de nombreuses communautés côtières.

*Mots clés: Captures accessoires, Crevettes, Valeur-ajoutée, Revenue, Marché, Moyens d'existence, Tanzanie*

## 1. INTRODUCTION

Shrimps are a valuable marine resource, not just for domestic use but also for export. Artisanal fishermen as well as industrial fisheries catch shrimp in Tanzania. The artisanal fishery operates from small dugout canoes and small-planked boats, which are powered mainly by pedals, poles, oars and sails. Few fishers also use boats with outboard engines. Industrial or commercial prawn fishery in Tanzania is based on four major species: the white shrimp (*Fenneropenaeus indicus* (formerly known as *Penaeus indicus*), giant black shrimp (*P. monodon*), tiger shrimp (*P. semisucatus*) and brown shrimp (*Metapenaeus monoceros*), with white shrimp making up the bulk of the catch (Bwathondi *et al.,* 2002, Haule, 2001). The industrial prawn fishery is associated with high catch of non-target fin fish species, including snapper, kingfish, shark, rays, small clupeids, lobsters, sardines and sea cucumbers. Most of these are consumed locally while a few high valued species such as sharks, kingfish and lobsters are exported. Taking into consideration the fishing costs it has been difficult for industrial prawn fishing vessels to keep by-catch on board, most prefer to discard it at sea although fisheries regulations of Tanzania require all by-catch to be landed at a specified site. However, fishing vessels conducting daily fishing close to major cities could easily land by-catch at justifiable cost. The current study on how by-catch utilization can be maximized and hence compensate for fishing cost while at the same time elevating fishers' income. Moreover, maximizing utilization of by-catch would ensure food security in the country and hence fulfil Millennium development goals. Suluda (1997) pointed out limited chill or cold storage capacity and low price paid for by-catch as set-backs for by-catch utilization. The benefit accrued from by-catch has failed to override costs for storing and transportation. The main challenge facing by-catch utilization in Tanzania is that most fish caught as by-catch is low priced when sold raw and hence provision of means to add value would shift the curve to low cost and high benefit. However, by-catch has significant importance to the country in terms of human consumption (Mgawe, 1999). Therefore, it is important to find out how by-catch may be utilized efficiently to cover the cost incurred during fishing and contribute to income generation in fishing communities. This study makes comparative advantages of prices over value addition to low value fish by-catch. The scope examined possible useful products of by-catch which would be an alternative source of income for small-scale fishers. The study identified preferable products for high economic gain by price comparison of various processed products

## 2. STUDY OBJECTIVES

The present study aimed at analyzing the comparative advantages of various specialty products produced from by-catch obtained from industrial fishing. The specific objectives included:

- Quantify the by-catch composition obtained from prawn fishing vessels
- Investigate distribution channels of by-catch from the landing site
- Make a comparison of market prices between fish sold fresh without any processing with specialty products.

## 3. METHODOLOGY

Data were collected by industrial stern trawler (27m), visits to various landing sites and markets, through key-informant interviews and discussion. About 30 sampling trips were conducted twice a month during full and new moon from March 2010 to February 2011. The fish caught during fishing trips were sorted out into prawns, small and medium size clupeids, high value fin fish such as tuna and tuna like species, and other high value fish such as squid, sharks and rays. After sorting the weight of each group of fish was measured. On landing the prawns were directly sold to wholesale traders and other high value species of fish were sold at retail price.

Medium and small size clupeids were then divided into two parts of 200 kg each, one group sold as low priced fish (un-processed). The remainder were dried, smoked, or made into specialty products then sold. The market price of various fish species and products were determined. The study further, determined the price of fish and specialty products with reference to distance. It focused on tracing fish distribution channels from the coastline (landing site) to inland markets.

## 4. RESULTS AND DISCUSSION

The results from the fishing vessel indicated the average catch of 49.3 kg of big fish, 1698 kg of small fish, mainly clupeids, and 107.03 kg of prawns (target species) per fishing vessels (ref figure below) . This result indicates the relation of small to large fish being negatively skewed with average weight of 156g that means most of fish caught were within this weight. Results indicate major species of big fish were sharks, rays, emperors, groupers and mackerels dominating the catch by over 95 percent of big fish composition, while the prawns were dominated by white shrimp (*Fenneropenaeus indicus*) - over 67 percent. Other species included giant black shrimp (*Penaeus monodon*) and brown shrimp (*Metapenaeus monoceros*).

Jiddawi (2002) identified 500 species of fish being utilized for food with reef fishes being the most important category, including emperors, snappers, sweetlips, parrotfish, surgeonfish, rabbitfish, groupers and goatfish. In this study it was found that some of these coral reef species are being caught along with shrimps in the trawl nets.

By-catch over 75 percent making a ratio of 1:4 is relatively high in prawn trawling; such by-catch requires a clear study on how such catch can be utilized to yield maximum profits that would cover the costs incurred for conducting fishing operations. These results concur with the findings of Pelgröm and Sulemane (1982) referring to a shrimp to by-catch ratio of 1:3 (catch rates greater than 50 kg/h), Gislason (1985) found 1:3.8, Anon. (1994) found 1:5 with 89 percent of the by-catch discarded, and Pacule and Baltazar (1995) found 1:4.3.

In spite of many fishing companies wishing to discard by-catch at sea to reduce cost, the need to utilize it remains paramount as far as food security is concerned.

Results generally indicate that processed products have competitive prices to un-processed fresh fish as shown in the Table 1 below.

**Table 1. Fish species composition, processing methods and respective price for specialty products**

| Scientific name | English name | Swahili name | Value (Tsh)/kg | | | |
|---|---|---|---|---|---|---|
| | | | Fresh | Fillets | Smoked | Sun dried |
| *Caranx tille* | Tille Trevally | Kolekole | 2,600 | 8,000 | 5,000 | 4,500 |
| *Pomadasys multimaculatum* | Cock grunter | Karamamba | 3,600 | 5,000 | 6,000 | 4,500 |
| *Sphyraena jello* | Pickhaudle barracuda | Msusa | 1,000 | - | 3,200 | 1,000 |
| *Epinephelus tauvina* | Grouper | Chewa | 1,500 | - | 3,600 | 1,800 |
| *Carcharinus sp.* | Shark | Papa | 1,800 | - | 6,000 | 1,500 |
| *Scomberomorus* spp. | Mackerels | Vibua | **1,200** | - | **6,600** | **5,600** |
| *Arius africanus* | Sea Catfish | Hongwe | 1,200 | - | 4,500 | 1000 |
| *Anguilla mossambica* | African longfin eel | Mkunga | 1,400 | - | - | 1,600 |
| *Hydrocynus vittatus* | Tiger fish | Kange | 1,700 | - | 3,500 | 2,000 |
| *Sardina* spp. | Sardines | Salidini | **750** | - | **4,800** | **3,800** |
| *Taeniura* spp. | Rays | Taa | **1,200** | - | **6,000** | **1,600** |
| *Octopus* spp. | Octopus | Pweza | **6,700** | - | - | **4,500** |
| *Metanephrops* spp. *Nephropsis* spp. | Lobster | Kamba koche | 15,000 | - | - | - |

*Source:* Field data.

This study also focused on tracing fish distribution channels from the coastline (landing site) to inland markets. It was found that fish caught from both artisanal and industrial fisheries are either exported or consumed within the country. Tracing the distribution of fish in relation to distance from the coast was a major area of this study.

It was generally found that most high value species (over 90 percent) from industrial fisheries are exported mainly to European or Chinese markets. Only low value species such as groupers, jacks and clupeids are sold in the country, at prices equal to those of artisanal products. It was established that over 90 percent of most high value species including prawns, squid, octopus and crabs from artisanal fish species are sold within the country, only 10 percent, mainly including species such as prawns, octopus, lobsters, crabs and squids are exported. Moreover, it was also observed that about 60 percent of fish from artisanal fishers are transported to Dar es Salaam Ferry market where they get relatively high prices due to high demand caused by rapid population growth of the city. The remaining 40 percent were transported locally with no formal distribution channels. The distance fish could go was mainly limited by the handling techniques used, icing of fish being the main method employed. Processed products were found as far as 1000 km from the coast, being distributed by small traders, while fresh fish from the same traders were limited to within a short distance of about 100 km from the coastline. Fresh fish encountered beyond 100 km were mainly transported using insulated specialized vans, owned by commercial fish dealers. Traditional fish processing such as sun drying, salting and smoking are still popular in fishing villages and among Tanzanian consumers and most of products from artisanal fishers are processed using these techniques. Sardines are mostly sun dried while the table size bony fish are smoked and cartilaginous fish such as rays and sharks are salt dried. Over 86 percent of products of artisanal fishers encountered at the market were dried and smoked fish. Common fish species of commercial important are shown in the Table above.

Various common traditional processing methods encountered during field work include but are not limited to drying, smoking, salting, frying and fermenting or different combinations of these, with sun-drying being widely applied in processing of small clupeids and smoking for large clupeids. In most cases sun-drying was a response to encounter deterioration of fish caught when the market did not meet fishers' expected prices. For instance, during the new moon when small clupeid catches were high fishers preferred processing their catch so as they to access distant markets where over 50 percent of catch was processed. The number was as low as around 10 percent during full-moon, when most of the small scale purse seine fleet is not operating. It was apparent that in this period the by-catch caught by industrial prawn trawlers had sufficient market to justify fishing and transportation costs. Therefore, the need to process by-catch as a means of adding value to earn more income had no sense during full-moon. The price offered at the market was as high as three times the normal market price, which was competitive enough to cover fishing costs. However, customer preferences remained the controlling factor for pricing on different fish species and products.

Customer preferences on specialty products made out of by-catch were rated. Generally, salted and smoked products were rated high especially away from the coast, while along the coast salted fermented fish, locally known as nguru and ng`onda, were ranked high. Generally, most coastal communities preferred fresh fish to processed fish. Preference of processed fish products increased with increasing distance from the coast, probably because in most cases they eat more processed than fresh fish. Field observations indicated over 40 percent of fish products encountered at the markets were smoked fish, 52 percent sun dried small clupeids, 5 percent fresh fish and 3 percent other fish products. Fermented fish (nguru and ng`onda) had high customer preferences in coastal urban markets as compared to a high preference for dried dagaa and smoked large clupeids in rural markets away from the coastline.

This study observed high prices for processed fish in urban areas compared to low prices in rural areas along the coastline and vice versa away from the coast. Such observations are common in African countries. Njai (2000) observed that the price of smoked fish in rural markets was higher than in urban markets in Gambia. Such prices are determined by transport and other costs incurred by the trader during distribution from the processing site to the marketing points in the rural areas.

Observations from the field indicated no proper packing materials were used and this resulted in post harvest loss and low prices for locally packed products. With the exception of fresh products, which are commonly carried in ice boxes, other products were packed in sacks of 20 kg to 100 kg, special baskets locally known as *matenga* made out of bamboo trees and other locally made materials. Packaging problems are found to be related to market prices and quality of products, thus poor packaging materials are associated with low prices and high loss of processed products.

## 5. CONCLUSION AND RECOMMENDATIONS

The findings from the current study have shown that it is possible to maximize the use of by-catch and hence ensure food security and enhance coastal communities' livelihoods. From this study it could be concluded that by-catch utilization can be maximized by employing various processing techniques to make specialty products that meet market demand. Low priced fish that dominate the prawn fishery can easily be converted into high value fish by making specialty products. Market price of processed fish has a direct relationship with customers' preferences, hence market feedback to processors will increase profit from the fish business and hence counter balance costs incurred for handling by-catch. Improvement would include increased awareness through training for fishermen and processors on improved handling techniques that can result in high quality fish.

Furthermore, the study recommends Tanzania encourage by-catch utilization by making laws prohibiting discarding at sea and installing fish handling and processing units along the coast at specified landing sites. This would include improving market distribution systems. Further research on finding possibilities for exporting specialty products to European markets is recommended; currently specialty products from artisanal fishing industry are sold within great lakes countries.

## 6. ACKNOWLEDGEMENT

Mr Yahya Mgawe is appreciated for all the support and encouragement he gave to me on conducting this study. I owe a great appreciation to the Principal and my fellow staff members of Mbegani Fisheries Development Centre for their day to day encouragement. I also thank Mr Anold Mbunda and Mukama Ndaro for their support during data collection and field visits; in spite of all harsh conditions met they were always there to provide a hand when needed.

## 7. REFERENCES

**Anon.** 1994. Projecto de recolha da fauna acompanhante em Quelimane: Análise da informação recolhida pelos fiscais. Instituto de Desenvolvimento de Pesca de Pequena Escala (IDPPE), Maputo, pp. 6–15 (in Portuguese).

**Bwathondi, P.O.J., Chande, A.I., Mhitu, H.A., Kulekana, J.J., Mwakosya, C.A., Shayo, S.D. & Bayona, J.D.R.** 2002. Investigation of the abundance and distribution of prawns at Bagamoyo and Rufiji delta. *Tanzania Fisheries Research Institute* p. 56

**Gislason, H**. 1985. A short note on available information about demersal fish on the shallow water part of Sofala Bank. RIP 13, IIP, Maputo, pp. 83–95.

**Haule, W.V.** 2001. Reducing the impact of tropical shrimp trawling fisheries on living marine resources through the adoption of environmentally friendly techniques and practices in Tanzania. *In* Tropical shrimp fisheries and their impact on living resources. *FAO Technical Paper* 500, pp 216–233.

**Jiddawi, N. S.** 2001. Marine Fisheries in Tanzania. *In* Present State of Marine Science. Tanzania. TCMP Working Documents. Ed. Ngusaru, A.

**Mgawe, Y.** 1999. Utilization of by-catch from the fishing industry in Africa, an overview. *In* Clucas, I.J. Mgawe, Y. and Teutscher, F. (eds). *Report and proceedings of FAO/DFID Expert consultation on by-catch utilization in Tropical fisheries*. Beijing, China, Sept 1998. Natural Resources Institute, Chatham, UK.

**Njai, S**. 2000. Traditional Fish Processing and Marketing of the Gambia. University of Iceland, UNFP

**Pacule, H. & Baltazar, L.** 1995. A fauna acompanhante de camarão no Banco de Sofala. Análise preliminar (Shrimp by-catch at Sofala Bank), pp. 79–95. RIP No. 20 (1), Instituto de investigação Pesqueira, Maputo (in Portuguese with English abstract).

**Pelgröm, H. & Sulemane, M**. 1982. Regional and country developments – Mozambique, pp. 139–140. In *Fish by-catch...Bonus from the Sea*. Ottawa: FAO/IDRC.

**Suluda, S.I.** 1997. Strategy for improving utilisation of shrimp by-catch. A case study on Sofala Bank shrimp fishery, Mozambique. MSc dissertation, University of Hull, UK. 140pp.

# MICRONUTRIENT ENRICHMENT OF MEALS FED TO PUPILS USING HIGHLY NUTRITIOUS AND LOW-COST UNDERUTILIZED FISH UNDER THE SCHOOL FEEDING PROGRAMME IN GHANA

## [ENRICHISSEMENT EN MICRONUTRIMENTS DES REPAS D' ELEVES NOURRIS EN UTILISANT DU POISSON HAUTEMENT NUTRITIF, A FAIBLE COUT ET SOUS-UTILISÉ SOUS LE PROGRAMME DE CANTINE SCOLAIRE AU GHANA

by/par

Margaret A. Ottah Atikpo[1], L.D. Abbey, M. Glover-Amengor, L. Lawer, J. Ayin and J. Toppe

### Abstract

Four underutilized fish species, namely Woevi, or one-man-thousand, (*Sierathrissa leonensis*), Flying gurnard (*Dactylopterus volitans*), common bogue (*Boops boops*) and anchovies (*Anchoa guineensis*); as well as tuna frames obtained as factory remnants, were either solar-dried or mechanically dried and milled into dry powder. The powders were analysed for their proximate and mineral composition, biochemical and microbiological status, in addition to sensory and shelf life tests. Characterisation of the fish species showed that the selected fish are of high nutritional value in either human food supplements or formulations, with high protein content and optimal amino acid profile, abundance of polyunsaturated fatty acids, and high in micronutrients, particularly minerals. It showed the potential of these products for food supplementation in the school feeding programme, although generally the products might be regarded as fish for the poor. The sensory evaluation showed that the school children rated all the foods fortified with the fish powders well on the positive side of the hedonic scale. All the foods were accepted by the children, in particular banku with anchovies and okro stew, rice with tuna frames stew and rice with flying gurnard stew.

*Key words: Fish powder, School feeding, Supplements, Micronutrients*

### Résumé

Quatres espèces de poissons sous-utilisés, appelé Woevi, ou one-man-thousand (*Sierathrissa leonensis*), Poule de mer (*Dactylopterus volitans*), bogue commun (*Boops boops*) et les anchois (*Anchoa guineensis*); ainsi que des restes des industries de thon , ont été soit séchés au soleil ou mécaniquement séchés et broyés en poudre. Les poudres ont été analysées pour leur composition organique et minérale, du statut biochimique et microbiologique, en plus des tests sensoriels et de durée de vie. La caractérisation des espèces de poisson a montré que les poissons sélectionnés sont d'une haute valeur nutritionnelle dans chacun des compléments ou formulations alimentaires, avec une haute teneur en protéines et un profil optimal en acide aminé, l'abondance des acides gras polyinsaturés, et riche en micronutriments, en particulier les minéraux. Il a été démontré le potentiel de ces produits pour les rations de l'alimentation fortifiée dans le programme de cantine scolaire, bien que généralement les produits pourraient être considérés comme poisson pour pauvres. L'évaluation sensorielle a montré que les écoliers ont donné un score positif sur l'échelle hédonique à tous les aliments enrichis avec les poudres de poissons. Tous les aliments ont été acceptés par les enfants, en particulier le banku avec les anchois et la sauce gombo avec restes de thon et riz avec soupe locale particulière.

*Mots clés: Poudre de poisson, Alimentation scolaire, Suppléments, Micro-éléments nutritifs*

## 1. INTRODUCTION AND LITERATURE REVIEW

Fish is of importance to the diet in the developing world. In about 30 low-income food-deficit countries in Africa and Asia, more than 1/3 of their daily intake of animal proteins comes from fish. Fish contains 70–80 percent water, 15–24 percent protein, 1–2 percent minerals and 0.1–22 percent fat (Clucas, 1985) high in fat-soluble vitamins like Vitamin A and D (Putro, 1990).

Fish contains macronutrients (proteins and fats) and micronutrients (vitamins and minerals) necessary for good nutrition, thus contributing effectively to food and nutrition security as an accompaniment to rice-based diets in

---

[1] CSIR-Food Research Institute, PO Box M20, Accra, Ghana. magatik@yahoo.co.uk

Asia; and maize and cassava-based diets in Africa. In Africa, population rise has caused an increasing demand for fish products resulting in increased focus on processing and use of underutilized fish species to combat malnutrition. Deficiencies of micronutrients such as vitamin A, iron and iodine are of public health significance in Africa as deficiencies may have serious health impacts, such as blindness, poor learning capabilities, poor growth and increased morbidity and mortality rates. Mainstreaming nutrition issues using a food based approach can help alleviating problems of malnutrition in urban and rural households in Ghana.

Fish proteins are easily digested and of high biological value. The fat content of fish depends on species and the season in which it is caught. Fat from fish is the most important source of the polyunsaturated omega 3 fatty acids (PUFAs) namely EPA (eicosapentaenoic acid) and DHA (docosahexaenoic acid), essential for brain development of children, are also associated with protection against cardiovascular diseases.

Fish micronutrients are made up of both vitamins and minerals. Fish can be a rich source of vitamins, vitamin A and D from fatty species, as well as thiamine, riboflavin and niacin (vitamins B1, B2 and B3). Minerals in fish such as iron, calcium, zinc, iodine (from marine fish), phosphorus, and selenium are particularly present in the bones (Toppe *et al.*, 2007). These minerals are highly bioavailable in fish. Fish is a very good source of minerals, in particular if eaten whole.

Fish should be an integral component of the diet in preventing malnutrition. The development of affordable and acceptable fishery products for school children would enhance micronutrient intake, especially small-sized fish that are consumed whole. Although most of the minerals are found in high amounts in fish bones, consumption of bones of larger fish is rarely practiced. Increased use of seafood, including bones, could contribute significantly to increasing the micronutrients level and thus reducing protein malnutrition.

## 2. MATERIALS AND METHODS

### Materials

Fish species used in this study were the West African pygmy herring (*Sierathrissa leonensis*), a *Clupeid* fish species (subfamily *Pellonulinae*) found in African inland waters and locally known as woevi, or one-man-thousand; flying gurnard (*Dactylopterus volitans*) (*Linnaeus*) otherwise known as *Cephalocanthus volitans* (local name in Accra is pampansre); Common bogue (*Boops boops*) known locally as otoe kpakpa (Anon, 1994); and anchovies (*Anchoa guineensis*). Tuna frames as by-product from processing tuna were obtained from Pioneer Fish Processing Limited at Tema. The one-man-thousand, was harvested in the Volta lake reservoir (fresh water), purchased at Kpong in the Eastern Region and conveyed to the laboratory on ice. The other freshly harvested fish species (all marine) were also immediately held on ice at 0 °C and conveyed from the Tema Fishing Harbour to the laboratory where they were frozen at -20 °C until analysis.

### Methods

The fish were cut into thin fillets and strips and dried to a low moisture level by a GTZ solar dryer and a FRI gas-fuelled mechanical dryer. A hammer mill was then used to mill it into fine powder, bones included. The powders were placed in sterile polyethylene bags, sealed aseptically, labelled and stored for analyses, including shelf life studies.

Protein, moisture, total fat, fatty acids, ash, phosphorus, calcium and iron were determined by the methods of AOAC (1990). Microbiological analyses were carried out by means of International standardized procedures including aerobic plate count in cfu/g at 30 °C/72h (NMKL 86, 2006), *E. coli* count in cfu/g (NMKL 125, 2005) and *Salmonella* spp./25g (NMKL 71, 1990).

Four local foods were selected and used as vectors for inclusion of the milled fish products for acceptability tests on school children. Each of the fish powders was tested and mixed into each of the local dishes below:

1. Okro stew served with banku;
2. Aprapransa;
3. Jollof rice;
4. Plain rice served with tomato stew.

Consumer acceptability was investigated using the four varieties of fish powder (woevi, flying gurnard, anchovies and tuna frames) in the different foods, a total of 16 samples. One thousand four hundred and sixty four (1464) school children were recruited for the test on the premises of the school. The children were divided into four groups; each group tested four different diets, all with different fish powders. Each portion included 10 g of fish powder. The school children were asked to rate on ballot sheets how much they liked each sample in terms of the following attributes: appearance, colour, aroma, texture, taste and overall acceptability. Rating was done on a 9-point hedonic scale with anchors 1-dislike extremely and 9-like extremely. Provision was also made on the ballot sheets for further comments from the school children for liking or disliking the samples. The children ate a meal per day during their lunch time in accordance with the school feeding programme.

All data obtained from the school children on the rating of sensory attributes of the foods were analysed using SPSS version 16. The means were tested for significance using one way analysis of variance (ANOVA) and Tukey's post test to determine significant differences between individual fish variety in each group and between the four food groups. Mean differences were considered significant at $p < 0.05$.

## 3. RESULTS AND DISCUSSION

### Chemical composition of fish species

Moisture content (Table 1) for both the flying gurnard and common bogue were within 70–80 percent moisture as expected (Clucas, 1985). The protein content of both fish species were above the levels of 16 percent reported for pelagic fish (Windsor and Barlow, 1981), and within 15–24 percent reported (Clucas, 1985). This suggests that the fish is a good source of protein and may be used in fish protein concentrate production or in food supplements (Windsor and Barlow, 1981). Proximate and chemical composition of a number of species have been reported by a number of workers (Paetow et al., 1966; Podsevalov and Perova 1975; Smith et al., 1980; Bykov 1985) on commercially important species, in contrast to this study on commercially lesser known but economically viable species in Ghana.

The low fat content of both flying gurnard and common bogue were below the range of 1–8 percent reported for other pelagic species (King and Poulter, 1985; Bykov 1985) and so may not be suitable for fish oil production (Urdahl, 1992) but might still provide a significant amount of the essential fatty acids (Table 3). The low fat content of the fish, as suggested by Talabi et al. (1980) is an indication of the fact that most of the lipids are present as phospholipids which are a rich source of polyunsaturated fatty acids (PUFA). In species of higher fat percentage (1–3.8 percent) Talabi et al. (1980) reported comparatively lower percentages of PUFA.

The ash content for both fish species was higher than the range of 0.5–1.8 percent wet weight for most other fish species (Sidwell, 1981). The high calcium content of both fish (especially for common bogue) renders them useful even for compromised individuals to boost calcium supplementation when such fish are consumed, of particular importance to growing children and elderly women (osteoporosis). Values of 580 mg/100 g calcium have been reported for whole sardines, anchovy and other fish species (Da Costa and Stern 1956; Sidwell, 1981), in accordance with the levels of 2000–3000 mg/100 g in the dried fish powders (Table 5). The iron content of the common bogue was more than 30 mg/100 g, about 50 percent higher than for flying gurnard. Values for other fish species (not dried) are reported within a wide range of 0.8–373 mg/100 g (Egass and Braekkan 1977; Sidwell, 1981; Teeny et al., 1984). The iron content of the common bogue was much higher than the reported range of 2–10 mg/kg from the Pacific coast of the USA (Gordon and Roberts 1977; Teeny et al., 1984). The iron data of the common bogue includes bones, which might explain the much higher level of iron.

The nutritional composition of the fish species indicate that they could be used for the development of infant foods and other food products.

**Table 1. Proximate and chemical compositions of fresh flying gurnard and common bogue (g/100 g)**

| Parameter | *Flying Gurnard | *Common Bogue |
|---|---|---|
| Moisture (wet weight basis) | 74.0 ± 2.5 | 71.2 ± 2.5 |
| Protein (N x 6.25) | 22.3 ± 3.4 | 18.3 ± 3.2 |
| Fat | 0.7 ± 0.1 | 5.3 ± 1.5 |
| Ash | 3.3 ± 0.7 | 3.3 ± 0.7 |

* *Values are means of three determinations ± standard deviation.*

**Amino acid composition of fish species**

The major amino acids (Table 2) present in the flying gurnard are glutamic acid and alanine; and in the common bogue is tyrosine. The flying gurnard is also a good source of lysine and leucine while threonine, alanine and leucine are observed for the common bogue. The percentage sulphur amino acid, methionine in the flying gurnard compares favourably to other species (Garrow and James, 1993; Batista *et al.*, 2001; Iwasaki and Harada, 1985). The overall profile of the essential amino acids of both fish species appear to suggest that they have a high class protein comparable to that of the mammalian meat which contains high levels of lysine and histidine (FAO, 1962; Garrow and James, 1993; Friedman, 1996). The fish may therefore be a good source of protein supplement in infants' diets and diets of school children involved in the school feeding programme.

**Table 2. Amino acid composition of salt soluble proteins of total protein in flying gurnard and common bogue (ug/mg total protein)**

| Amino acids | * Flying Gurnard | *Common Bogue |
|---|---|---|
| Aspartic | 6.05 ± 0.45 | 6.25 ± 0.21 |
| Glutamic | 9.9 ± 0.22 | 4.39 ± 0. 05 |
| Hydroxy Proline | 1.39 ± 0.11 | 1.24 ± 0.10 |
| Serine | 4.76 ± 0.13 | 2.20 ± 0.18 |
| Glycine | 6.98 ± 0.22 | 1.38 ± 0.35 |
| Histidine | 0.71 ± 0.07 | 2.74 ± 0.26 |
| Arginine | 5.59 ± 0.04 | 0.66 ± 0.12 |
| Threonine | 4.88 ± 0.08 | 7.32 ± 0.01 |
| Alanine | 12.07 ± 0.17 | 7.54 ± 0.44 |
| Proline | 3.46 ± 0.39 | 2.39 ± 0.56 |
| Tyrosine | 4.48 ± 0.05 | 8.35 ± 0. 31 |
| Valine | 5.9 ± 0.03 | 5.43 ± 0.02 |
| Methionine | 3.21 ± 0.34 | 1.20 ± 0.40 |
| Cysteine | 1.0 ± 0.06 | 1.80 ± 0.12 |
| Isoleucine | 4.31 ± 0.02 | 4.85 ± 0.01 |
| Leucine | 8.27 ± 0.10 | 7.29 ± 0.23 |
| Phenylalanine | 3.34 ± 0.01 | 4.80 ± 0.16 |
| Tryptophane | 5.34 ± 0.09 | 6.04 ± 0.14 |
| Lysine | 8.31 ± 0.64 | 5.26 ± 0.37 |

* *Values are means of three determinations ± standard deviation.*

**Fatty acids profile of fish species**

From the fatty acids profiles (Table 3) the flying gurnard showed comparatively higher relative values of Poly Unsaturated Fatty Acids (PUFA) as compared to the common bogue. However, the ω3 content, predominantly EPA and DHA, represented about 50 percent of total fat in both species. Total fat content in common bogue was more than 7 times higher (Table 1), making common bogue a seven times better source of the ω3 fatty acids.

**Table 3. Fatty acid profile of the flying gurnard and common bogue (g/100 g total fat)**

| Fatty acid | *Flying Gurnard | Common Bogue |
|---|---|---|
| ∑ Saturated | 20.2 ± 1.4 | 18.4 ± 1.2 |
| ∑ Mono unsaturated | 12.6 ± 1.0 | 7.1 ± 0.6 |
| ∑ Poly unsaturated | 63.6 ± 0.7 | 38.9 ± 0.4 |
| Total ω3 content | 51 ± 3.7 | 48.13 ± 1.3 |

* *Values are means of three determinations ± standard deviation.*

Similar observations were made by Pozo *et al,* (1992) and Batista *et al.* (2001) in their studies on pelagic fish. The abundance of ω3 fatty acids suggests an additional advantage for the use of the fish in the formulation of infant foods as they help in the healthy growth and development of the brain, the nervous system and functioning of the eye (Bjerve *et al.,* 1992; Koletzko, 1992; Newman *et al.,* 1993; Agostoni *et al.,* 1995; Cockburn, 1997).

**Consequence of drying methods for fish powder production**

Solar and mechanically dried flying gurnard and anchovies processed into fish powder (Table 4) contained high levels of protein, adequate enough for protein supplementation of meals for school children involved in the school feeding programme in Ghana. Especially so is fish powder obtained from mechanically dried flying gurnard. The fat content of the flying gurnard was higher than that of the anchovies; however, its oleic acid was comparatively lower than observed in the anchovies. Thus the keeping quality of fish powder prepared with flying gurnard could be higher.

**Table 4. Proximate analysis of solar and mechanically dried fish powder (g/100 g)**

| Parameter | Flying Gurnard | | Anchovies | |
|---|---|---|---|---|
| | Solar dried | Mechanical dried | Solar dried | Mechanical dried |
| Moisture | 6.0 | 5.0 | 18.0 | 17.7 |
| Fat | 12.9 | 13.6 | 3.9 | 4.2 |
| Protein | 64.2 | 74.3 | 57.0 | 66.2 |
| Free fatty acids (oleic)* | 29.2 | 5.3 | 64.1 | 48.7 |

*g free fatty acids/100 g total fat.*

The moisture content of the flying gurnard was very low and indicative of good keeping properties during storage of the powder for use as supplement for the local foods. The low moisture content of the powder would inhibit the growth of many opportunistic mesophilic and pathogenic microorganisms. Both drying methods used seem to dry the fish adequately. However, solar drying resulted in a much higher level of free fatty acids in the flying gurnard due to oxidation processes. In the case of common bogue, both drying methods resulted in high free fatty acids content.

**Micronutrient content of fish and recommended daily allowances**

Table 5 shows the mean values of the minerals in the fish powder while Table 6 shows the percentage Recommended Daily Allowances (RDAs) in 20 g of fish powder for the various micronutrients. The micronutrient content of the dried fish powder showed highest level of zinc in Tuna, highest Iron content in flying gurnard, high calcium content in all the fish species but, especially in the flying gurnard (Table 5). Also magnesium content was highest in the anchovies, followed by woevi. With respect to selenium, the highest content was in woevi.

**Table 5. Micronutrient content of dried fish powder from the fish species**

| Fish type | Mean Values | | | | |
|---|---|---|---|---|---|
| | Zn mg/100 g | Fe mg/100 g | Ca mg/100 g | Mg mg/100 g | Se µg/100 g |
| Woevi | 29.0 | 17.2 | 2994.0 | 852.0 | 184.0 |
| Flying Gurnard | 7.0 | 31.6 | 3448.0 | 345.0 | 168.0 |
| Tuna | 36.9 | 20.2 | 1786.0 | 667.0 | 111.0 |
| Anchovies | 10.1 | 26.1 | 2852.0 | 1095.0 | 174.0 |

**Table 6. Percentage recommended daily allowance in 20 g fish powder**

| Minerals mg/100 g | RDA | | | | |
|---|---|---|---|---|---|
| | RDA mg/day | Percentage in 20 g Fish Powder (%) | | | |
| | | Woevi | Flying Gurnard | Tuna | Anchovies |
| Zn | 14 | 41 | 10 | 53 | 14 |
| Fe | 18 | 19 | 35 | 22 | 29 |
| Ca | 1200 | 50 | 57 | 30 | 48 |
| Mg | 370 | 46 | 19 | 36 | 59 |

**Microbiological analysis of dried fish powder**

Microbial examination of a processed food product provides information that serves as the most important criterion for judging the success of the process using the effectiveness of the production controls as well as the microbiological stability and safety of the food. It was observed that due to the effectiveness of the drying process, all the samples had low and acceptable bacterial and fungal loads. Significantly, pathogens were absent from the fish, indicating the wholesomeness of the dried fish and fitness for human consumption. The absence of *Escherichia coli* in the dried fish products showed that there was no faecal contamination of the fish. Coliforms, other than *Escherichia coli* are a good indicator of unsatisfactory processing techniques or sanitation procedures during handling. The absence of these organisms therefore showed that proper and hygienic procedures were used during the drying process employing both mechanical and solar dryers.

## Sensory evaluation of fish powder in foods fed to school children

*Appearance*: The appearances of all the foods were liked with mean score above 7.0 (Table 7). However, apranpransa with tuna frames, rice with tuna frames stew, rice with anchovies stew, banku with flying gurnard okro stew, apranpransa with flying gurnard, banku with woevi okro stew, jollof with woevi stew, jollof with tuna frames, banku with anchovies okro stew and rice with flying gurnard stew were rated highest on appearance with a mean score above 8.0. On the 9-point hedonic scale, a score above 8.0 implies that the appearances of all those foods were liked very much. ANOVA indicated that the mean rating of 7.49 for rice with woevi stew was significantly lower than that of the other foods. Nevertheless, on the hedonic scale, 7.49 correspond to moderate liking, thus showing that the sample was acceptable in appearance.

*Colour*: As with appearance, the school children liked the colour of all the meals, giving a rating of above 7.0 which indicated a moderate liking on the hedonic scale (Table 7).

*Aroma*: Rice with woevi stew had a mean score of 6.9 which means it was liked slightly. On the whole, the aromas of all the meals were acceptable as is evident by the fact that none of them was rated below 6 (like slightly) on the hedonic scale (Table 7).

*Texture*: Only apranpransa had the texture rated and were rated on the positive side of the hedonic scale (Table 7). Apranpransa with woevi had the lowest rating with mean score of 7.14 which on the hedonic scale means liked moderately and texture of apranpransa with Flying Gurnard was liked very much (8.17).

*Mouthfeel:* The school children did not really like how the rice with woevi stew felt in their mouth (6.7), however they liked how the other foods felt. The other foods had mean rating above 7.3 which is moderately liked on the hedonic scale. Banku with anchovies okro stew was liked very much (8.3). According to ANOVA, the foods were not significantly different from each other except rice with woevi which was rated low (Table 7).

*Taste:* In terms of taste, only rice with woevi stew was slightly liked. All the foods were rated high with a mean score above 7.4; especially banku with anchovies okro stew was liked very much. Turkey's post test showed that the taste of the foods were liked very much and were not significantly different from each other (Table 7).

*Overall Acceptability:* For overall acceptability, the school children rated all the foods on the positive side of the hedonic scale (Table 7). All the foods were accepted by the children especially banku with anchovies okro stew (8.4). According to the analysis of variance, Tukey's test, the foods were not significantly different from each other.

**Table 7. Means (±SD) and significance[1] for consumer acceptance testing of underutilized fish incorporated in fish samples**

| Food Sample | Sensory Attributes[2] | | | | | | |
|---|---|---|---|---|---|---|---|
| | Appearance | Color | Aroma | Texture | Mouthfeel | Taste | Overall Acceptability |
| Rice with woevi stew | 7.5 ± 2.3$^a$ | 7.0 ± 2.5$^a$ | 7.0 ± 2.3$^a$ | 0.0 ± 0.0 | 6.7 ± 2.2$^a$ | 6.8 ± 2.3$^a$ | 6.9 ± 2.2$^a$ |
| Jollof with anchovies | 7.6 ± 1.6$^a$ | 7.6 ± 1.7$^a$ | 7.3 ± 1.6$^a$ | 0.0 ± 0.0 | 7.4 ± 1.3$^{ab}$ | 7.5 ± 1.4$^{ab}$ | 7.5 ± 1.2$^{ab}$ |
| Banku with tuna f. okro stew | 7.9 ± 1.6$^{ab}$ | 7.0 ± 1.9$^a$ | 7.5 ± 1.7$^{abcd}$ | 0.0 ± 0.0 | 7.3 ± 1.6$^{ab}$ | 7.5 ± 1.6$^{bc}$ | 7.5 ± 1.4$^{ab}$ |
| Jollof with flying g. | 7.9 ± 1.6$^{abc}$ | 7.9 ± 1.6$^b$ | 7.9 ± 1.3$^{bcd}$ | 0.0 ± 0.0 | 7.5 ± 1.3$^{bcd}$ | 7.7 ± 1.3$^{bcd}$ | 7.8 ± 1.1$^{abcde}$ |
| Aprapransa with anchovies | 7.9 ± 1.7$^{abc}$ | 7.8 ± 1.6$^b$ | 7.8 ± 1.8$^{bcd}$ | 0.0 ± 0.0 | 7.9 ± 1.9$^{bcdef}$ | 7.7 ± 1.8$^{bcd}$ | 7.8 ± 1.7$^{bcde}$ |
| Aprapransa with woevi | 8.0 ± 1.7$^{abc}$ | 7.4 ± 1.8$^{ab}$ | 7.5 ± 1.6$^{abcd}$ | 0.0 ± 0.0 | 7.5 ± 1.6$^{bc}$ | 7.5 ± 1.8$^{ab}$ | 7.5 ± 1.8$^{ab}$ |
| Aprapransa with tuna f. | 8.1 ± 1.5$^{abc}$ | 7.7 ± 1.4$^{ab}$ | 7.6 ± 1.5$^{abcd}$ | 7.4 ± 1.4$^b$ | 7.8 ± 1.3$^{bcdef}$ | 7.4 ± 1.5$^{ab}$ | 7.7 ± 1.3$^{bc}$ |
| Rice with tuna frames stew | 8.2 ± 0.9$^{abc}$ | 8.2 ± 0.8 | 8.1 ± 1.0$^{cd}$ | 7.1 ± 1.6$^b$ | 8.0 ± 1.0$^{bcdef}$ | 8.0 ± 1.2$^{bcd}$ | 8.4 ± 1.1$^{bcde}$ |
| Rice with anchovies | 8.2 ± 1.3$^{abc}$ | 7.5 ± 1.5$^{ab}$ | 7.6 ± 1.6$^{abcd}$ | 7.8 ± 1.7$^c$ | 7.6 ± 1.3$^{bcde}$ | 7.7 ± 1.5$^{bcd}$ | 7.7 ± 1.1$^{bcde}$ |
| Banku and flying g. okro stew | 8.2 ± 0.9$^{abc}$ | 7.7 ± 1.2$^{ab}$ | 7.5 ± 1.7$^{abcd}$ | 0.0 ± 0.0 | 7.9 ± 1.1$^{bcdef}$ | 8.1 ± 0.9$^{bcd}$ | 7.9 ± 1.0$^{bcde}$ |
| Aprapransa with flying gurnard | 8.3 ± 0.8$^{bc}$ | 8.3 ± 0.7$^{cdef}$ | 8.3 ± 0.6$^{ed}$ | 8.2 ± 0.8$^d$ | 8.2 ± 0.8$^{def}$ | 8.3 ± 0.6$^d$ | 8.2 ± 0.7$^{cde}$ |
| Banku and woevi okro stew | 8.4 ± 0.9$^{bc}$ | 8.2 ± 0.9$^{cdef}$ | 8.1 ± 0.9$^{cd}$ | 0.0 ± 0.0 | 8.2 ± 1.0$^{cdef}$ | 8.2 ± 0.8$^{cd}$ | 8.2 ± 0.8$^{cde}$ |
| Jollof with woevi stew | 8.5 ± 0.8$^{bc}$ | 8.5 ± 0.7$^{def}$ | 8.2 ± 0.9$^{cd}$ | 0.0 ± 0.0 | 7.5 ± 1.1$^{bcd}$ | 8.1 ± 1.2$^{bcd}$ | 8.0 ± 0.6$^{bcde}$ |
| Jollof with tuna f. | 8.6 ± 0.6$^{bc}$ | 8.6 ± 0.6$^{def}$ | 8.3 ± 0.7$^{de}$ | 0.0 ± 0.0 | 7.9 ± 0.9$^{bcdef}$ | 7.9 ± 0.8$^{bcd}$ | 8.2 ± 0.6$^{cde}$ |
| Banku with anchovies okro stew | 8.6 ± 0.5$^{bc}$ | 8.3 ± 0.5$^{def}$ | 8.2 ± 0.8$^{de}$ | 0.0 ± 0.0 | 8.3 ± 0.7$^f$ | 8.3 ± 0.7$^d$ | 8.4 ± 0.6$^e$ |
| Rice with flying g. stew | 8.6 ± 0.7$^c$ | 8.3 ± 0.9$^{def}$ | 8.2 ± 1.1$^{de}$ | 0.0 ± 0.0 | 8.3 ± 0.9$^{ef}$ | 7.7 ± 1.3$^{bcd}$ | 8.4 ± 0.9$^{de}$ |

*[1] values in the same column with different superscripts are significantly different at p<0.05.*
*[2] sensory attributes were evaluated on a 9-point hedonic scale as follows: 1- dislike extremely/ 9- like extremely.*

## 4. CONCLUSION AND RECOMMENDATIONS

This study shows the potential of using low cost, but highly nutritious fisheries resources found locally in combating malnutrition. The combination of low cost, high nutritional value, simple technology and acceptability among the children testing the products is unique. Characterisation of the fish species showed that the selected fish are of high nutritional significance in either human food supplements or formulations, as it has high protein content, good general amino profile and abundance of polyunsaturated fatty acids. As for all the tested fish products, the high nutritional content of one-man-thousand was notable; although generally regarded as fish for the poor, it was well accepted in the tested diets for food supplementation in the school feeding programme. The other products tested were also of high nutritional value and could contribute significantly in the fight against malnutrition. The high level of acceptance among the children, who ate and evaluated the different products, showed that the products tested were not only highly nutritious, but also highly accepted by schoolchildren. Of particular interest was the high acceptance of the powder made of tuna frames, since this is a product available in high quantities and at low cost. The product based on tuna frames was highly accepted, and could also open up the possibility of using this highly nutritious ingredient as a supplement in traditional foods.

The study revealed the potential use of fish powder in local foods in combating malnutrition, which should be further explored. Cost analyses and optimization of the technology should be further developed. The technology should be extended and transferred to local producers at artisanal level and eventually for the industrial level of fish powder production to meet the needs of Ghanaian households in the rural and urban communities in order for better applicability. With such approach, the fish powder will be extensively utilised in both homes and schools in Ghana. Research into the health and nutritional benefits of using this product in local diets should be further encouraged, and the use of this product in local foods could be promoted. Getting fish products into the menu of existing school feeding programmes would be key to the success of introducing this highly nutritious product.

## 5. REFERENCES

**Agostoni, C., Tryan, S., Bellu, E., Riva, E. & Giovanni, M.** 1995. Neuro-development quotient of healthy term infants at 4 months and feeding practice: The role of long chain polyunsaturated fatty acids. *Paediatric Research* 38, pp. 262–266.

**Anonymous.** 1994. Fisheries Research and Utilization Branch Report. Ministry of Food and Agriculture, Ghana.

**AOAC.** 1990. Official Methods of Analysis (15th edn.). Association of Official Analytical Chemists, Washington, DC.

**Batista, I., Pires, C., Bandarra, N.M. & Gonçalves, A.** 2001. Chemical characterization and preparation of salted minces from bigeye grunt and longfin bonefish. *Journal of Food Biochemistry*. 25 (6), pp. 527–540.

**Bjerve, K.S., Thoresen, L., Bønaa, K., Vik, T., Johnsen, H. & Brubakk, A.M.** 1992. Clinical studies with alpha-linolenic and long chain n-3 fatty acids. *Nutrition* 8, pp.130–135.

**Bykov, V.P.** 1985. Marine Fishes. Russian Translation Series 7 A.A. Balkema, Rotterdam.

**Clucas, I.J.** 1985. Fish handling, preservation and processing in the tropics. Part 1. Report of the Tropical Products Institute (TPI), No. G144 p. 9.

**Cockburn, F.** 1997. Proceedings of a Conference on Food, Children and Health held at The Royal Society of Medicine. Wimpole Street, London.

**Da Costa, A. & Stern, J.A.** 1956. The calcium and phosphorus contents of some foreign and domestic canned sardines. *Food Research*, 21, pp. 242–249.

**Egass, E. & Braekkan, O.R.** 1977. The selenium content in some Norwegian fish products. *Fiskeridirektoratets Skrifter, Serie Ernaering* 1, pp. 87–91.

**FAO.** 1962. Fish in Nutrition. Papers presented at the Washington conference. Eds Eirik Heen & Rudolf Kreuzer Fishing News (Books) Ltd. London.

**FAO.** 1995. Quality and quality changes in fresh fish, *FAO Fisheries Technical Paper* 348, pp. 93–117

**Friedman. M.** 1996. Nutritional value of proteins from different food sources. A review. *J. Agric. Food Chem.* 44, pp 6–49.

**Garrow, J.S. & James, W.P.T.** 1993. Human Nutrition and Dietetics. Churchill, Livingstone, London.

**Gordon, D.T. & Roberts, G.L.** 1977. Mineral and proximate composition of Pacific Coast fish. *J Agric. Food Chem.* 25, pp. 1262–1268.

**Gormley, T.R., Walshe, T., Hussey, K. & Butler, F.** 2002. The effects of fluctuating vs constant frozen storage temperature regimes on some quality parameters of selected food products. *Lebensmittel-Wissenschaft und-Technologie* 35, pp 190–199.

**Huss, H.H.** 1988. Le poisson frais, sa qualité et altération de qualité. Manuel prepare pour le programme de perfectionnement FAO/DANIDA sur la technologie du poisson. *FAO Fisheries Series* No. 29. Rome. Italy.

**Iwasaki, M. & Harada, R.** 1985. Proximate and amino acid composition of the roe and muscle of selected marine species. *J. Food Sci.* 50, pp. 1585–1587.

**Julshamn, K., Haugnes, J. & Utne, F.** 1978. The contents of 14 major and minor elements (minerals) in Norwegian fish species and fish products, determined by atomic absorption spectrophotometry. *Fiskeridirektoratets Skrifter, Serie Ernaering* 1, pp.117–135.

**King, D.R. & Poulter, R.G.** 1985. Frozen storage of Indian mackerel (*Rastrelliger kanagurta*) and big eye (*Priacanthus hamrus*). *Trop. Sci.* 25, pp.79–90.

**Koletzko, B.** 1992. Fats for the Brain. *European Journal of Clinical Nutrition* 46 (suppl 1), pp. 551–562.

**Newman, W. P., Middaugh, J. P., Propst, M. T. & Rogers, D. R.** 1993. Arteriosclerosis in Alaska nativesand non-natives. *The Lancet* 342 pp. 1056–1057.

**NMKL 71.** 1990. Nordic Committee on Food Analysis. Salmonella Bacteria: Detection in Foods, 2nd edition.

**NMKL 125.** 2005. Nordic Committee on Food Analysis. Enterobacteriaceae: Detection in Foods.

**NMKL 86.** 2006. Nordic Committee on Food Analysis. Aerobic Bacteria: Detection in Foods.

**Paetow, A., Schober B. & Papenfuss, H.J.** 1966. The chemical composition and organoleptic quality of fish from West African fishing grounds. *Fisherei-Forschung Wissenschaftliche* 4, pp. 99–101.

**Podsevalov, V.N. & Perova, L.A.** 1975. Fat content of mackerel and scad flesh in relation to month of catching. *Rybnoe Khozyaistvo* 3, p. 72.

**Pozo, R., Perez, V. & Saitua, E.** 1992. Total lipids and omega-3- fatty acids from seven species of pelagic fish. In Pelagic fish: The Resource and its exploitation, J. R. Burt, R. Hardy and K. J. Whittle, (Eds.), pp. 142–147. Fishing News Books.London, UK.

**Putro, S.** 1990. Setting up a processing line. In: *Infofish International*. 1/90, pp. 44–48.

**Sidwell, V.D.** 1981. Chemical and Nutritional Composition of Finfishes, Whales, Crustaceans, Mollusks and their Products. Technical Memorandum, National Oceanic and Atmospheric Administration, National Marine Fisheries Service, US Department of Commerce, Washington, DC, p. 432.

**Sinhuber, R.O. & Yu, T.C.** 1958. 2-Thiobabituric acid method for the measurement of rancidity in fishery products: II. The quantitative determination of malonaldehyde. *Food Technol.* 12 (1), pp. 9–12.

**Smith, J.G.M., McGill, A.S., Thomson, A.B. & Hardy, R.** 1980. Preliminary investigation into the chill and frozen storage characteristics of scad (Trachurus trachurus) and its acceptability for human consumption. *In*: Advances in Fish Science and Technology, ed. Connell J.J. pp. 303–307. Fishing News Books, Oxford.

**Talabi, S.O., Fetuga, B.L. & Ologhobo, A.** 1980. Utilization of big-eye fish, *Brachydeuterus auritus* meal and fish protein concentrate production: A preliminary evaluation. *Advances in Fishery Science and Tech.* pp. 335–338.

**Teeny, F.M., Gauglitz, E.J. Jr., Hall, A.S. & Houle, C.R.** 1984. Mineral composition of the edible muscle tissue of seven species of fish from the Northeast Pacific. *J. Agric. Food Chem.* 32, pp. 852–855.

**Toppe, J., Albrektsen, S., Hope, B. & Aksnes, A.** 2007. Chemical composition, mineral content and amino acid and lipid profiles in bones from various fish species. *Comp. Biochem. Physiol.* 146B, pp. 395–401.

**Urdahl, N.** 1992. By-products from pelagic fish. In Pelagic fish: The Resource and its exploitation, J. R. Burt, R. Hardy and K. J. Whittle, (Ed.) p. 142–147. Fishing News Books.

**Windsor, M. & Barlow, S.M.** 1981. Introduction to fishery by-products. Fishing News Books, Oxford.

# MARINADES DE COQUILLAGES ET FRUITS DE MER DES ILES DU SALOUM ET DE FADIOUTH AU SENEGAL

## *[MARINADES OF SHELLFISH AND SEAFOOD OF THE SALOUM ISLANDS AND FADIOUTH IN SENEGAL]*

by/par

Momar Yacinthe Diop[1]

### Résumé

Ces technologies parmi lesquelles les marinades de coquillages et fruits de mer, constituent des innovations sur un marché qu'il faut animer en permanence par de nouvelles recettes et/ou de nouveaux conditionnements.

Les coquillages et fruit de mer sont des aliments très fragiles. La méthode traditionnelle de conservation consiste pour la plupart à cuire et ensuite sécher les produits, avec une mobilisation en temps et ressources sans commune mesure avec le faible rendement et la rémunération y afférente.

L'appropriation par les communautés organisées de techniques de conservation et de valorisation des coquillages et fruits de mer, respectueuses de l'environnement va permettre à ces dernières de fabriquer des produits finis sains, de bonne qualité et mieux rémunérateurs. Il s'agit de produits transformés sous un label qualité, et surtout des produits à valeur ajoutée. La douceur du goût, le fondant et l'utilisation facile de ces produits répondent bien à l'évolution de la demande actuelle.

Les consommateurs apprécient bien ce nouveau produit qui se vend très bien dans les foires, marchés traditionnels ou hebdomadaires (1000 à1500F le bocal de 325 ml. Les femmes qui s'adonnent à cette activité ont vu leur revenu augmenter très substantiellement (1000 à 6000F CFA) par bassine de 40 kg de matière première.

Les résultats se traduisent aussi par le renforcement des pratiques locales de gestion développées par les femmes pour garantir l'exploitation durable de la ressource. Cette nouvelle technologie est potentiellement génératrice de beaucoup d'emplois et présente des atouts économiques majeurs qui méritent beaucoup plus d'attention. L'expérience a toutefois montré qu'il reste encore beaucoup d'efforts à fournir pour positionner ce produit d'une valeur commerciale très importante.

*Mots clés: Marinades, Communautés de pêche, Coquillages, Conservation, Valeur ajoutée, Îles, Marché*

### Abstract

These technologies including pickles of shellfish and seafood are innovations on a market that should constantly evolve with new recipes or new packages.

Shellfish and seafood are very sensitive food. The traditional preservation method is mostly cooking and then drying the products, with a mobilization in time and resources incommensurable with the subsequent low yield and income.

Ownership by organized communities of environment-smart preservation and utilization techniques of shellfish and seafood, will allow the latter to manufacture finished products that are healthy, of good quality and better returns. These are processed products under quality label and mostly value-added products. The softness of taste, fondant and the easy use of these products meet the evolution of the current demand.

Consumers appreciate well this new product well sold in, fairs, traditional or weekly markets (1000 to 1500F CFA per 325 ml jar) Women who engage in this activity saw their income increase very substantially (1000 to 6000F CFA) per 40 kg pan of raw material.

[1] Chef Atelier Poisson et Produits Halieutiques, Institut de Technologie Alimentaire, BP2765, Route Des Pères Maristes, Hann Dakar, Sénégal. momaryacinthe@yahoo.fr

Results also reflect the strengthening of local management practices developed by women to ensure the sustainable exploitation of the resource. This new technology is potentially generating many jobs and has major economic assets that deserve much more attention. Experience has however shown that much more effort remains to better place on the market this product of a very important commercial value.

*Key words: Marinades, Shellfish, Fishing communities, Preservation, Added value, Islands, Market*

## 1. INTRODUCTION

Le marché est axé sur une meilleure qualité et sécurité sanitaire des produits, qu'ils soient frais, congelés ou transformés. Une évolution de taille par rapport aux décennies précédentes est que la demande en poisson n'est plus uniquement tributaire de l'augmentation de consommation de protéines par habitant, mais aussi des préférences des consommateurs plus exigeants en matière de sécurité sanitaire, de qualité gustative et de convenance d'utilisation de produits.

Il convient alors pour les professionnels du secteur de la pêche, de répondre à l'attente des consommateurs avec des produits naturels, bons pour la santé, savoureux et pratiques, dans un marché où le développement passe par l'innovation. L'innovation est un élément indispensable sur un marché qu'il faut animer en permanence par de nouvelles recettes ou de nouveaux conditionnements.

Les coquillages constituent une civilisation dans les îles du Saloum et de Fadiouth. Ils jouent un rôle crucial dans l'alimentation humaine et constituent la principale source de revenus des femmes. L'exploitation des coquillages est le seul secteur de la pêche où les femmes contrôlent toute la chaîne de production. Elles procèdent à la collecte, transformation et vente des produits finis.

L'insuffisance des équipements, l'hygiène environnementale, la qualité des produits finis, les problèmes d'emballage et de conditionnement font que ces produits transformés sont le plus souvent de moindre qualité et consommés au niveau national.
Il s'agit maintenant d'innover le secteur par la mise au point de nouveaux produits qui répondent au goût et besoin des populations locales, mais aussi qui peuvent être exportés au profit des producteurs.

L'Atelier Poisson et Produits Halieutiques de l'Institut de Technologie Alimentaire (ITA) a eu à développer ces dernières années des technologies très appropriées dans le domaine de la valorisation du poisson et des coquillages. Ces technologies parmi lesquelles les marinades de coquillages et fruits de mer, constituent des innovations majeures.
L'objectif principal des activités de marinage est le renforcement de la capacité d'utilisation responsable des ressources halieutiques en vue d'une meilleure sécurité alimentaire et pour des revenus accrus des communautés de pêche.

L'appropriation par ces communautés organisées de ces techniques de conservation et de valorisation des coquillages et fruits de mer, respectueuses de l'environnement vont permettre à ces dernières de fabriquer des produits finis sains, de bonne qualité et mieux rémunérateurs. Il s'agit de produits transformés sous un label qualité, et surtout des produits à valeur ajoutée. La douceur du goût, le fondant et l'utilisation facile de ces produits répondent bien à l'évolution de la demande actuelle.

## 2. MATERIEL ET METHODE

### MATERIEL

Il est constitué par les coquillages, animaux à corps mous protégés le plus souvent par une coquille et qui vivent sur le fond enfoui ou en surface. Les espèces ciblées dans ce document sont les arches et les huitres:

- Les arches (Arca senilis) sont des mollusques bivalves. Ce sont des animaux filtreurs qui se nourrissent de petites particules en suspension dans l'eau. Elles colonisent tous les types de fonds meubles sableux ou vaseux. L'espèce peut atteindre 30 ans d'âge pour une taille maximale de 75 mm.

- L'huître est un mollusque bivalve, c'est-à-dire un animal emprisonné entre deux valves calcaires aussi dures que sa chair est "molle", qu'il peut à peine entrouvrir pour filtrer l'eau qui lui apporte sa nourriture principale : le plancton. Car l'huître est un organisme filtreur, microphage et omnivore. Elle se nourrit de diatomées (microalgues) dont la "navicule bleue" (qui lui donne sa couleur vert pâle, des spores d'algues), des organismes microscopiques et débris divers.

Elle appartient à la famille des Ostréidés qui contient plusieurs genres (Voir Annexe 1).

## METHODE

Traditionnellement les technologies utilisées par les femmes transformatrices pour le traitement des mollusques sont la fermentation pour le cymbium et le murex, la cuisson pour les huîtres et les coques. Les produits sont ensuite séchés.

«Le marinage», nous paraît la technologie la mieux adaptée au contexte local, vue la simplicité de sa mise en œuvre.

Les femmes ont été initiées aux procédés de fabrication de marinades.

### *Marinage*

Le marinage est l'opération qui consiste à immerger des animaux marins ou parties d'animaux marins dans une marinade chauffée ou non, pendant un temps suffisant pour substituer une partie de leur eau de constitution par du vinaigre ou par un acide organique.

Le marinage confère au produit des qualités de saveur particulières (goût acide) et lui assure une certaine durée de conservation.

L'objectif du marinage est de réduire l'activité bactérienne responsable de l'altération du produit (poisson, mollusque), notamment les germes pathogènes, en procédant à une acidification du produit par le vinaigre ou un acide organique autorisé.

Il existe plusieurs types de marinades. Dans cette expérience, nous avons procédé au marinage à chaud. L'abaissement du pH a un effet de ralentissement sur les micro-organismes responsables de l'altération du produit, ce qui permet d'augmenter leur durée de conservation. C'est le principe de ce procédé qui est utilisé (pH < 4,5), permettant d'inhiber les bactéries responsables des toxi-infections (Staphylocoques, Salmonella, Clostridium perfringens et botulinum) comme on peut le constater dans les résultats des analyses microbiologiques (Annexe 2).

Le vinaigre qui représente le bain de conditionnement, est utilisé dilué (degré alcoolique 8°). L'ajout de légumes et autres ingrédients va nécessiter un bain dit de macération.

Le but recherché est la conservation des denrées en vue d'en différer la consommation: Les micro-organismes étant le principal obstacle, il s'agit soit de les éliminer, soit de les rendre inopérants. Le traitement thermique par pasteurisation, vise à allonger la durée limite de consommation (DLC) tout en préservant les qualités de goût du produit. L'action du pH reste prépondérante.
- 
### *Quelques précautions pour la préparation d'une marinade:*

- Une nouvelle utilisation du même bain de macération augmente les risques de prolifération bactérienne.
- Dans la préparation, le vinaigre doit être limpide, sans odeur de moisi, et sa teneur en acide lactique clairement indiquée. L'eau utilisée a une grande importance pour la conservation pour les qualités organoleptiques du produit. Il est déconseillé aux femmes de se servir d'eau d'origine douteuse. Une eau même potable peut ne pas convenir en dilution dans une marinade (présence de calcaire, etc.).
- Le sel devra être à usage alimentaire et contenir la plus petite quantité possible de chlorures de calcium et de magnésium : le premier confère un goût amer. L'addition de végétaux comme les oignons peut modifier l'équilibre eau/acide de l'ensemble. Il convient de laisser macérer les végétaux au préalable pour éviter cet inconvénient. Les oignons peuvent, par suite de fermentation, conférer un goût désagréable et même altérer la semi-conserve. Pour éviter cet inconvénient, on peut les faire tremper deux à trois jours dans une solution constituée de 2L de vinaigre et 150 g de sel par kg d'oignons.

- Certains légumes (carotte, poivron…) et épices (piment) sont tranchés et cuits à la vapeur.
- ns le conditionnement :
  - La concentration en acide doit être surveillée et le pH maintenu au-dessous de 4,5. L'effet inhibiteur est d'autant plus efficace que la teneur en sel est proche de 5 pour cent et que le pH est bas.
  - Le remplissage doit être fait à refus : le contact du produit avec l'air peut entraîner la formation de moisissures. Certaines de ces moisissures peuvent utiliser l'acide acétique augmentant ainsi le pH et, par voie de conséquence, permettre la croissance des bactéries déjà présentes.
  - Le rapport produit/liquide doit être respecté. En aucun cas, il ne faut changer ce rapport sous peine de voir le pH modifié.

**Figure 1. Etapes de fabrication de marinades d'arches**

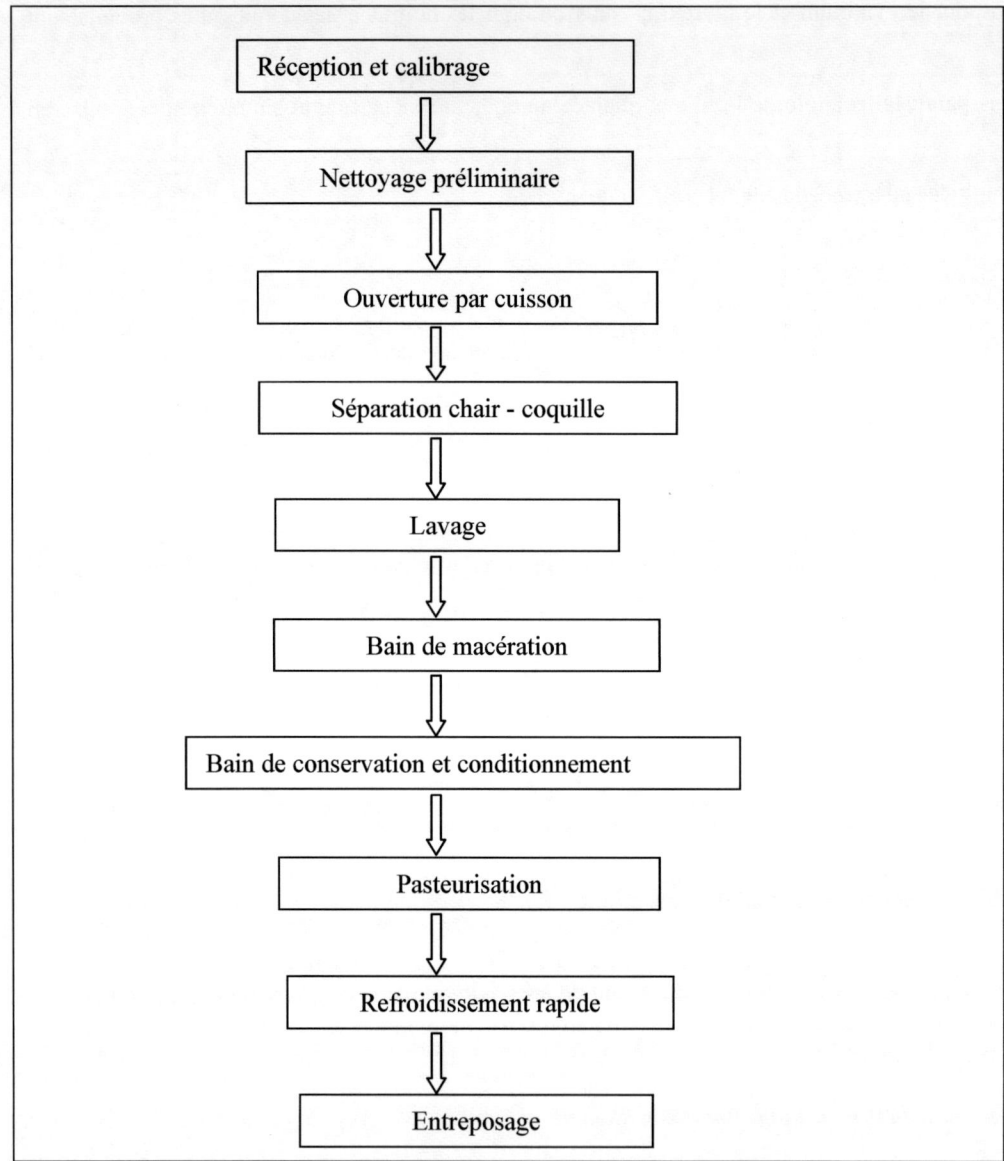

## 3. ETUDE COMPARATIVE DES DEUX MODES DE VALORISATION

Avec le repos biologique, la campagne annuelle pour l'exploitation des arches dure six mois et celle des huitres, quatre mois. Dans les îles du Saloum, il y a un phénomène de remontée des eaux suivie d'une marrée basse appelée « wâmé », que l'on observe trois fois dans le mois. Pendant cette période dite « wâmé » les femmes peuvent effectuer trois sorties dans le mois pour récolter des arches et cueillir des huitres.

## Arches

Pour chaque sortie, elles sont au nombre de vingt cinq (25) et louent une pirogue à raison de 300 F CFA/personne. Chacune pourra récolter tout au plus trois (03) bassines, les arches étant difficiles à exploiter. Ce qui fait que chaque femme pourra avoir 9 bassines d'arches par mois, soit 54 bassines au total pour toute la campagne.

Chaque bassine d'arches pèse 40 kg.

## Huitres

Pour chaque sortie, elles sont au nombre de dix (10) et louent une pirogue à raison de 500 F CFA/personne. Chaque femme pourra cueillir jusqu'à dix bassines d'huitres et quelques fois plus. Les huitres sont plus faciles à exploiter, mais elles prennent du volume dans la pirogue, ce qui explique la limitation du nombre de femmes à dix. Ce qui fait que chaque femme pourra obtenir 30 bassines d'huitres par mois, soit 120 bassines au total pour toute la campagne.

Chaque bassine d'huitres pèse 20 kg.

## Bois

Les femmes peuvent trouver du bois mort dans la mangrove, mais elles préfèrent acheter le tas assez volumineux vendu à trois mille F CFA qu'elles peuvent utiliser pendant trois mois.

## Main d'œuvre

Les femmes transformatrices ne payent pas la main d'œuvre. Elles vivent généralement dans de grandes concessions avec plusieurs membres d'une même famille. C'est cette famille qui constitue la main d'œuvre. Cela permet de préserver les bénéfices au sein de la famille et de mieux contrôler la vente. Ce sont des micros entreprises familiales.

## Valorisation des arches

Revenu mensuel d'une femme transformatrice.

### Méthode artisanale

- Coût location pirogue par mois : 900 F CFA
- Bassine
  - o Durée de vie : 1 an
  - o *Amortissement mensuel 1 : prix de l'élément/durée de vie = 1000/12= 85F CFA*
- Marmite de cuisson
  - o Durée de vie : 5 ans
  - o *Amortissement mensuel 2 : 15 000/60=250*
- Emballage (sac de jute) : 100 F CFA
- Bois : 1000 F CFA

Total des charges = 900 + 85+ 250+ 100+ 1000 = 2335 F CFA

| Poids matière première (kg) | Poids arches décoquillées (kg) | Poids arches Séchées (kg) | Total des charges (F CFA) | Prix de vente (F CFA) |
|---|---|---|---|---|
| 360 | 22,5 | 10 | 2335 | 10 000 |

Vente mensuelle = poids arches séchées x prix au kilogramme = 10 x 1000 F= 10 000F CFA
**Revenu mensuel** = vente mensuelle – total des charges = 10 000 – 2335 = **7665 F CFA**

**Méthode améliorée (marinage)**
- Coût location pirogue par mois : 900 F CFA
- Bassine
    - Durée de vie : 1 an
    - *Amortissement mensuel 1 : prix de l'élément/durée de vie = 1000/12= 85F CFA*
- Marmite de cuisson
    - Durée de vie : 5 ans
    - *Amortissement mensuel 2 : 15 000/60=250*
- Emballage (bocaux en verre) : 24 300 F CFA
- Vinaigre : 3200 F CFA
- Ingrédients (sel, légumes, etc.…): 4 000 F CFA
- Bois : 1000 F CFA

Total des charges = 900 + 85+ 250 + 24 300 + 3200 +4000 + 1000 = 33 820 F CFA

| Poids matière première (kg) | Poids arches décoquillées (kg) | Marinades arches en bocaux | Total des charges (F CFA) | Prix de vente (F CFA) |
|---|---|---|---|---|
| 360 | 22,5 | 108 | 33 820 | 108 000 |

Vente mensuelle = nombre de bocaux x prix d'un bocal = 108 x 1000 F= 108 000 F CFA
**Revenu mensuel = 108 000 – 33 820 = 74 180 F CFA**

*Valorisation des huitres*

**Méthode artisanale**
- Coût location pirogue par mois : 900 F CFA
- Bassine
    - Durée de vie : 1 an
    - *Amortissement mensuel 1 : prix de l'élément/durée de vie = 1000/12= 85F CFA*
- Marmite de cuisson
    - Durée de vie : 5 ans
    - *Amortissement mensuel 2 : 15 000/60=250*
- Emballage (sac de jute) : 100 F CFA
- Bois : 1000 F CFA
- Eau : 250 F CFA

Total des charges = 900 + 85+ 250+ 100+ 1000 + 250 = 2585 F CFA

| Poids matière première (kg) | Poids huitres décoquillées (kg) | Poids huitres Séchées (kg) | Total des charges (F CFA) | Prix de vente (F CFA) |
|---|---|---|---|---|
| 600 | 37,5 | 18,750 | 2585 | 65 625 |

Vente mensuelle = poids huitres séchées x prix au kilogramme = 18,75 x 3500 F= 65 625F CFA
**Revenu mensuel : 65 625 – 2585 = 63 040 F CFA**

**Méthode améliorée (marinage)**
- Coût location pirogue par mois : 900 F CFA
- Bassine
    - Durée de vie : 1 an
    - *Amortissement mensuel 1 : prix de l'élément/durée de vie = 1000/12= 85F CFA*
- Marmite de cuisson
    - Durée de vie : 5 ans
    - *Amortissement mensuel 2 : 15 000/60=250*
- Emballage (bocaux en verre) : 33 750 F CFA
- Vinaigre : 6400 F CFA

- Ingrédients (sel, légumes, etc.…): 8 000 F CFA
- Bois : 2000 F CFA
- Eau : 250 F CFA

Total des charges = 900 + 85+ 250 + 33 750 + 6400 +8000 + 2000 + 250 = 51 720 F CFA

| Poids matière première (kg) | Poids huitres décoquillées (kg) | Marinades huitres en bocaux | Total des charges (F CFA) | Prix de vente (F CFA) |
|---|---|---|---|---|
| 600 | 37,5 | 150 | 51 720 | 180 000 |

**Revenu mensuel : 180 000– 51 720 = 128 280 F CFA**

## 4. DISCUSSIONS

Cette nouvelle technologie est très rentable dans l'ensemble. Avec la méthode artisanale, le rendement est faible et le prix de vente mensuel très bas (7665 F pour les arches et 65625 F pour les huitres avec des charges respectives de 2335 F et 2585 F).

Pour la même quantité de matière première, avec le marinage le prix de vente mensuel est assez conséquent (108 000 F pour les arches et 180 000 F pour les huitres avec des charges respectives de 32 820 F et 51 720 F).

La technique du marinage permet de présenter d'excellents produits prêts à être consommés sans aucune autre forme de préparation. Par ailleurs, la durée limite de consommation est de six mois ce qui laisse une marge suffisante au réseau de distribution pour en assurer la vente.

Ces coquillages consommés par la majorité de la population sénégalaise de manière traditionnelle (séché ou fermenté), une fois valorisés sous forme de marinades commencent à bien pénétrer le marché.

Les marinades de coquillages et fruits de mer sont très bien vendus au niveau des foires et d'autres événements (veillées religieuses, séances de lutte traditionnelle). Les femmes arrivent à écouler en moyenne plus de 100 bocaux par jour.

La filière de la collecte, transformation et vente de coquillages implique beaucoup de femmes qui ont en charge de larges familles. Pendant la période active, elle permet aux actrices de bénéficier de revenus qui permettront de subvenir à leurs dépenses quotidiennes. En adoptant adéquatement le marinage, elles vont non seulement régler les dépenses immédiates mais en plus épargner pour les besoins durant la période inactive (repos biologique).

Quelques femmes ont rencontré un cas d'altération sur des huitres, quelques heures après leur préparation.

Cette forme d'altération est d'origine enzymatique. Peu de temps après la mise en marinade, le liquide de couverture prend une apparence opalescente, en raison du passage en solution sous forme colloïdale de quantités considérables de glycogène provenant de la chair. La teneur en glycogène rapportée à la chair humide peut atteindre 7 pour cent (Houwing). Cette teneur élevée est à l'origine d'une forme d'altération qui se caractérise par une clarification du liquide de couverture accompagnée d'une forte chute du pH. On peut observer sur le fond du pot un dépôt gris-blanchâtre.

La cause de cette altération réside dans le fait qu'une partie des enzymes glycogénolytiques des huitres peut échapper à la destruction lors du traitement thermique d'ouverture/cuisson. Ces enzymes entrent en activité après la mise en marinade pour former du glucose à partir du glycogène en présence. La marinade devient alors un excellent milieu nutritif pour la multiplication des bactéries lactiques qui transforment le glucose en acide lactique. Il a été recommandé aux femmes pour corriger cette anomalie, une forte cuisson à la vapeur après décoquillage.

Nous avons vu que cette activité de récolte de coquillages est physiquement très exigeante. Passé un certain âge, les femmes n'ont plus de condition physique pour pratiquer cette activité. Avec l'augmentation substantielle des revenus par l'application du marinage, les femmes pourront limiter cette activité diminuant la charge de la collecte, ce qui aura des conséquences bénéfiques sur leur santé.

Plusieurs améliorations pourront aider les actrices à mieux gérer cette filière, qui sont regroupées en deux aspects à savoir :

- Aspect gestion biologique
  - L'application du repos biologique devrait être accompagnée d'une aide des autorités par la mise en place de projets de développement ;
  - Fixer et faire respecter la taille  minimale des captures ;
  - Vulgariser les informations sur les périodes de reproduction ;
  - Réglementer les périodes de capture.
- Aspect gestion économique
  - La formation en  BPH et BPF permettra d'obtenir des produits artisanaux de meilleure qualité mieux conditionnés, et vendus au meilleur prix ;
  - Promotion des investissements en matériel par l'Etat ;
  - Formation en assurance qualité.

## 5. CONCLUSION ET RECOMMANDATIONS

Aujourd'hui plusieurs indices nous permettent d'affirmer que les produits de marinade de coquillages commencent à bien s'implanter dans le paysage national et même international :

- Acceptabilité des produits marinés par les consommateurs (tests sensoriels)
- Vente importante des produits marinés au niveau des foires et marchés,
- Formation et renforcement de capacité au niveau de plusieurs sites de production
- Plusieurs demandes de formation par les femmes des îles qui n'ont pas encore accès à la technologie  de marinage de coquillages et fruits de mer,
- Formation de femmes Gambiennes à la technologie par leurs homologues de l'île de Moundé. Cette formation s'est tenue dans plusieurs départements de la Gambie ;

Cependant nous avons noté des points faibles au niveau de la filière :

- Données biologiques et écologiques peu connues par les femmes (taille des prises) ;
- Insuffisance d'équipements adéquats et d'infrastructures, en effet les femmes utilisent comme emballage des caisses en carton de fortune pour les produits finis
- Insuffisance des moyens de transport : les femmes acheminent les produits finis par autobus vers les marchés, ce qui entraine quelquefois des pertes ;

Il s'agit maintenant de:

- Sensibiliser les femmes sur les paramètres biologiques et écologiques, la taille des captures,
- Former toutes les femmes des îles qui n'ont pas encore accès à cette nouvelle technologie,
- Etablir des règles de gestion de cette filière,
- Renforcer la surveillance au niveau des sites de prélèvement,
- Encourager le secteur privé à investir dans la filière,
- Continuer à assurer la promotion des produits par leur commercialisation au niveau des grandes surfaces, national et international.

## 6. BIBLIOGRAPHIE

**Diop, M.Y.** 2009. Rapport de formation des femmes « Suivi sur les modes de traitement et de  valorisation des mollusques dans les îles de Dionewar,Falia, Djirnda, Fadiouth, Ndangane Sambou, Bassoul,Thialane et Moumdé », 22 p.

**Equipe RAP-UMR LEMAR.** 2011, Programme femmes et coquillages « rapport final », 10 p.

**Guérin, C**. 2005. Les produits transformés : panés, marinades, surimi, tartinables, Agrocampus, Rennes, France, 4-9 p.

**Knockaert, C.** 1989. Les marinades des produits de la mer (IFREMER, Collection Valorisation des Produits de la Mer- Service de la Documentation et des Publications (SDP)), 8 ;15-18 p.

**Malang, S.** 2005, Le marinage du poisson, EISMV/HIDAO, 4 p.

**PHOTOS DES COQUILLAGES OBJET DU PROGRAMME DE VALORISATION**

**Figure 1.** *Cymbium sp* **(Volute)**

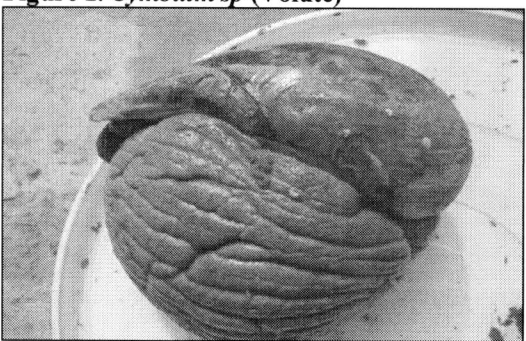

***Source:*** Guide de suivi bioécologique des coquillages exploités dans les iles de Niodior, Dionewar, Falia et de Fadiouth.

**Figure 2.** *Murex cornutus* **(Rocher)**

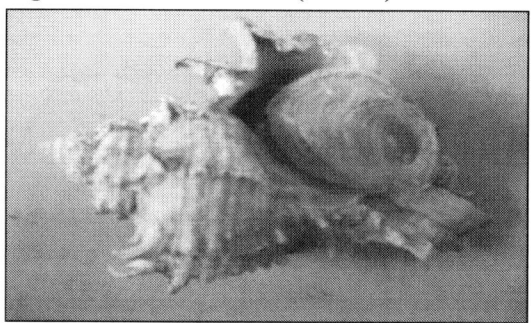

***Source:*** Guide de suivi bioécologique des coquillages exploités dans les iles de Niodior, Dionewar, Falia et de Fadiouth.

**Figure 3.** *Murex duplex* **(Rocher)**

***Source:*** Guide de suivi bioécologique des coquillages exploités dans les iles de Niodior, Dionewar, Falia et de Fadiouth.

**RÉSULTATS D'ANALYSES MICROBIOLOGIQUES**

| PARAMETRES | UNITE | COQUES FRAICHES | MARINADE DE COQUES |
|---|---|---|---|
| Flore aérobie totale | UFC/g | $1,5 \times 10^4$ | < 10 |
| Coliformes fécaux | UFC/g | < 10 | < 10 |
| Clostridium sulfito-réducteurs | UFC/g | $2 \times 10^1$ | 0 |
| Staphylocoques pathogènes | UFC/g | < 100 | < 100 |
| Salmonelles | UFC/25g | Abs | ABS |

| PARAMETRES | UNITE | Semi conserve d'Arches | Semi conserve de Murex | Semi conserve de Crevettes |
|---|---|---|---|---|
| Flore aérobie totale | UFC/g | $1,8 \times 10^4$ | $3,7 \times 10^2$ | $1,3 \times 10^3$ |
| Coliformes fécaux | UFC/g | < 10 | < 10 | < 10 |
| Clostridium sulfito-réducteurs | UFC/g | < 10 | < 10 | < 10 |
| Staphylocoques pathogènes | UFC/g | < 100 | < 100 | < 100 |
| Salmonelles | UFC/25g | Abs | Abs | Abs |

**DIAGRAMME DE PRODUCTION**

Le décoquillage des coques fait subir au produit une cuisson suffisante pour rendre inutile un traitement de macération indispensable dans le cas du marinage à froid.

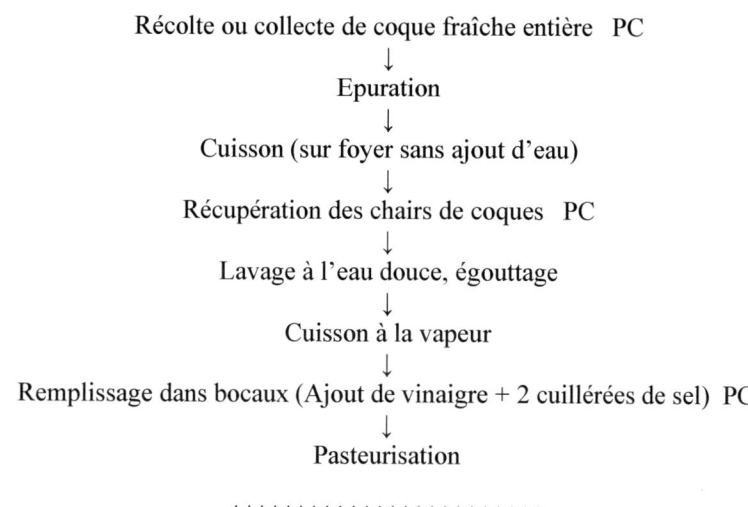

**Processus de fabrication d'arche mariné (pagne)**

Récolte ou collecte de coque fraîche entière   PC
↓
Epuration
↓
Cuisson (sur foyer sans ajout d'eau)
↓
Récupération des chairs de coques   PC
↓
Lavage à l'eau douce, égouttage
↓
Cuisson à la vapeur
↓
Remplissage dans bocaux (Ajout de vinaigre + 2 cuillérées de sel)  PC
↓
Pasteurisation

\*\*\*\*\*\*\*\*\*\*\*\*\*\*\*\*\*\*\*\*\*\*\*

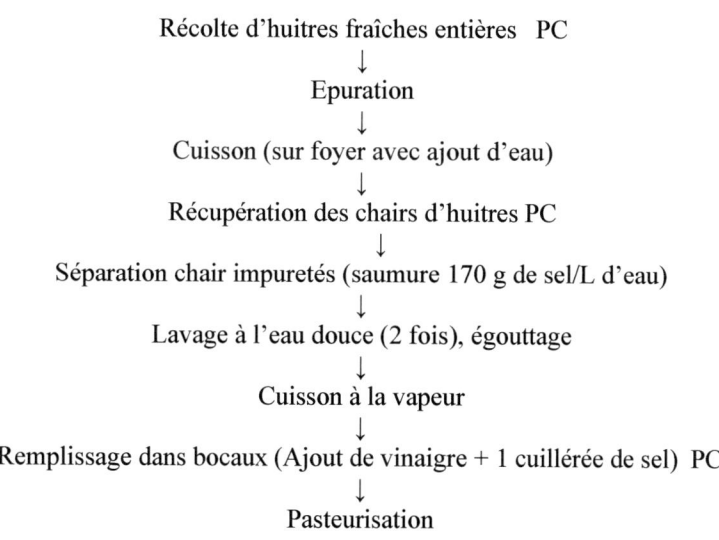

**Processus de fabrication d'huitre mariné (yokhoss)**

Récolte d'huitres fraîches entières   PC
↓
Epuration
↓
Cuisson (sur foyer avec ajout d'eau)
↓
Récupération des chairs d'huitres PC
↓
Séparation chair impuretés (saumure 170 g de sel/L d'eau)
↓
Lavage à l'eau douce (2 fois), égouttage
↓
Cuisson à la vapeur
↓
Remplissage dans bocaux (Ajout de vinaigre + 1 cuillérée de sel)  PC
↓
Pasteurisation

NB : L'épuration s'effectue pendant 24h dans une bassine remplie d'eau de mer à laquelle sont ajoutées 3 gouttes d'eau de javel.
PC : Point critique

**Processus de fabrication du cymbium mariné (yeet)**

Récolte de cymbium frais PC
↓
Epuration
↓
Apprêt (casse, éviscération)
↓
Blanchiment, grattage peau, Coupure en petits bouts PC
↓
Cuisson à la vapeur
↓
Mise en boîte
↓
Remplissage dans bocaux (Ajout de vinaigre + 1 cuillérée de sel)  PC
↓
Pasteurisation


\*\*\*\*\*\*\*\*\*\*\*\*\*\*\*\*\*\*\*\*\*\*\*


**Processus de fabrication du murex mariné (touffa)**

Récolte de Touffa
↓
Epuration
↓
Apprêt (casse, éviscération)
↓
Blanchiment, grattage peau PC
↓
Cuisson à la vapeur
↓
Mise en boîte
↓
Remplissage dans bocaux (Ajout de vinaigre + 1 cuillérée de sel)  PC
↓
Pasteurisation

# CAMDEBOO SATELLITE AQUACULTURE PROJECT: CANNING CATFISH FOR THE FOOD SERVICE INDUSTRY

## [PROJET AQUACULTURE SATELLITE : POISSON CHAT MIS EN CONSERVE POUR L'INDUSTRIE DU SERVICE DE L'ALIMENTAION]

by/par

Liesl de la Harpe[1]

### Abstract

Due to an 80 percent drop in the South African pilchard quota since 2004, the production of alternate sources of fish for canning presents a viable and market driven business opportunity, particularly as approximately 89 percent of South Africans rely on pilchards for some of their protein and micronutrient intake. In order to satisfy current demand for canned fish 1,500 satellite farms (namely Catfish) would need to be established.

This paper takes stock of the work completed and planned initiatives by the Camdeboo Satellite Aquaculture Project (CSAP) The CSAP Model consists of one central management farm surrounded by 50 identical satellite farming systems within 150 km radius, each cluster has the potential to provide a nutritional food source to 900,000 people. So far several phases have been successfully concluded, including the training and skills development of some key players, the product development and market acceptance assessment, the bankable feasibility study and ongoing incubation phase. The future will focus on strategic marketing initiatives to better position the product, a model replication and utilization of by-products from processing.

*Key words: Camdeboo, South Africa, Catfish farming, Canning, Pouch, Food service*

### Résumé

Du fait de la chute du quota du pilchard depuis 2004, la production de sources alternatives de poisson pour la conserverie représente une perspective d'affaire commerciale opportune et viable, surtout que 89 pour cent des Sud Africains dépendent du pilchard pour certains de leur protéine et micronutriments. Pour satisfaire la présente demande pour la conserve de poisson 1.500 fermes satellites (notamment de poisson-chat) devront être établies.

Ce document rend compte du travail mené et les initiatives programmées par le Camdeboo Satellite Aquaculture Project (CSAP). Le Modèle CSAP consiste en une ferme central gérée entourée par 50 systèmes satellites identiques de culture dans un rayon de 150 km, chaque groupe a un potentiel de produire une source d'aliment nutritifs à 900.000 personnes. Jusqu'ici différentes phases ont été conduites avec succès, y compris la formation et développement de compétence de certains des acteurs, le développement de produit et l'évaluation de l'acceptation par le marché, l'étude de faisabilité et la phase d'incubation en cours. Le future sera centré sur les initiatives de marketing stratégique pour mieux positionner le produit, la réplication du modèle et l'utilisation des coproduits de la transformation.

*Mots clés: Camdeboo, Afrique du Sud, Culture du poisson-chat, Conserverie, Poche pour emballage, Service de l'alimentation*

## 1. INTRODUCTION

Climate change, over-fishing and pollution are rapidly depleting fish stocks worldwide, with far reaching implications on food security and aquatic biodiversity. To protect water resources and ensure sustainable fishing production, quotas and restrictions on ocean fishing have been implemented. Due to an 80 percent drop in the South African pilchard quota since 2004, the production of alternate sources of fish for canning presents a viable and market driven business opportunity, particularly as approximately 89 percent of South Africans rely on pilchards for some of their protein intake.

---

[1] Blue Karoo Trust, PO Box 534, Wolwas Road 6280, Graaff-Reinet, South Africa. liesl@blue-karoo.co.za

The Camdeboo Satellite Aquaculture Project (CSAP) is an innovative and highly replicable model which presents a comprehensive solution to address fish supply shortfalls, create sustainable self-employment opportunities for rural women, promote pro-poor economic growth, and encourage social equity in the Camdeboo. The Project thus has a high level of support at local, provincial and national government levels as well as from the private sector.

The CSAP concept centres around the establishment of aquaculture clusters which comprise a central management farm and a network of satellite farming systems. Each aquaculture cluster is designed to produce 1,166 tonnes of fish per month and provide employment to 500 people, primarily rural women. The aquaculture clusters benefit both through economies of scale as a result of their collaboration as well as from the support, training and mentorship provided by the central structure.

Due to the fact that the Camdeboo is a dry region with distinct rainfall patterns, farmers have developed a practical and highly suitable irrigation and storage infrastructure which is currently underutilized. The Project will promote collaboration between unemployed rural women, many of whom reside on isolated farms, and water resource owners to utilize these resources in order to establish commercially viable and sustainable fish farms. This collaboration will centre on the water being borrowed from the daily farming activities for use in the recirculating aquaculture system for a predetermined period and then returned, for use as irrigation water, in an enriched form due to high nutrient levels in the effluent water.

The availability of these underutilized resources provides an enormous local competitive advantage, and thus an opportunity to develop a new economic activity and income stream based on the region's physical and social characteristics.

In order for the women to operate the aquaculture production systems (farm the fish) effectively and to participate as "partners" in a commercially viable business, an intensive training plan has been developed, and is currently being piloted with our first group of trainees. Participants are gaining a sense of purpose and personal fulfillment as well as the opportunity to become self-reliant and empowered members of the community.
The freshwater fish produced by the aquaculture clusters will be processed and packaged in 3 kg retort pouches and A10 cans in order to increase shelf life and sold at an affordable price in order to fulfill the growing gap caused by the reduction in the annual pilchard quota.

## 2. OVERALL DEVELOPMENTAL OBJECTIVE OF THE PROJECT

To reduce the number of households living below the poverty line by creating sustainable self-employment of rural women and pro-poor economic growth.

To produce an affordable, nutritious food source for low income groups as an alternative to, and thereby reducing reliance on, overexploited wild fish stocks.

## 3. RELEVANCE OF THE PROJECT TO MILLENNIUM DEVELOMENT GOALS

The Project is directly relevant to the following Millennium Development Goals (MDG):

MDG 1: Eradicate extreme poverty and hunger
MDG 3: Promote gender equality and empower women
MDG 7: Ensure environmental sustainability

## 4. PROJECT FRAMEWORK

The CSAP project framework is broken into the following five phases:

### Design (concluded)

With financial support from Thina Sinako (a joint venture between the European Union and Eastern Cape Provincial Treasury) the design of the Project took place between November 2007 and June 2008. The primary objective during this phase was to determine in principle the feasibility of the Project as a viable and sustainable commercial venture. Primary activities included environmental authorization, project design, BBBEE assessment, feasibility study, supplier and input assessment, compilation of legal documentation, formulation of a training plan and the identification and appraisal of associates, sub-contractors, stakeholders and implementing partners.

### Product development and market acceptance (concluded)

Following the design of the Project, a decision was taken to develop the product and ensure its marketability prior to implementing the initiative. Financial support from IDC and DBSA/Cacadu LEDI culminated in the development of three "canned freshwater fish in tomato sauce" products. These products were subjected to an independent Market Acceptance Survey (funded by ECDC and the UK Department for Environment, Food and Rural Affairs (DEFRA) to establish whether the canned fish products were acceptable to the bulk catering target market, as well as how these products compared to pilchards. Based on results from the survey, catfish in tomato sauce was selected as the best suited product for the CSAP.

### Bankable feasibility study (concluded)

This phase involved updating and finalizing information for the farming and factory business plans as well as the packaging of the selected product in bulk packaging in order to reduce costs. The finalized documentation was submitted to the relevant PSC Members for evaluation in terms of the following criteria: Financial, Technical, Operational, Marketing, Social, Environment.

### Incubation (active)

The next step in the development of CSAP is the two-year Incubation Phase which will focus on the establishment of the first central management farm as well as the processing facility. The central management farm will provide training, employment, support and mentorship to 62 local community members, the majority of whom will be rural women. Through the collaboration with various SMME service providers, approximately 40 additional jobs will be sustained through the farming and processing activities, and an additional 40 rural women will have access to a skills development opportunity. The processing facility will be established to receive process, package and market the product – providing permanent employment to approximately 15 to 25 people.

The success of the Incubation Phase will give commercial lenders confidence in the concept of farming freshwater fish for canning, which is essential to the delivery of the next phase of the Project.

### Commercialisation (future)

This phase will commence in April 2014 with the aim of duplicating the aquaculture production systems refined during the Incubation Phase of CSAP, and in so doing, launch 50 satellite farms around the central farm, thus establishing the first cluster. Thereafter, this "cluster model" will be duplicated in similar rural areas, such as mining and farming communities, as well as around canneries which are not operating at full capacity. In order to satisfy current demand for canned fish in South Africa, 1,500 satellite farms would need to be established. This translates to 30 clusters, at least 15,000 direct jobs (the majority of which will be self-employment for rural women), and a nutritional food source for 27,000,000 people during the Commercial Phase of this initiative. The potential for duplication of this initiative throughout Africa, and even globally, will be explored during the Commercial Phase, as well as the development of additional value-added products such as fish fingers, fish cakes and other sauce variations for canned fish.

## 5. PRODUCT DEVELOPMENT

### Primary Product

The CSAP have developed a sterile, shelf-stable fish product which is an affordable source of protein and micronutrients. The product, which has been named Karoo Catch, is a minced catfish in tomato sauce which is packaged in 3 kg retort pouches and A10 cans. Although cans are conventionally used in this industry, the equivalent cost to produce a final product in a can, is R0,70 per kg more expensive than using a pouch. Based on the cost comparison, cans are not being considered for the CSAP at this stage.

Production of the CSAP preserved fish product includes the following activities:

- Fish are harvested at 1 kg, electrocuted and transported to the processing facility
- Fish are then brined in a salt water and ice solution in order to remove slime
- Once placed on the draining table, fish are decapitated just behind the skull, and the tail is removed
- Remaining viscera are manually removed and the fish are gently rinsed
- Fish are then minced, inclusive of bones and skin
- The minced fish is then mixed together with tomato sauce, water and vitacel, and heated to 85 °C
- The pouch / can is hot filled and sealed
- The pouch / can is then placed in the retort and sterilized at 121 °C for a set period of time.

### By-products

Heads, tails and viscera (offal or waste) will account for 32 percent of the raw material processed in the factory. These by-products present a potential income source for the CSAP through the:

- Production of animal feed, other than catfish feed.
- Manufacturing of fish meal (ingredient for animal feed or used for soil improvement).
- Usage as a raw material with other biomass for the manufacturing of compost through aerobic digestion.
- Usage for the generation of energy through anaerobic digestion together with other bio-degradable wastes.

All of these options will be explored in order to utilize these by-products to generate additional revenue for the CSAP.

## 6. MARKETINGING STRATEGY

The freshwater fish produced by the aquaculture clusters will be processed and packaged and sold locally at an affordable price to bulk markets including caterers, public sector kitchens (prisons, hospitals, schools, etc.) and feeding schemes (school feeding schemes, other community feeding schemes, etc.). The intention is not to compete with established brands in formal markets but rather to provide a sustainable and cost-effective bulk source of protein and essential micronutrients directly to kitchens.

### Primary Target Market: National School Nutrition Programme (NSNP)

During 2009/10, the NSNP fed 1,480,907 learners in the Eastern Cape Province (71 percent of all learners in the province) daily, thus requiring 5,065 tonnes of tinned fish per annum in order to provide each learner with a weekly allowance of 90g of this nutritional food source.

Nationally, the South African NSNP requirement in 2009/10 financial year was 22,499 tonnes per annum. At present, a large percentage of this requirement is being met through the importing of canned fish due to local demand being far higher than local supply. The sharp increase in import quantities however attracts another problem to South Africa namely substandard quality. Nutritional Value analysis conducted on pilchards canned in certain other countries, for example, revealed variations between the data specified on the label and that found when testing the actual contents of the can.

### Market Acceptance

Selection of the species, product and packaging for the BKT's product has been guided by formal interaction with the target market through two separate surveys.

The first market acceptance survey was conducted early in 2011 in order to establish which of the canned fish in tomato sauce products, and therefore fish species, was most acceptable to the target market, as well as how these products compared with canned pilchards. Over 1,000 one-on-one surveys (inclusive of product tasting) were conducted at schools, prisons, hospitals, military and the public over a two month period. The outcome of the survey was that the large catfish was selected as the product of choice, and scored in a very similar range to pilchards.

During the second survey, five bulk catering networks (FoodBank PE, Nelson Mandela Bay Municipality "War On Hunger Campaign", prisons, schools and hospitals) were approached in October 2011 in order to determine the market acceptance of the developed product (i.e. minced whole catfish in tomato sauce) and to select the packaging option of choice. Seventeen organizations were each given a 3 kg tin or pouch and were requested to prepare a meal, as they would usually, in their own environment. The meal was then tasted by various end users and a one-on-one survey conducted to determine various product attributes and packaging preferences. The overall feedback indicated that the product was highly acceptable in terms of the overall presentation, quality and taste. There was no preference regarding packaging (i.e. 3 kg tin or 3 kg pouch). However, it was found that, despite the pouch being unfamiliar to all participants, all that used it selected the pouch as their option of choice. The pouch has thus been selected as the packaging option with which to proceed based on its ease of use, cost-effectiveness and practicality.

## *Positioning*

A Proudly South African product that is high in protein and essential micronutrients, competitively priced and creates new jobs in the local market.

## *Competition and Pricing*

Karoo Catch will be marketed as a nutritional alternative for products high in protein, and as such needs to be competitive with other canned fish products, more specifically canned pilchards. Successful penetration into the market will be determined, to a large extent, by the pricing, quality, as well as the supply consistency of the product in comparison with similar products.

In addition, CSAP's product has the following undisputed competitive advantages over its main competitors:

- The bulk 3 kg pouches eliminate the need to open an equivalent of just over seven cans to gain the same volume of product.
- There is no need for a can opener – which can be a challenge in many rural areas.
- The product is "ready to use" and there is no need to remove bones, scales or eggs as in canned pilchards.
- The nutritional value of the product is consistent (due to it being a farmed product), compared to pilchards which are harvested and imported from across the world which results in inconsistent nutritional content.
- And very importantly, the production of a farmed fish is consistent, predictable and reliable compared to the wild fish stocks which are seasonal, and even then, not always available.

## *Branding*

The brand name, logo design and all aspects of communication must be consistent and drive a message to the consumer that communicates that the product is a safe and a high quality product.

| | |
|---|---|
| Name: | Karoo Catch |
| Tagline: | Fish–a-Women Empowering Themselves |
| Brand Personality: | Katie is a typical hardworking and caring farm worker that has now chosen to empower herself through training and skills development and has taken ownership of her future. She will have her own recipe book and contributions will be made to it by her customers. |
| Endorsements: | Applications for the following endorsements are in process - Proudly South African, SABS, Heart and Stroke Foundation, Halaal and Kosher |

## 7. DEVELOPMENTAL IMPACT OF THE PROJECT

### *Human resource development, job creation and food security*

During the Incubation Phase, the first central farm will be established and training, employment, support and mentorship provided to 62 local community members, the majority of whom will be rural women. An additional 40 rural women will have access to a skills development opportunity. Through the collaboration with various SMME service providers, approximately 40 additional jobs will be sustained through the farming and processing activities of the Incubation Phase alone. Priority will be given to local SMME service providers wherever possible throughout the development of the Project. The planned output during this Phase is 792 tonnes of fish per year which will meet the nutritional requirements of 15.6 percent of the NSNP in the Eastern Cape.

During the Commercial Phase, 50 satellites will be established around the central farm, providing the same training and employment opportunities to an additional 438 local community members and a food source to 900,000 people.

Further down the line, and in order to satisfy current SA demand for tinned fish, 30 clusters (i.e. 30 central farms and 1,500 satellite farms) would need to be established. This translates into the creation of at least 15,000 direct jobs (the majority of which will be self-employment for rural women) and an affordable source of protein and nutrients to approximately 27,000,000 people.

## Model replication

Success of the Incubation Phase of the CASP will enable BKT to prove the business concept and thus give commercial lenders confidence in the concept of producing freshwater fish for canning. Commercial lenders are in place, pending the successful implementation of the Incubation Phase. The Commercial Phase will see the launch of 50 satellite farms around the central farm, thus establishing the first cluster. Thereafter, this "cluster model" will be duplicated in similar rural areas, such as mining and farming communities, as well as around canneries which are not operating at full capacity due to the reduction in fishing quotas.

## Macro-economic indicators (in terms of Employment and GDP)

It is envisaged that, once the CSAP initiative is fully implemented (factory, central farm and 50 satellite farms), approximately 500 direct permanent and sustainable job opportunities would be created in the Camdeboo Local Municipality. Based on general accepted industry ratios, a total of 2,751 new job opportunities can thus be expected as a result of the implementation of just one cluster.

| Variable: Labour | Base |
|---|---|
| Annual production value from project (R million) | 358.2 |
| Direct (A) labour impact | 500 |
| Indirect (B) labour impact | 1 342 |
| Direct + indirect (A+B) labour impact | 1 842 |
| Induced (C) labour impact | 909 |
| Total (A+B+C) labour impact | 2 751 |

Similarly, the direct and indirect effects that this Project can potentially have on the GDP of the Camdeboo are outlined in the table below. These figures are based on the anticipated turnover of the BKT processing facility (selling the final product), and includes the facility's own contribution to GDP.

| Variable: GDP | Base |
|---|---|
| Annual production value from project (R million) | 358.2 |
| Direct (A) GDP impact | 34.2 |
| Indirect (B) GDP impact | 225.6 |
| Direct + indirect (A+B) GDP impact | 259.8 |
| Induced (C) GDP impact | 168.2 |
| Total (A+B+C) GDP impact | 428 |

## 8. SUSTAINABILITY

### Financial/Economic Sustainability

Thorough financial analysis over a 10 year period indicates that, through the cooperation between the various separate entities, each entity and the venture as a whole can be economically viable. The central farm will achieve financial sustainability though the income generated by the three production systems, the hatcheries as well as the small percentage of sales which each satellite farm will pay to the central farm for support services.

### Institutional Sustainability

The central farm will be managed as a fully fledged business. It will operate as the hub of the entire cluster, and a base from which the initiative can grow, supplying broodstock, providing training, extensions services, negotiating bulk purchasing and supply agreements, marketing the end products, and managing the activities of the various entities. In addition, the central farm will supply each satellite farm with the system design, facilitation of system construction, business plan and projected financials, and operating manuals at no cost. The assistance which each satellite farm will receive on setup and the provision of ongoing support services from the central farm will equip the individual entities for success.

### *Environmental Sustainability*

Climate change is inevitable. The ocean temperatures are rising, and the water is acidifying, upsetting the ecological balance of the oceans, and impacting negatively on fish species. The impacts of changes in fish yield, distribution, catch variability and seasonality of production have an enormous effect on the growing number of people worldwide who are food insecure. Continued reliance on the wild stocks is this not sustainable.

CSAP offers an alternative, yet sustainable and predictable, method of fish production, within a controlled environment, which could potentially even contribute to carbon sequestration to some extent by providing raw material for biofuels which may be used to generate heat for the system as an alternative to electricity, whilst reducing reliance on wild fish stocks. This enhances the value of aquaculture as an important source of animal protein with a smaller carbon footprint, relevant potential for additional mitigation of carbon release into the atmosphere, and thus a low carbon developmental path.

## 9. CONCLUSION

The use of recirculating aquaculture is a highly effective method of breeding fish in captivity with minimal water usage. Given the rapid depletion of ocean fish and the simultaneous demand for the product due to population growth and growing awareness that fish is a healthy source of protein and essential micronutrients, the CSAP presents a viable, environmentally sound, sustainable and cost effective method of bridging the gap between supply and demand.

Due to an 80 percent drop in the South African pilchard quota since 2004, the production of alternate sources of fish for preservation in cans and or pouches presents a viable and market driven business opportunity, particularly as approximately 89 percent of South Africans rely on pilchards for some of their protein intake. Based on the CSAP Model, which consists of one central management farm surrounded by 50 identical satellite farming systems within 150km radius, each cluster has the potential to provide a nutritional food source to 900,000 people.

The job creation potential of the CSAP will make a valuable contribution towards solving one of South Africa's greatest challenges – unemployment. During the Incubation Phase alone, which involves the establishment of the first central management farm, the Project will provide employment to 62 currently unemployed people – the majority of whom will be rural women, whilst sustaining approximately 40 jobs within the SMME sectors, and providing a training opportunity for an additional 40 local community members who will be future satellite farmers.

The Commercial Phase, which is set to commence in April 2014, will see the duplication of the refined aquaculture production systems, will see the launch of 50 satellite farms around the central farm creating skills development and employment opportunities for an additional 438 people. In order to supplement the gap caused by the drastic reduction in the pilchard quota, and thus satisfy local demand for canned fish, 1,500 satellite farms would need to be established, which translates to 30 clusters, and the creation of at least 15,000 direct jobs (the majority of which will be for rural women).

It is understood that the Project is taking on an enormous challenge – the growing of fish on land for canning in bulk packaging, and producing a quality product which is approved by the South African National Regulator for Compulsory Specifications, is not done anywhere in the world. However, the drivers have the institutional capacity to implement the Project and, together with the project partners and project steering committee, have the knowledge, experience and expertise to develop and implement the policies, procedures and regulations which are required to oversee the implementation of the Project.

# TECHNOLOGIE AMÉLIORÉE D'EXTRACTION DE L'HUILE DE *BRYCINUS LEUCISCUS* ET VALORISATION DU TOURTEAU OBTENU APRÈS EXTRACTION

## *[IMPROVED* BRYCINUS LEUCISCUS *OIL EXTRACTION TECHNOLOGY AND UTILIZATION OF THE CAKE OBTAINED AFTER EXTRACTION]*

by/par

Oumou Traore Cissé[1]

**Résumé**

*Brycinus leuciscus* est un poisson de petite taille, très abondant dans les pêcheries du Delta Central du Niger, de Sélingué et de Manantali. Le séchage et l'extraction de son huile pour la friture et la préparation de sauce, constituent les principaux moyens de valorisation. Dans la méthode traditionnelle d'extraction il est utilisé une marmite pour faire bouillir le poisson et la récupération de l'huile est faite à la surface du récipient. Le rendement en huile est très faible et sa qualité est mauvaise. La présente recherche a été initiée pour améliorer ces paramètres et mieux exploiter cette ressource très périssable.

L'objectif est d'améliorer la technique d'extraction de l'huile de *Brycinus leuciscus*. Cet objectif a été poursuivi avec la valorisation des sous-produits issus de cette transformation et qui renforce le rendement habituellement atteint par la méthode traditionnelle d'extraction de l'huile.

Des différentes méthodes d'extraction développées, il ressort de l'analyse des résultats que la presse hydraulique a un rendement moyen (24 pour cent), plus élevé que celui de l'extraction en système d'immersion et celui de la presse simple. L'huile obtenue est de meilleure qualité physico-chimique et organoleptique.

Le cube d'assaisonnement obtenu à partir du tourteau a été bien apprécié par le panel de dégustation et la technologie est utilisée par les transformatrices de Mopti pour une commercialisation en vue de l'assaisonnement des sauces accompagnant le riz et la pâte de farine de mil.

En termes de rentabilité économique, la marge bénéficiaire de la presse hydraulique est de 29 pour cent.

***Mots clés: Brycinus leuciscus, Valorisation, Huile, Cubes, Presse, Sous-produits***

**Abstract**

*Brycinus leuciscus* is a small size fish, very abundant in Central Delta of Niger, Sélingué and Manantali's fisheries. Drying and extraction of its oil for deep frying and sauce preparation constitute the principal means of utilization. The traditional extraction method uses a pot for the boiling the fish and the recovery of oil is done on the surface of the container. The oil yield is very weak and its quality is bad.

This research was initiated in order to improve these parameters and better utilize this perishable resource. The general objective is to improve the technique of oil extraction of *Brycinus leuciscus*. This goal went further with the value-addition to the by-products resulting from this processing, which reinforces the output usually reached by the traditional method of oil extraction.

Of the different extraction methods developed, the analysis of the results shows that the hydraulic press has an average yield (24 percent), higher than the immersion system of extraction and the simple press. The oil obtained has a better physicochemical and organoleptic quality.

The seasoning cube obtained from the cake was well appreciated by the sensorial panel and the technology is used by women processors of Mopti for marketing as flavouring for sauces accompanying rice and millet paste. In terms of economic profitability, the margin of the hydraulic press is 29 percent.

***Key words: Brycinus leuciscus, Utilization, Oil, Cube, Press, Byproducts***

[1] Institut Economie Rurale, IEK, BP 258, Rue Mohamed V, Bamako, Mali. oumouni2006@yahoo.fr

## 1. INTRODUCTION

De tous les poissons de la zone d'inondation du fleuve Niger, *Brycinus leuciscus* est l'espèce la plus abondante (Daget, 1952). C'est un characidae de petite taille, du genre Alestes (maximum 100 mm de longueur standard), Figure 1.

**Figure 1. *Brycinus leuciscus* comme matière première**

Son aire de répartition comprend les bassins de la Gambie, du Sénégal, du Niger et de la Bénoué. La pêche de l'espèce est très limitée et s'effectue de novembre à février. *Brycinus Leuciscus* a une courte durée de vie. Le nombre d'individus atteignant la deuxième année est relativement faible (Niaré, Benech, 1993).

*Brycinus leuciscus* fait l'objet de pêche spéciale avec des filets très spécifiques, de faible maille. Très grasse, l'huile extraite constitue la principale matière grasse utilisée dans les préparations culinaires des zones productrices. L'huile est très prisée dans le Delta Central du Niger et représente une spéculation importante pour les communautés de pêcheurs (Quensière, 1994). Compte tenu de la mauvaise qualité physico-chimique et organoleptique de l'huile issue de *Brycinus leuciscus* fermenté, elle est moins chère que l'huile végétale (750 contre 1 000 FCFA). Avec la technique améliorée, elle coûte 1 250 FCFA.

Le tourteau obtenu après l'extraction de l'huile de *Brycinus leuciscus* est vendu au marché comme aliments volaille.

La technique traditionnelle d'extraction à partir du poisson fermenté, le conditionnement dans des barriques métalliques et les mauvaises conditions de conservation, influencent négativement la qualité de l'huile: une odeur très piquante, un goût de rancidité oxydante et une courte durée de conservation.

Pour lever ces contraintes, le projet de recherche intitulé « Amélioration de la technologie d'extraction d'huile de *Brycinus Leuciscus* » a été exécuté, sur demande des utilisateurs des résultats de la recherche du Centre Régional de Recherche Agronomique de Mopti.

Des résultats de cette recherche, il ressort une bonne qualité chimique et organoleptique, avec la méthode améliorée d'extraction d'huile à partir du poisson frais. Ces résultats ont fait l'objet de présentation à la réunion des Experts FAO sur la Technologie, l'Utilisation et la Commercialisation du poisson, tenue à Bagamoyo en Tanzanie en 2005 et les principales recommandations ont été la poursuite de l'étude par la détermination des caractéristiques chimiques et nutritionnelles de l'huile et la mise au point d'un extracteur pour le pressage de l'huile de *Brycinus Leuciscus*.

Les activités de recherche se sont alors poursuivies sur la détermination de la composition en acides gras de l'huile de *Brycinus leuciscus*. La méthodologie était la suivante :

L'extraction des lipides totaux a été effectuée sur les matières premières (têtes et troncs) et sur *Brycinus leuciscus* entier frais, *Brycinus leuciscus* entier fermenté, Têtes de *Brycinus leuciscus* frais et Tronc de *Brycinus leuciscus* frais.

Le présent rapport donne les résultats de recherche sur la composition de l'huile et les caractéristiques de l'extracteur.

L'objectif général est de contribuer à l'amélioration de la qualité de l'huile de *Brycinus leuciscus* pour une large consommation et une diversification de son utilisation, afin de valoriser cette espèce sous utilisée, mais abondante dans le Delta Central du Niger au Mali.

Les objectifs spécifiques sont:

- Développer un extracteur hydraulique pour l'huile de *Brycinus leuciscus,*
- Conduire des essais comparatifs avec les autres méthodes d'extraction
- Développer un cube d'assaisonnement à partir de tourteau de *Brycinus leuciscus.*

## 2. MATERIEL ET METHODES

Pour la mise au point de la presse hydraulique, les tests avec la presse simple et le système d'extraction à immersion ont été effectués. Suivant les résultats des différents tests (presse et système à immersion), une sélection de caractéristiques techniques a été faite pour concevoir la presse hydraulique de type bain- marie.

En termes de valorisation du tourteau issu de l'extraction de l'huile de *Brycinus leuciscus,*
des cubes d'assaisonnement ont été produits.

### Les tests d'extraction effectués

### *Matériel*

Le matériel était composé de:

- un échantillon de *Brycinus leuciscus* frais,
- une presse simple,
- un fourneau à charbon de bois,
- un système d'extraction à immersion,
- une bouteille de gaz de 6 kg,
- la presse hydraulique,
- une bouteille de gaz de 12 kg.

### *Méthodes*

Deux essais ont été effectués dans une entreprise de construction d'équipements, dans la recherche pour l'amélioration de la technologie d'extraction d'huile de *Brycinus leuciscus*.

Un kg de poisson était utilisé à chaque essai. La presse est constituée d'un cylindre en inox, un panier cylindrique perforé dans lequel le poisson est logé et une manivelle pour le presser.. Le poisson contenu dans la presse, était chauffé à l'aide d'un fourneau à charbon de bois, entre 70 et 80 °C et la température était mesurée avec un thermomètre. A la fin de l'extraction, le poisson était pressé et l'huile était récupérée dans un récipient, à travers un bec verseur. Le rendement d'extraction et les caractéristiques organoleptiques de l'huile, ont été déterminés (voir figures 2, 3, 4 et 5).

**Figure 2. Poisson dans la presse manuelle**

**Figure 3. Huile après extraction**

**Figure 4. Huile de Brycinus leuciscus**

**Figure 5. Tourteau issu de l'extraction de l'huile**

Par rapport au système d'extraction à immersion, trois essais ont été effectués, avec 1 kg de *Brycinus leuciscus* en deux répétitions. Le poisson contenu dans le passoir est plongé dans l'eau jusqu'à immersion. L'ensemble est chauffé à l'aide du gaz. La température de l'eau variait de 80 à 90 °C. La durée de l'extraction était de 30 minutes, 45 minutes et 1 heure pour chaque essai.

Deux essais ont été effectués avec la presse hydraulique. L'appareil est en acier inoxydable de type bain-marie, composé d'une cage cylindrique pour le chargement du poisson frais. Sa contenance est de 60 litres d'eau. L'alimentation se fait au gaz.

Le poisson frais est chargé dans la cage qui est plongée dans le bain-marie et le tout porté à environ 90 °C, pendant une heure trente minutes pour la première extraction. Lorsque le poisson est suffisamment cuit, les trois quart d'eau du bain-marie sont soutirés dans un récipient et le reste contenant l'huile est collecté dans un autre. *Brycinus leuciscus* encore chaud, contenu dans la cage est pressé pour récupérer le reste d'huile. Le mélange est laissé au repos pendant 5 à 10 minutes.

Après la première extraction qui dure une heure trente minutes, les suivantes durent trente (30) minutes, car l'eau du bain-marie déjà chaude, est encore réutilisée.

La capacité de chargement de la presse étant de 10 à 15 kg, la capacité journalière pour 8 heures de fonctionnement varie de 160 à 240 kg de poisson frais (voir figures 6 et 7).

**Figure 6. Presse hydraulique**

**Figure 7. Huile issue de l'extraction de *Brycinus leuciscus***

Après extraction, les échantillons d'huile ont été récupérés, séchés pour éliminer l'eau résiduelle, refroidis et filtrés à l'aide d'un tamis nylon et pesés. L'huile est conditionnée et conservée dans des bidons en plastique. Le rendement d'extraction et les caractéristiques organoleptiques de l'huile, ont été déterminés.

**Développement de cube d'assaisonnement à partir de tourteau de Brycinus leuciscus**

*Matériel*

- Tourteau
- Ingrédients
- Mixer
- Séchoir
- Tamis nylon
- Moule

*Méthode*

Le tourteau et les ingrédients broyés ont été séchés dans un séchoir, entre 45 et 55 °C pendant 12 à 14 heures. Le mélange a été broyé dans un mixer et tamisé avec un tamis nylon. La formulation des cubes a été faite dans un moule et les cubes ont été séchés à l'étuve et conditionnés dans des sachets en plastique pour la conservation (voir figure 8).

**Figure 8. Cubes d'assaisonnement**

Un panel de dégustation a été constitué pour apprécier la qualité organoleptique du produit. L'analyse se révèle comme une méthode sensible pour détecter la présence de composés volatils issus de l'oxydation des lipides. Un classement d'échantillons d'huile a été demandé à un jury, composé de 15 personnes, sur l'odeur, le goût et la couleur. L'échelle de notation était la suivante:

*Couleur*

Jaune clair = 5
Jaune or = 4
Brun = 3
Sombre = 2

*Goût et odeur*

Très bon = 5
Bon = 4
Passable = 3
Mauvais = 2

## 3. RESULTATS OBTENUS

*Composition de l'huile de Brycinus leuciscus en acides gras saturés et insaturés*

L'huile de *Brycinus leuciscus* contient des acides gras saturés, mono et poly insaturés à des proportions différentes (Graphique 1) et le profil est donné dans le Tableau 1.

**Graphique 1. Acides gras de l'huile de Brycinus leuciscus**

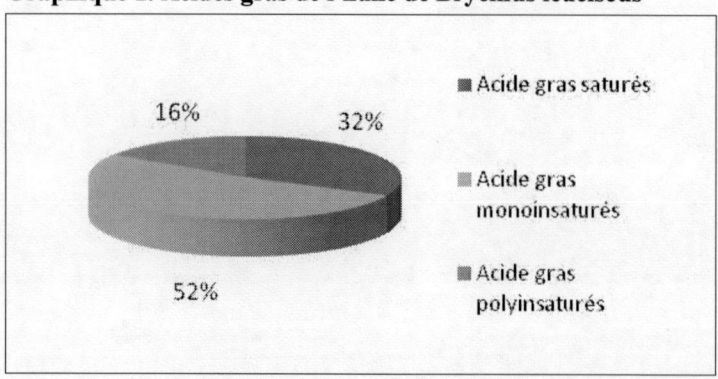

**Tableau 1. Composition en acides gras des échantillons d'huile de Brycinus leuciscus**

| Acides gras saturés | Acides gras mono insaturés | Acides gras poly insaturés |
|---|---|---|
| Myristique 14 : 0 | | |
| Palmitique 16 : 0 | Palmitoléïque 16 : 1 (n-7) | |
| Margarique 17 : 0 | | |
| Stéarique 18 : 0 | Oléïque 18 : 1 (n-9) | Linoléïque 18 : 2 (n-6) |
| | | Linolénique 18 : 3 (n-3) |
| Arachidique 20 : 0 | Gadoléïque 20 : 1 ( n-9) | Arachidonique 20 : 4 (n-6) |
| | | EPA, Ecosapentahenoïque 20: 5 (n-3) |
| | | DHA, Docosahexahenoïque 22: 6 (n-3) |

*Note :* L'huile de *Brycinus leuciscus* contient des acides gras saturés, mono et poly insaturés.

### Rendement d'extraction de l'huile de Brycinus leuciscus

L'extraction avec la presse a donné un rendement moyen en huile de 18 pour cent. Pour l'extraction en système immersion, le taux maximum d'huile est obtenu 30 minutes après que la température ait atteint 80 à 90 °C. Les rendements moyens d'extraction et les caractéristiques de la presse hydraulique sont donnés respectivement dans les Tableaux 2 et 3.

**Tableau 2. Rendements moyens avec l'extraction en système d'immersion**

| Caractéristiques % | Extraction en système à immersion (bain-marie) | | | | | |
|---|---|---|---|---|---|---|
| | Essai 1 | | | Essai 2 | | |
| | 30 min | 45 min | 60 min | 30 min | 45 min | 60 min |
| Rendement      1 | 22,5 | 22,0 | 21 ,9 | 21,8 | 21,7 | 21, 8 |
| 2 | 22,8 | 21,9 | 21,9 | 22,6 | 22,3 | 22,5 |
| 3 | 22,0 | 21,9 | 21,8 | 21,9 | 21,8 | 21,9 |
| Rendement moyen | 22,4 | 21,93 | 21,86 | 22,1 | 21,93 | 22,06 |

**Tableau 3. Données de la presse hydraulique pour 8 heures de travail par jour**

| Variable | Quantité | Unité | Prix unitaire(FCFA) | Montant en FCFA /jour |
|---|---|---|---|---|
| Capacité journalière | 240 | Kg | 200 | 48 000 |
| Main d'œuvre | 1 | H/J | 1 250 | 1 250 |
| Gaz 6 kg | 1 | Bouteille | 2 250 | 2 250 |
| Amortissement journalier de la machine | 1 | | 208 | 208 |
| Total charges | - | - | - | 51 708 |
| Quantité d'huile obtenue | 64 | Litre | 1 250 | 80 000 |
| Résidu (Tourteau) | 182 | Kg | 100 | 18 200 |
| Revenu total | | | | 98 200 |
| Bénéfice | | | | 46 492 |

Nombre de chargements par jour: 16, soit 240 kg de poisson.
Rendement en huile = 24%, soit 57,6 kg ou 64 litres.
Résidu (Tourteau) = 76%, soit 182 kg.

Production de cube d'assaisonnement à base de tourteau de Brycinus leuciscus

Le Tableau 4 donne les résultats de 15 participants du panel de dégustation par rapport au goût, à la couleur et à l'odeur du produit.

**Tableau 4. Evaluation organoleptique du cube d'assaisonnement**

| Caractéristiques | Très bon | Bon | Passable | % d'appréciation |
|---|---|---|---|---|
| Couleur | 70 | 30 | - | 100 |
| Goût | 80 | 20 | - | 100 |
| Odeur | 85 | 15 | - | 100 |
| Appréciation générale | 78,33 | 21,66 | | |

## 4. DISCUSSION

L'huile de *Brycinus leuciscus* contient les acides gras saturés représentant 32 pour cent et les insaturés 68 pour cent, avec une proportion importante d'acides gras mono insaturés (52 pour cent). Les acides gras Oméga 3 dans l'huile de *Brycinus leuciscus* sont : l'acide eicosapentanoïque (EPA), l'acide docosahexanoïque (DHA) et l'acide alpha linolénique, Oméga 6, r l'acide arachidonique et l'acide linoléique. L'huile de Brycinus leuciscus contient des acides gras insaturés essentiels pouvant être utilisés dans la nutrition.

Du point de vue caractéristique organoleptique, l'huile obtenue avec la presse a une couleur sombre, due à la température trop élevée de la presse, très proche du foyer d'alimentation du feu.

Les rendements moyens en huile, varient de 21,86 à 22,4 pour cent. La durée n'a pas influencé le taux d'extraction. Ce qui signifie qu'à 30 min après avoir atteint 80 à 90 °C, le taux maximum d'huile est obtenu.

Le rendement est meilleur à celui de la presse, malgré qu'une certaine quantité d'huile reste dans le tourteau. Les échantillons d'huile ont une bonne qualité organoleptique : couleur jaune or à jaune crème et bonne odeur.

Le rendement moyen d'extraction est de 24 pour cent. Les qualités physico-chimiques et organoleptiques sont bonnes: acidité inférieure à 1 pour cent, couleur jaune or, bonne odeur et goût agréable.

Pour une exploitation à grande échelle, une presse hydraulique d'une capacité de 100 kg est recommandée pour augmenter la quantité d'huile produite.

Les caractéristiques techniques obtenues ont permis de déterminer la rentabilité de la presse hydraulique. La marge bénéficiaire est relativement bonne.

L'extracteur est actuellement utilisé par le Programme de Compétitivité et Diversification Agricole de Mopti, au Centre de Démonstration, de Diffusion et de Prestation sur les technologies de transformation par les transformatrices de poisson.

Le cube d'assaisonnement a été bien apprécié par le panel et les quelques ménagères auxquelles, le produit a été remis pour tester dans la préparation de la sauce gombo pour la pâte de mil. Les utilisatrices ont fait comme commentaires, la forte ressemblance avec les cubes industriels Jumbo, vendus sur les marchés.

Le tourteau est utilisé pour assaisonner les sauces, comme complément pour la fortification des farines infantiles.

Le renforcement de capacité des transformatrices de poisson à la production du cube d'assaisonnement à base de poisson, permet d'augmenter les revenus et de diversifier l'utilisation de *Brycinus leuciscus*, de même que les enquêtes de marché et de consommation.

Aussi, la méfiance des consommateurs vis-à-vis de l'utilisation des cubes de fabrication industrielle décriés à tort ou à raison, pour des potentiels effets sur la santé, constitue une opportunité à la production, à la diffusion et à la consommation de cubes à base de tourteau de poisson. Ceci représentant un substitut à l'utilisation des cubes Maggi ou Jumbo, etc.

## 5. CONCLUSION ET RECOMMANDATIONS

Les rendements moyens en huile, avec le système à immersion, varient de 21,86 à 22,4 pour cent.

Le rendement moyen d'extraction avec la presse hydraulique est de 24 pour cent."L'huile de *Brycinus leuciscus* contient des acides gras polyinsaturés bénéfiques pour la nutrition et la santé.

La presse hydraulique pour l'huile de *Brycinus leuciscus* a un meilleur rendement et une bonne qualité physico-chimique et organoleptique.

La technologie a été adoptée par les transformatrices de poisson et fait l'objet d'une exploitation familiale et au niveau associative. Elle a été conçue et adaptée au niveau de la production des transformatrices de poisson. Une presse de grande capacité (100 kg) est envisageable pour augmenter la productivité.

Les tests de dégustation ont conclu une bonne appréciation du produit constitué par le cube d'assaisonnement par les consommateurs.

Les analyses sont en cours pour déterminer la qualité physico-chimique et nutritionnelle du cube d'assaisonnement à base de tourteau de *Brycinus leuciscus*.

La fiche technique de production de cube d'assaisonnement à base de *Brycinus leuciscus* et le référentiel technico- économique pour la presse hydraulique, sont disponibles.

En termes de recommandations, les actions suivantes sont à prévoir pour une meilleure valorisation de la technologie d'extraction de l'huile de *Brycinus leuciscus*:

- Les enquêtes de consommation et des études de conservation/ conditionnement et de marché du cube d'assaisonnement à base de *Brycinus leuciscus* ;
- La promotion et la diffusion des techniques d'extraction et de production de cube d'assaisonnement ; et
- La conception d'emballages appropriés.

## 6. REFERENCES

**Daget, J.** 1952. Biologie des poissons du Niger Moyen. Bulletin de l'Institut Français d'Afrique Noire. Tome XIV, Page 199–201. Janvier 1952.

**Niaré, T. & Benech V.** 1993 Modification de la croissance de Brycinus leuciscus suite aux changements hydro climatiques et halieutiques dans la plaine inondée du Delta Central du Niger.

**Quensière, J.** 1994. La Pêche dans le Delta Central du Niger. Approche pluridisciplinaire d'un système de production halieutique. IER-ORSTOM-KHARTALA.

**Traoré, O.** 2005. Amélioration de la qualité de l'huile de *Brycinus leuciscus*. Consultation des Experts FAO sur la technologie, l'utilisation et l'assurance qualité du poisson et des produits de la pêche, Bagamoyo, Tanzanie.

**Traoré, O.** 2010. Effet du mode d'extraction sur le rendement et les qualités nutritionnelles, physico-chimiques et sensorielles de l'huile de *Brycinus leuciscus*. Rapport de thèse, 95-98.

# EATING FISH: ENJOYING THE BENEFITS AND AVOIDING THE RISKS

## [MANGER DU POISSON: APPRÉCIER LES AVANTAGES ET ÉVITER LES RISQUES]

by/par

Jogeir Toppe[1]

**Abstract**

The risks of consuming potentially contaminated foods have traditionally received greater attention than the benefits of eating the same food. Fish and other seafood products are regarded as among the main food sources of some contaminants such as heavy metals (e.g. Hg) and dioxin-like compounds. However, there is now also a growing focus on the risks of *not* consuming certain foods, given their potential beneficial components. In order to address the growing concern linked to the risks of consuming fish, and the increasing focus on the role of fish in our diets, FAO and WHO recently held an expert consultation on the risks and benefits of fish consumption. The experts underlined the important role fish and fisheries products have in a healthy diet. It is beyond doubt that the consumption of fish plays an important role in the early development of the brain and neural system of children, and in preventing coronary heart diseases (CHD) among adults. As in most foods, contaminants can be present in seafood. However, in general it is widely accepted that the levels of contaminants are well below safe levels in practically all fish. However, for large predatory fish, or fish caught in polluted waters, the levels of contaminants might exceed safe levels. This paper is based on the report of the joint FAO/WHO Expert Consultation on the Risks and Benefits of Fish Consumption.

*Key words: Fish consumption, Benefits, Risks, Risk/benefit assessment, EPA, DHA, Mercury, Dioxin*

**Résumé**

Les risques de consommer des aliments potentiellement contaminés ont traditionnellement reçu plus d'attention que les avantages de manger la même nourriture. Le poisson et d'autres produits de la mer sont considérés comme étant parmi les aliments sources principales de certains contaminants comme les métaux lourds (par ex.: le mercure) et les composés apparentés à la dioxine.

Cependant, il y a aussi maintenant un intérêt croissant sur les risques de ne pas consommer certains aliments, compte tenu de leurs composantes bénéfiques potentielles. Afin de répondre à l'inquiétude croissante liée aux risques de la consommation de poisson et l'intérêt croissant sur le rôle du poisson dans notre régime alimentaire, la FAO et l'OMS ont récemment organisé une consultation d'experts sur les risques et les avantages de la consommation de poisson. Les experts ont souligné l'importance du rôle des produits de la pêche dans un régime alimentaire sain. Il n'y a aucun doute que la consommation de poisson joue un rôle important dans le développement précoce du cerveau et du système nerveux des enfants et dans la prévention des maladies cardio-vasculaires (MCV) chez les adultes. Comme dans la plupart des aliments, les contaminants peuvent être présents dans les fruits de mer. Cependant, en général il est largement admis que les niveaux de contaminants soient bien en-dessous des niveaux sécurisés dans pratiquement tous les poissons. Toutefois, pour les grands poissons prédateurs, ou les poissons capturés dans les eaux polluées, les niveaux des contaminants pourraient dépasser les niveaux de sécurité. Ce document est basé sur le rapport de la Consultation d'Expert FAO/OMS sur les Risques et les Avantages de la Consommation de Poisson.

*Mots clés: Consommation de poisson, Avantages, Risques, Evaluation des risques/avantages, EPA, DHA, Mercure, Dioxine*

## 1. INTRODUCTION

It is generally accepted that the consumption of fish and other seafood has well-established nutritional and health benefits. However, fish products can also be harmful since they might accumulate contaminants present in the aquatic environment. Some contaminants are a result of human activities, but some are beyond the control of humans. Consumers of fisheries products will always be exposed to both contaminants and nutrients, present in fish as in all other foods. There has been a growing need for more knowledge on the risks and benefits of fish

---

[1] Fisheries Officer, Food and Agriculture Organization, Viale delle Terme di Caracalla, 00153 Rome, Italy. Jogeir.toppe@fao.org

consumption in order to give good advice to consumers on how to enjoy the health benefits of seafood consumption while minimizing the concurrent negative consequences. This paper is based on the report of the joint FAO/WHO Expert Consultation on the Risks and Benefits of Fish Consumption (FAO/WHO, 2011).

The risks of consuming potentially contaminated foods have traditionally received greater attention than the benefits of eating the same food. However, there is now a growing focus on the risks of not consuming certain foods, and among them fish products, given their potential beneficial components. Nutritional benefits derive not only from the long-chain polyunsaturated fatty acids (LCPUFAs) – docosahexaenoic acid (DHA) and eicosapentaenoic acid (EPA) – but also from amino acids, micronutrients (vitamins, minerals) and possibly from other nutrients (e.g. taurine), all found in fish.

The fact that fish consumption helps prevent coronary heart disease (CHD) has been well known for some time. There is now an increasing focus on fish as a source of DHA and iodine, which are essential for the early development of the brain and neural system. These nutrients are almost exclusively found in foods from the aquatic environment. The role of fish in mitigating mental disorders, such as depression and dementia, is also receiving increased attention from scientists.

However, the presence of contaminants in some fish and fish products and other foods is of increasing concern to consumers. Some fish products are known to contain contaminants such as methyl mercury (mercury in its most toxic form) and dioxins (all dioxin-like compounds). In general, it is believed that the levels of such contaminants in seafood are well below the maximum levels established for their safe intake. Nevertheless, in fish caught in polluted waters or in large, long-lived predator species, the levels of contaminants might exceed the levels regarded as safe for consumption.

It is well known that ingested mercury might have a negative impact on the development of the neural system of children and that some fish species can be the main source of mercury in many diets. Fish can also be a source of dioxins in populations that consume fish frequently. However, the occurrence of dioxins among individuals in these populations is generally not higher than in populations having low fish consumption (Sasamato et al., 2006; Mazet et al., 2005; Schecter et al., 2001; Tsutsumi et al., 2001). Therefore, reducing the consumption of fish might reduce the exposure to mercury in human diets, but the exposure to dioxins will probably be the same for individuals even if they significantly reduce their consumption of fish.

When consumption of a food can be associated with both potential health risks and benefits, risk managers try to identify an intake level that minimizes risks and maximizes benefits. It is particularly important to establish such levels when consumption levels are close to levels that should not be exceeded (European Food Safety Authority, 2007).

Advice on limiting the consumption of fish for vulnerable groups, such as children and pregnant women, is being given by many public health authorities. While the intention is only to limit consumption of products believed to have elevated levels of contaminants, the effect in some cases has been a significant reduction in seafood consumption. However, a reduction in seafood consumption could result in a diet that might not ensure an optimal intake of essential nutrients. Both children and adults run this risk. As LCPUFAs are essential in the early development of the brain and neural system in children, advice aiming to limit the consumption of contaminated fish must be couched in such terms that not all fish is given a "bad name". Similarly, as seafood consumption reduces cardiovascular diseases among the adult population, messages intended to reduce the exposure of fish products to contaminants should go hand in hand with the promotion of safe fish products.

## 2. METHODOLOGY

Addressing the risks and benefits of fish consumption is a complex and resource-demanding scientific task that includes: (i) an assessment of the health risks associated with the consumption of fish and other seafood; (ii) an assessment of the health benefits of fish and other seafood consumption; and (iii) a subsequent comparison of the health risks and health benefits.

Most informed observers would probably agree that the solution to this issue consists of sound, science-based advice that weighs the benefits and costs for human health of consuming fish. Although much work has been done in this field, the subject is not exhausted and conclusions reached to date have not obtained universal endorsement.

Some studies (van Kreijl *et al.*, 2006; Mozaffarian and Rimm 2006) have tried to balance the positive and negative sides of consuming foods of high nutritional value but that are also a source of contaminants. However, to date, the procedures used have been controversial, and experts (European Food Safety Authority, 2007) in this field maintain that new procedures need to be developed in order to carry out quantitative assessments of the risks and benefits to human health of consuming fish and other seafood. Once the methodology has been developed, the required data will need to be obtained. The new procedures should make it possible to compare nutritional benefits with the possibility of adverse effects while accounting for the uncertainties – this should be possible for all groups in the population. In addition, scientists should be able to make quantitative comparisons of the human health risks and benefits of seafood consumption.

### *Recent actions*

In order to assist governments in giving advice to vulnerable population groups on the potential risks and benefits of consuming fish and seafood, the Codex Alimentarius Commission requested FAO and the World Health Organization (WHO) to hold an expert consultation on health risks associated with mercury and dioxins in fish and the health benefits of fish consumption.

The joint FAO/WHO Expert Consultation on the Risks and Benefits of Fish Consumption was held from 25 to 29 January 2010 at FAO Headquarters, Rome, Italy (FAO/WHO, 2011). Seventeen experts in nutrition, toxicology and risk-benefit assessment discussed the risks and benefits of fish consumption.

## 3. RESULTS AND DISCUSSION

The experts agreed that consumption of fish provides energy, protein and a range of essential nutrients, and that eating fish is part of the cultural traditions of many peoples. In some populations, fish and fishery products are a major source of food and essential nutrients, and there may be no alternative and affordable food sources for these nutrients.

Among the general adult population, consumption of fish, and in particular oily fish, lowers the risk of CHD mortality. There is an absence of probable, or convincing, evidence of mercury causing CHD. Although there is a risk that dioxins may cause cancer, the risk is comparatively small and seems to be outweighed by reduced CHD mortality for those who eat fish. By weighing the benefits of LCPUFAs against the risks of mercury for women of childbearing age, it is established that fish in the diet in most circumstances lowers the risk that women give birth to children with suboptimal development of the brain and neural system that may follow on not eating fish (FAO/WHO, 2011), as shown in Table 1 below.

Although a quantification of the risks and benefits of eating fish is complicated and difficult to visualize, the table below (Table 1) below shows an effort to quantify the effects on child IQ as a result of the fish consumed by the mother during and after pregnancy. This quantitative assessment was based on several assumptions: Fish serving size was estimated to be 100 g. Ratio of DHA to EPA + DHA was assumed to be 0.67. Maternal body weight was assumed to be 60 kg. The estimate of IQ points gained from DHA exposure using the coefficient of 4 IQ points for 100 mg of DHA intake. The maximum positive effect from DHA was estimated at 5.8 points. IQ points lost from methylmercury exposure, was calculated using the central estimate of −0.18 and the higher and more conservative value calculated using the upper-bound estimate of −0.7 IQ points lost per microgram of mercury in maternal hair per gram maternal hair. Table 1a–d shows the effects on child IQ as a result of the child's mother consuming one, two, four or seven servings of fish per week with different EPA plus DHA and methylmercury concentrations in the fish consumed (FAO/WHO, 2011).

**Table 1. Estimated changes in child IQ resulting from the child's mother having consumed fish with different methylmercury and EPA plus DHA contents at one, two, four and seven servings per week***

(a) One serving per week

| | | EPA + DHA | | | |
|---|---|---|---|---|---|
| | | $x \leq 3$ mg/g | $3 < x \leq 8$ mg/g | $8 < x \leq 15$ mg/g | $x > 15$ mg/g |
| Methylmercury | $x \leq 0.1$ µg/g | −0.1 +0.8 | −0.1 +2.1 | −0.1 +4.4 | −0.1 +5.8 |
| | $0.1 < x \leq 0.5$ µg/g | −0.5 +0.8 | −0.5 +2.1 | −0.5 +4.4 | −0.5 +5.8 |
| | $0.5 < x \leq 1$ µg/g | −1.2 +0.8 | −1.2 +2.1 | −1.2 +4.4 | −1.2 +5.8 |
| | $x > 1$ µg/g | −2.3 +0.8 | −2.3 +2.1 | −2.3 +4.4 | −2.3 +5.8 |

(b) Two servings per week

| | | EPA + DHA | | | |
|---|---|---|---|---|---|
| | | $x \leq 3$ mg/g | $3 < x \leq 8$ mg/g | $8 < x \leq 15$ mg/g | $x > 15$ mg/g |
| Methylmercury | $x \leq 0.1$ µg/g | −0.2 +1.5 | −0.2 +4.2 | −0.2 +5.8 | −0.2 +5.8 |
| | $0.1 < x \leq 0.5$ µg/g | −0.9 +1.5 | −0.9 +4.2 | −0.9 +5.8 | −0.9 +5.8 |
| | $0.5 < x \leq 1$ µg/g | −2.3 +1.5 | −2.3 +4.2 | −2.3 +5.8 | −2.3 +5.8 |
| | $x > 1$ µg/g | −4.7 +1.5 | −4.7 +4.2 | −4.7 +5.8 | −4.7 +5.8 |

(c) Four servings per week

| | | EPA + DHA | | | |
|---|---|---|---|---|---|
| | | $x \leq 3$ mg/g | $3 < x \leq 8$ mg/g | $8 < x \leq 15$ mg/g | $x > 15$ mg/g |
| Methylmercury | $x \leq 0.1$ µg/g | −0.3 +3.1 | −0.3 +5.8 | −0.3 +5.8 | −0.3 +5.8 |
| | $0.1 < x \leq 0.5$ µg/g | −1.9 +3.1 | −1.9 +5.8 | −1.9 +5.8 | −1.9 +5.8 |
| | $0.5 < x \leq 1$ µg/g | −4.7 +3.1 | −4.7 +5.8 | −4.7 +5.8 | −4.7 +5.8 |
| | $x > 1$ µg/g | −9.3 +3.1 | −9.3 +5.8 | −9.3 +5.8 | −9.3 +5.8 |

(d) Seven servings per week

| | | EPA + DHA | | | |
|---|---|---|---|---|---|
| | | $x \leq 3$ mg/g | $3 < x \leq 8$ mg/g | $8 < x \leq 15$ mg/g | $x > 15$ mg/g |
| Methylmercury | $x \leq 0.1$ µg/g | −0.5 +5.4 | −0.5 +5.8 | −0.5 +5.8 | −0.5 +5.8 |
| | $0.1 < x \leq 0.5$ µg/g | −3.3 +5.4 | −3.3 +5.8 | −3.3 +5.8 | −3.3 +5.8 |
| | $0.5 < x \leq 1$ µg/g | −8.2 +5.4 | −8.2 +5.8 | −8.2 +5.8 | −8.2 +5.8 |
| | $x > 1$ µg/g | −16.3 +5.4 | −16.3 +5.8 | −16.3 +5.8 | −16.3 +5.8 |

*\* The numbers in the upper row in each cell are estimates of IQ points lost from methylmercury exposure, using the upper-bound estimate of −0.7. The number in the lower row in each cell is the estimate of IQ points gained from DHA exposure. Shaded cells represent the estimates where the net effect on child IQ, using the upper-bound estimate for methylmercury, is negative.*

***Source:*** FAO/WHO, 2011.

The results from the expert consultation clearly showed the link between EPA and DHA in the diet of the mother and an increase in IQ of the child. On the other side, it also showed the link between methylmercury ingestion and a decrease in IQ. Most fish contain both mercury and EPA/DHA, but the study shows that in most cases the net effect is positive. Based on available databases (FAO/WHO, 2011) on the nutrient and contaminant content of seafood, Table 2 was developed. This table gives an indication of which species of fish or shellfish might fit into the categories used in Table 1, giving an indication of which species might have a net negative impact on the neurodevelopment of a child if consumed frequently. The tables do not take into account fish from contaminated areas, and is only indicative. Similar tables should be made based on local/regional data for local/regional advice.

**Table 2. Classification of the content of EPA plus DHA by total mercury content in finfish and shellfish species, based on data from North America, Northern Europe and Japan**

| | | EPA + DHA | | | |
|---|---|---|---|---|---|
| | | $x \leq 3$ mg/g | $3 < x \leq 8$ mg/g | $8 < x \leq 15$ mg/g | $x > 15$ mg/g |
| Mercury | $x \leq 0.1$ µg/g | Butterfish, Catfish, Atlantic cod, Pacific cod, Atlantic croaker, Haddock, Pike, European plaice, Pollock, Saithe, Sole, Tilapia, Clams, Cockle, Crawfish, Cuttlefish, Oysters, Periwinkle, Scallops, Scampi, Sea urchin, Whelk | Flatfish, John Dory, Ocean and Mullet perch, Sweetfish, Wolf fish, Mussels, Squid | Redfish, Atlantic salmon (wild), Pacific salmon (wild), Smelt, Spider crab, Swimcrab | Anchovy, Herring, Mackerel, Rainbow trout, Atlantic (farmed) salmon, Sardines, Sprat, Atlantic cod (liver), Saithe (liver), Crab (brown meat) |
| | $0.1 < x \leq 0.5$ µg/g | Anglerfish, Catshark, Dab, Grenadier, Grouper, Gurnard, Hake, Ling, Lingcod, Scorpionfish, Nile perch, Pout, skate/ray, Porgy snapper, Sheepshead snapper, Yellowfin tuna, Tusk, Whiting, Lobster, American lobster | Freshwater bass, Carp, Freshwater perch, Scorpion fish, Tuna, Albacore tuna, Crab, Norway lobster, Spiny lobsters, | Saltwater bass, Bluefish, Goatfish, Atlantic halibut (farmed), Greenland halibut, Horse mackerel, Spanish mackerel, Seabass, Seabream, Atlantic tilefish, Skipjack tuna | Eel, Pacific mackerel, Sablefish |
| | $0.5 < x \leq 1$ µg/g | Marlin, Orange roughy, Bigeye tuna | King mackerel, Shark | Alfonsino | Pacific bluefin tuna |
| | $x > 1$ µg/g | | Swordfish | | |

***Source:*** FAO/WHO, 2011.

At levels of maternal dioxin intake (from fish and other dietary sources) that do not exceed the established long-term tolerable intakes of dioxins, the risk of suboptimal neural development is negligible (FAO/WHO, 2011). If the maternal dioxin intake (from fish and other dietary sources) exceeds the established long-term tolerable intakes of dioxins, this risk may no longer be negligible. Among infants, young children and adolescents, evidence is insufficient to derive a quantitative framework of health risks and benefits. However, healthy dietary patterns that include fish established early in life influence dietary habits and health during adult life.

### Future perspectives

Although there is no association between resource sustainability and health, the issue of sustainability must be considered if proven health benefits lead to greatly increased demand for seafood. With the known wide range of benefits from seafood consumption, it is pertinent to consider whether increased production is possible. For the last 20 years, global landings from capture fisheries have been stagnant at around 89–93 million tonnes. Even with the widespread failure to manage fishery resources properly, which has resulted in a situation where some 28 percent of stocks are overexploited, there is general scientific agreement that significantly more cannot be produced from wild fish populations.

However, total global fish production has continued to rise, amounting to about 140 million tonnes in 2007 (FAO, 2011). The balance is made up by production from aquaculture, which now amounts to 50 million tonnes, accounting for 44 percent of all fish for human consumption.

Global fish consumption has gradually increased, regardless of the increasing world population, and stood at 17.0 kg of fish (live weight equivalent) per capita per year in 2007 (FAO, 2011). A widespread recognition of the benefits of seafood consumption would inevitably lead to additional demand. If the recommendations of authorities in the United Kingdom of two meals of 140 g of fish per week (Scientific Advisory Committee on Nutrition/Committee on Toxicity, 2004) were to become true, then annual per capita consumption would have to rise to 23.3 kg This translates into an additional production of 42.2 million tonnes for 2007, rising to 84 million tonnes in 2050.

Aquaculturists are optimistic that far more fish can be produced, but there are issues of nutritional quality using land-based feeds. It would be necessary to incorporate LC n-3 into PUFAs the feeds. Intensive research is required on how this could be achieved, including on production from hydrocarbons by yeast fermentation, extraction from algal sources (Mata et al., 2010) and/or genetic modification of plants to become LC n-3 PUFA producers. However, for now and probably for the new decade, the source of LC n-3 PUFAs will remain marine capture fisheries.

Increased focus on the health benefits of consuming fishery products is providing more and more evidence of the positive effects seafood might have on mental health. Mental illness and depression are increasing globally. Some experts predict that they will become a major burden in terms of global health, especially in the developed world (Hibbeln and Davis, 2009). In 2004, mental health overtook heart disease as the leading health problem in Europe and was estimated to cost €386 billion a year (Andlin-Sobocki et al., 2005). More recent studies suggest that consumption of seafood and in particular long chain n-3 polyunsaturated fatty acids (LC n-3 PUFAs) may also have a positive impact on dementia (Morris et al., 2005) and Alzheimer's disease, with the most promising evidence for the benefits on mood/depression (Peet and Stokes, 2005). However, such benefits should be considered as emerging, as they are not as well established as reductions in CHD deaths and improved early neurodevelopment.

## 4. CONCLUSIONS AND RECOMMENDATIONS

To minimize risks in target populations, the expert consultation recommended that states should acknowledge that fish is an important food source containing energy, protein and a range of essential nutrients as well as being part of the cultural traditions of many peoples. States should therefore emphasize: (i) that fish consumption reduces CHD mortality in the adult population; and (ii) that fish consumption improves the neurodevelopment of foetuses/infants and is therefore important for women of childbearing age, pregnant women, and nursing mothers. In order to provide sound advice to different population groups, it will also be important to develop, maintain and/or improve regional databases of the specific nutrients and contaminants in the fish available for consumption. Risk management and communication strategies that aim to minimize risks and maximize benefits from eating fish should be developed and evaluated.

The demand for fisheries products is increasing, and will most likely continue to grow as people become more concerned about their health. The increasing demand for fish will mainly be met by increased aquaculture production, although more focus on reducing post harvest losses and diverting more of the industrial fish for direct human consumption will become more and more important. The nutritional qualities of fish as food are obvious, not only as a source of proteins and essential fatty acids, but also as an optimal source of micronutrients deficient in many diets worldwide. Fish could and should play a bigger role in combating malnutrition in the less privileged regions of this world. However, promoting increased consumption of fisheries product must go hand in hand with a sustainable production and harvest of this unique source of nutrients.

## 5. REFERENCES

**Andlin-Sobocki, P., Jönsson, B., Wittchen, H.-U. & Olesen, J.** 2005. Costs of disorders of the brain in Europe. *European Journal of Neurology*, 12(1), pp. 1–27.

**European Food Safety Authority**. 2007. Risk-benefit analysis of foods: methods and approaches. *Summary Report EFSA Scientific Colloquium* 6. Parma, Italy.

**FAO**. 2009. FAOSTAT statistical database. Rome. (available at http://faostat.fao.org/)

**FAO**. 2011. The State of World Fisheries and Aquaculture 2010, 218p. (available at www.fao.org/docrep/013/i1820e/i1820e00.htm)

**FAO/WHO**. 2011. Joint FAO/WHO Expert Consultation on the Risks and Benefits of Fish Consumption. *FAO Fisheries and Aquaculture Report* No. 978.

**Hibbeln, J.R. & Davis, J.M**. 2009. Considerations regarding neuropsychiatric nutritional requirements for intakes of omega-3 highly unsaturated fatty acids. *Prostaglandins, Leukotrienes and Essential Fatty Acid*s, 81(2), pp. 179–186.

**Kreijl C.F. van, Knaap, A.G.A.C. & Van Raaij, J.M.A**. 2006. Our food, our health, Healthy diet and safe food in the Netherlands. National Institute for Public Health and the Environment. Bilthoven, The Netherlands.

**Mazet A., Keck G. & Berny, P**. 2005. Concentrations of PCBs, organochlorine pesticides and heavy metals (lead, cadmium, and copper) in fish from the Drôme river: Potential effects on otters (Lutra lutra), *Chemosphere*, 61(6), pp. 810–816.

**Mata, T.M., Martins, A.A. & Caetano, N.S**. 2010. Microalgae for biodiesel production and other applications: a review. *Renewable and Sustainable Energy Reviews*, 14, pp. 217–232.

**Morris, M.C., Evans, D.A., Tangney, C.C., Bienias, J.L. & Wilson, R.S**. 2005. Fish consumption and cognitive decline with age in a large community study. *Archives of Neurology*, 62(12), pp. 1849–1853.

**Mozaffarian, D. & Rimm, E.B**. 2006. Fish intake, contaminants, and human health - Evaluating the risks and the benefits. *Jama-Journal of the American Medical Association* 296(15), pp. 1885–1899.

**Peet, M. & Stokes, C**. 2005. Omega-3 fatty acids in the treatment of psychiatric disorders. *Drugs*, 65(8), pp. 1051–1059.

**Plaza, M., Herrero, M., Cifuentes, A. & Ibáñez, E**. 2009. Innovative natural functional ingredients from microalgae. *Journal of Agricultural and Food Chemistry*, 57(16), pp. 7159–7170.

**Sasamoto T., Ushio F., Kikutani N., Saitoh Y., Yamaki Y. & Hashimoto, T**. 2006. Estimation of 1999–2004 dietary daily intake of PCDDs, PCDFs and dioxin-like PCBs by a total diet study in metropolitan Tokyo, Japan, *Chemosphere* **64**, pp. 634–641.

**Schecter, A., Cramer, P., Boggess, K., Stanley, J., Päpke, O., Olson, J., Silver, A. & Schmitz, M**. 2001. Intake of dioxins and related compounds from food in the U.S. population, *J Toxicol Environ Health A*. 63(1), pp. 1–18.

**Scientific Advisory Committee on Nutrition/Committee on Toxicity**. 2004. Advice on fish consumption: benefits and risks. The Stationery Office, Norwich, UK.

**Tsutsumi, T., Yanagi, T., Nakamura, M., Kono, Y., Uchibe, H., Iida, T., Hori, T., Nakagawa, R., Tobiishi, K., Matsuda, R., Sasaki, K. and Toyoda, M**. 2001. Update of daily intake of PCDDs, PCDFs, and dioxin-like PCBs from food in Japan. Chemosphere 45, pp. 1129–1137.

# PROCÉDÉ AMELIORÉ POUR LA RÉDUCTION DES PERTES APRES CAPTURE LIEES AU SECHAGE NATUREL PAR TEMPS DE PLUIE

## [IMPROVED PROCESS FOR THE REDUCTION OF POST-HARVEST LOSSES RELATED TO NATURAL DRYING DURING THE RAINY SEASON]

par/by

Oumoulkhairy Ndiaye[1]

**Résumé**

Le séchoir à charbon à ventilation électrique, d'origine cambodgienne et adapté en Indonésie a été expérimenté au Centre National de Formation des Techniciens des Pêches (CNFTPA) à Dakar au Sénégal, dans le cadre du programme coopératif de recherche des technologies améliorées pour la réduction des pertes post-capture coordonné la FAO.

Les contraintes liées à l'utilisation de ce type de séchoir en pêche artisanale ont conduit à la mise au point d'un système de séchoir mécanique composé de fourneau équipé d'une forge manuelle, à partir du four Parpaing ou Banda, 2 outils très connus des communautés de pêche. Ce séchoir a été conçu à partir du système de fumage FAO-Thiaroye.

Les résultats des essais sur des crevettes décortiquées ont été très concluants comparés au séchage naturel et cette méthode améliorée s'adapte à toutes les saisons, ce qui résout la problématique des fortes pertes après capture enregistrées en pêche artisanale en période d'hivernage, de forte humidité et par temps de froid.

*Mots clés: Pertes, Hivernage , Séchage, Forge manuelle, Séchoir, FAO-Thiaroye*

**Abstract**

A coal-fired dryer with electric ventilation, of Cambodian origin but adapted in Indonesia was tested at the National Training Centre of Fisheries Technicians (CNFTPA) in Dakar, Senegal, within the framework of the cooperative research programme in improved post-harvest reduction technologies coordinated by FAO.

The constraints to the use of this dryer in artisanal fisheries have led to the design of a mechanical drying system made up of a furnace equipped with a manual forge, built from Parpaing or Banda oven, both widely known in fisheries communities. This drier design was based onthe heating system of the FAO-Thiaroye smoking system.

The results from the trials on whole shrimps have been very conclusive compared to natural drying, and this improved method is all-weather user-friendly, which addresses the problem of high post-harvest losses during the rainy season, high humidity and cold weather.

*Key words: Losses, Rainy season, Manual forge, Drier, FAO-Thiaroye*

## 1. INTRODUCTION

Le séchage naturel au soleil fait partie des plus anciennes méthodes de conservation des aliments et est très utilisée en pêche artisanale sous les tropiques. Le principe est très simple et consiste à mettre le poisson au contact d'un courant d'air chaud et sec. L'air apporte au poisson la chaleur nécessaire à la vaporisation de l'eau qu'il contient. Le poisson se débarrasse de son humidité entraînant une perte de poids.

La méthode est peu onéreuse, mais demeure très longue et génératrice de pertes post-capture surtout en saison des pluies et période de fraîcheur.

Lorsque les conditions climatiques ne permettent pas d'avoir un séchage naturel facile, on peut agir soit sur la température, la vitesse de circulation de l'air et aussi sur l'épaisseur des produits à sécher en adoptant des procédés améliorés.

---

[1] CNFTPA, BP 2241, 10 km de Rufisque, Dakar, Senegal. oumoulinda@yahoo.fr

L'idée d'expérimenter le séchoir à charbon a été discutée durant la réunion des experts en technologie, utilisation et assurance qualité du poisson organisée par la FAO à Agadir du 24 au 28 Novembre 2008.

Un four de séchage d'origine Cambodgienne adapté en Indochine dans le cadre d'un projet d'urgence de la FAO et s'apparentant au four Parpaing/Chorkor avait été présenté à cet effet. L'apport en air s'effectue à l'aide d'un ventilateur dont l'alimentation en électricité pose un problème d'accessibilité dans les communautés de pêche.

Pour rendre ce séchoir à charbon d'utilisation pratique et performante, les experts ont échangé sur des possibilités de production d'air en intégrant un système de ventilation naturelle qui soit durable, écologique et économique. L'alimentation en air à travers l'apport en énergie hydraulique ou sous forme de poulie était une des éventualités mentionnées.

C'est dans ce contexte que le CNFTPA a mis au point un prototype de séchoir à charbon avec l'installation d'une forge pour la ventilation permettant ainsi de sécher des produits de pêche toute l'année sans se soucier des conditions climatiques.

Un dispositif de chauffage issu du système de fumage FAO-Thiaroye a été identifié pour constituer l'armature du séchoir mécanique à charbon, par la confection d'un fourneau équipé d'une forge.

La forge est un appareil non électrique destiné à projeter l'air avec force dans le fourneau contenant de la braise. Quatre (4) séries d'essais ont été réalisées pour maîtriser d'avantage les paramètres de séchage, apprécier la qualité organoleptique des produits et les performances techniques et économiques.

## 2. MATERIEL ET METHODE

### MATERIEL

**Figure 1. Four parpaing**

Le four parpaing rectangulaire construit dans une zone aérée, de longueur, de largeur et de hauteur 5 m X 1 m X 90 cm, respectivement. Il est divisé en deux compartiments muni chacun d'un foyer. La distance entre le cœur du feu et le grillage est de 70 cm.

Il est équipé d'un grillage amovible en matériaux inoxydables facilitant le nettoyage, résistant à la chaleur et muni de couvercle en métal. Sa capacité de chargement est de 400 kg de poisson.

**Figure 2. Fourneau du séchoir à charbon et la forge**

Les dimensions du fourneau à braise confectionné avec l'assistance d'un forgeron local sont de longueur, largeur et hauteur de 1mx1m x 24cm, respectivement. Ce fourneau est équipé de quatre (4) roulettes pour son déplacement, de tuyaux d'aération et d'une forge pour attiser le feu et projeter l'air à l'intérieur du four.

Le petit matériel (balance, thermomètre à sonde, hygromètre) a servi à suivre les paramètres techniques lors des essais de séchage.

**METHODE**

Des essais comparatifs du séchage naturel et du séchage amélioré avec le séchoir à charbon ont été effectués pour mieux apprécier les innovations apportées.

Des crevettes entières lavées, légèrement salées et égouttées ont été utilisées pour réaliser les essais de séchage.

Le principe consiste à étaler le produit à sécher sur une claie déposée sur le four parpaing, à chauffer l'air de séchage à l'aide d'un fourneau équipé d'une forge permettant ainsi d'augmenter la température et la masse d'air dans l'enceinte du four.

**Figure 3 et 4. Crevettes en séchage sur claie alimentée par un fourneau de braise**

En début de séchage, la quantité de braise est très réduite, le produit est mis à l'écart de la braise et la température est contrôlée de sorte à fluctuer entre 35 et 45 °C au maximum afin d'éviter d'une part la cuisson des crevettes et d'autre part le croûtage. La durée de cette étape est en moyenne d'une heure de temps.

Après cette étape, les crevettes sont retournées pour les rendre homogènes et la quantité de braise est augmentée de manière à élever la température aux environs de 60 °C et l'humidité relative entre 60 et 70 pour centpermettant ainsi d'accélérer le processus de séchage.

Le séchage naturel a été effectué sur une claie surélevée pour un cycle complet de séchage qui s'est achevé après 7 heures de temps à 30 °C avec une humidité relative de 85 percent. La claie est recouverte de moustiquaires afin d'éviter la contamination du produit par les mouches.

**3. RESULTATS ET DISCUSSION**

**PERFORMANCES TECHNIQUES**

Après deux heures de temps, le produit est totalement sec, de lustre attrayant, de couleur rougeâtre recherchée dans les crevettes séchées et le goût a été très apprécié par les panélistes du centre.

Le séchage naturel effectué au courant du mois d'août, en période hivernale a duré deux jours tandis que celui réalisé au four à charbon n'a duré que trois heures de temps.

Le séchoir à charbon a permis d'obtenir en trois heures de temps des crevettes séchées de très bonne qualité et l'opérateur est moins tributaire des aléas climatiques. Ce procédé offre des possibilités de sécher de grandes quantités de produits séchés même en temps de pluie.

Les problèmes de croûtage et de cuisson du produit sont maîtrisés avec une bonne aération et une réduction importante de la quantité de braise.

**Tableau 1. Paramètres de séchage**

| Paramètres | Séchoir à charbon | Séchage naturel en saison des pluies |
|---|---|---|
| Température en début de séchage | 35 °C | 30 °C |
| Humidité relative | 70% | 85% (séchage à proximité de la mer) |
| Distribution de l'air | Plus ou moins homogène | Hétérogène |
| Contrôle des opérations de production | Facile | Très contraignant en saison des pluies |
| Durée de séchage | 3 heures | 3 à 6 jours en saison des pluies |
| % de perte d'eau | 60% | 55% |
| Rendement | 40% | 45% |
| Qualité finale du produit | Très bonne et appréciée | Légèrement moisie en saison des pluies |
| Nombre de session de production | 4 (en moyenne par jour) | 2 (par semaine en saison des pluies) |
| Capacité | 400 kg | 20 claies de capacité 20 kg chacune |
| Occupation de l'espace | 5 m2 (peu d'espace) | 40 m2 (trop d'espace) |

## PERFORMANCES ECONOMIQUES

**Tableau 2. Amortissement du matériel**

| Matériels | Valeur en $ | Durée de vie | Amortisse-ment annuel en $ | Amortisse-ment mensuel en $ | Nombre moyen de sessions mensuel | Amortisse-ment par session de séchage en $ |
|---|---|---|---|---|---|---|
| Séchoir à charbon de capacité de 400 kg | 600 | 10 ans | 60 | 5 | 24 | 0,20 |
| Fourneau | 140 | 5 ans | 28 | 2,33 | 24 | 0,09 |
| Claie de séchage (20 claies de capacité 20 kg chacune) | 1000 | 3 ans | 333,33 | 27,77 | 8 | 3,47 |

Le coût moyen pour la confection du séchoir à charbon est d'environ de 600 $ pour le four de capacité 400 kg et de 140 $ pour le fourneau.

Vingt claies de séchage sont nécessaires pour arriver à sécher la même quantité pour le séchage naturel.

La claie confectionnée au CNFTPA avec des matériaux de récupération coûte 50 $. A capacité égale, le coût total des claies de séchage naturel revient à 1000 $.

La durée de vie est estimée au moins à 10 ans pour le four et 5 ans pour le fourneau. Ce dernier devra être repeint avec la peinture antirouille au moins tous les ans.

**Tableau 3. Compte d'exploitation pour une session de production**

| Charges pour une session de production | | |
|---|---|---|
| **Intrants** | **Séchoir à charbon** | **Séchage naturel** |
| Crevette | 100 kg x 3 $ | 100 kg x 3 $ |
| Sel | 1 sachet x 0,5 $ | 1 sachet x 0,5 $ |
| Charbon | 10 kg x 0,4 $ | - |
| Transport intrants | 4 $ | 4 $ |
| Main d'ouevre | 2 personne x 2$ | 2 personne x 2$ |
| Amortissement par session du matériel de séchage | 0,3 $ | 3,47 $ |
| Entretien matériel | 1 $ | 1 $ |
| Total charges | 313,8 $ | 312,97 $ |
| Prix de vente | 40 kg x 9 $ | 45 kg x 7,5 $ |
| Total prix de vente | 360 $ | 337,5 $ |
| Bénéfice net pour une session | 46,2 $ | 24,53 $ |
| Bénéfice moyen par semaine par exemple en raison d'une session par jour (séchoir à charbon) et 2 fois par semaines (séchage naturel), surtout en période d'hivernage ou parfois aucun séchage n'est possible sur plusieurs jours. | 46,2 x 7 jours = 323,4 $ | 24,53 x 2 jours= 49,06 $ |

Le coût du séchoir à charbon est amorti en 20 sessions de production tandis que pour le séchage naturel la durée dépasserait les 6 mois.

L'analyse économique comparée des deux types de séchage montre que le bénéfice généré des produits séchés pour 100 kg de crevettes fraîches est 1,9 fois plus important pour le séchoir à charbon que le séchage naturel en saison des pluies.

Les avantages significatifs de ce séchoir sont liés:

- à la réduction importante de la durée de séchage; trois heures pour le séchoir à charbon et 2 à 5 jours en saison pluvieuse;
- au nombre de sessions journalières, en moyenne 4 pour le séchoir à charbon et 1 à 2 fois par semaine pour le séchage naturel en saison des pluies;
- à la qualité du produit fini, meilleure pour le séchoir à charbon et légèrement moisi pour le séchage au naturel en saison des pluies;
- au fait qu'il peut aussi servir de four de fumage et de dispositif de stockage de produits finis jusqu'à la période de soudure.

## 4. CONCLUSION ET RECOMMANDATION

L'utilisation de procédés améliorés de séchage indépendant des aléas climatiques revêt un intérêt certain et voir même est indispensable pour réduire significativement les nombreuse pertes post-capture liées au séchage des produits de pêche et d'aquaculture en saison des pluies. Le séchoir mécanique constitue un système de séchage simple, efficace, adapté à tous les climats avec une occupation réduite de l'espace. Il apporte beaucoup de facilité dans la production de produits séchés avec un meilleur gain en revenu. Son adoption pourrait être d'un apport conséquent pour les transformatrices de poisson.

Tout en recommandant la documentation et dissémination pour le séchage des espèces pour lesquelles les essais ont été concluants, il serait important de mener des expérimentations sur d'autres produits de pêche qui sont en général soumis au séchage naturel.

## 5. BIBLIOGRAPHIE

**ALTERSIAL, ITA, GRET.** 1988. Étude technique et économique de l'amélioration des procédés traditionnels du traitement du poisson, document n°2, 70 p.

**BIT.** 1986. Le séchage solaire, méthodes pratiques de conservation des aliments.

**BIT, FAO–PNUD.** 1991. Transformation du poisson à petite échelle. Série technologique, Dossier technique n° 3, 106 p.

**Diei, Y. & Ndiaye, O.** 1998. Guide des bonnes pratiques de manutention et de transformation artisanale du poisson. Programme pour le Développement Intégré des Pêches Artisanale en Afrique de l'Ouest (DIPA). Cotonou, Bénin, 35p; DIPA/WP/129.

**FAO.** 2001. Septième consultation d'experts sur la technologie du poisson en Afrique, Saly-Mbour, République du Sénégal, 10-13 décembre 2001, rapport sur les Pêches n° 712.

**GERES.** 1998. Guide d'aide à la décision pour la réalisation d'unités artisanales de séchage.

**GRET.** 1994. Conserver et transformer le poisson, 295 p.

**Teutsher, F., Tall, A. & Jallow, A.** 1994. Rapport de l'atelier sur le thème "A la recherche des améliorations en Technologie du poisson en Afrique de l'Ouest". Pointe Noire, Congo 7-9 novembre 1994, DIPA/WP/66.

# PROXIMATE COMPOSITION AND LEVELS OF POLYCYCLIC AROMATIC HYDROCARBONS (PAHS) IN CATFISH (*CLARIAS GARIEPINUS*) USING DIFFERENT SMOKING SYSTEMS

## *[COMPOSITION IMMÉDIATE ET LES NIVEAUX D'HYDROCARBURES AROMATIQUES POLYCYCLIQUES(HAP) DANS LE POISSON-CHAT (CLARIAS GARIEPINUS) EN UTILISANT DIFFÉRENTS SYSTÈMES DE FUMAGE]*

by/par

G.R Akande[1], A.O Olusola; R.S Adeyemi, M.M. Salaudeen, and A.O. Abraham-Olukayode
(presented by Yvette Diei-Ouadi)

## Abstract

The proximate composition of smoked catfish *Clarias gariepinus*, showed a product with high levels of protein and lipid, as the moisture is reduced. Protein content had a range of 64.80–81.11 percent, lipid content 9–22.43 percent, moisture 8.33–25.33 percent while ash content ranged from 3.33–5.53 percent. The presence and concentration of seventeen polycyclic aromatic hydrocarbons (PAHs) were determined in hot smoked African catfish, *Clarias gariepinus*, a fresh water fish of economic value in Nigeria. The smoked fish were processed using traditional and improved smoking kilns and using firewood, charcoal and gas as sources of fuel energy. Skin on smoked fish flesh was pulverized and a 5 g sample (on dry basis) was used for each of the five samples. Soxhlet extraction of all samples was carried out with hexane: dichloromethane (3:1 v/v) at 50 °C for 6h and the aromatic containing portion of the eluant was subsequently analyzed using gas chromatography (GC/MS). The results showed that the limits of detection (LOD) for the individual seventeen PAHs found in the samples ranged from 2–142 µg/kg while benzo(a)pyrene (B(a)P) was not detected in three of the fish samples, except the two samples that were traditionally smoked with firewood where the concentration exceeded the maximum limit of 5 µg/kg set by the EU. The concentration level of the PAHs was also studied with respect to the emerging EU regulation effective 2012. The concentration (sum total) of benzo(*a*)pyrene, benzo(a)anthracene , benzo(b)fluoranthene and chrysene (PAH4) ranged from 16 to 366 µg/kg for all the samples studied. Firewood smoked fish from Borno State and gas smoked fish were both above the maximum level of 30 µg/kg PAH4 set by the EU. Results from this work confirm that the actual level of individual PAHs in fish products is dependent on variables such as the type of fuel energy used in the smoking process. Thus, the relatively high concentration of (B(a)P) beyond the EU limit of 5 µg/kg of smoked fish in the traditional smoked catfish using firewood may be attributed to the direct firing of smoking process compared to charcoal, charcoal/firewood and gas smoked fish in which benzo(a)pyrene were not detected.

*Keywords: PAHs, PAH4, Benzo(a)pyrene, Smoked fish, C. gariepinus, Improved smoking kiln, Traditional smoking kiln, Proximate composition*

## Résumé

La composition globale du poisson-chat fumé *Clarias gariepinus*, a montré un produit avec des niveaux élevés de protéines et de lipides, car l'humidité est réduite. La teneur en protéines varie entre 64,80-81,11 pour cent, les lipides de 9-22,43 pour cent, humidité entre 8,33-25,33 pour cent tandis que la teneur en cendre variait de 3,33-5,53 pour cent. La présence et la concentration des dix-sept hydrocarbures aromatiques polycycliques (HAPs) ont été déterminées dans le poisson-chat fumé africain, *Clarias gariepinus,* un poisson d'eau douce de valeur économique du Nigéria. Le poisson fumé a été transformé en utilisant des fumoirs traditionnels et améliorés avec le feu de bois, le charbon, le gaz comme sources de combustible. La peau sur la chair de poisson fumé a été pulvérisée et un échantillon de 5g (sur base sèche) a été utilisé pour chacun des cinq échantillons. L'extraction de Soxhlet de tous les échantillons a été réalisée avec de l'hexane: dichlorométhane (3:1 v/v) à 50 °C pendant 6 heures et les aromatiques contenant une portion de l'éluant ont été par la suite analysées à l'aide de la chromatographie en phase gazeuse (GC/MS). Les résultats ont montré que les limites de détection (LDD) pour les dix-sept HAPs trouvées dans les échantillons variaient de 2-142 µg/kg tandis que le benzo(a)pyrène (B(a)P) n'a été détecté que dans trois des échantillons de poissons, à l'exception des deux échantillons qui étaient traditionnellement fumés au bois où la concentration dépasse la limite maximale de 5 µg/kg fixée par l'Union Européenne.

---

[1] Department of Fish Technology, Nigerian Institute for Oceanography and Marine Research, 3, Wilmot Point Road, Bar Beach, Victoria Island, P.M.B. 12729, Lagos, Nigeria. akandegra@yahoo.com

Le niveau de concentration des HAPs a également été étudié en ce qui concerne la réglementation de l'UE prenant effet en 2012. La concentration (somme totale) de benzo(a)pyrène, benzo(a)anthracène, benzo(b)fluoranthène et chrysène (PAH4) varie de 16 à 366 µg/kg pour tous les échantillons étudiés. Le poisson fumé au feu de bois de l'Etat de Borno et le poisson fumé au gaz étaient tous les deux au-dessus du niveau maximal 30 µg/kg PAH4 défini par l'UE. Les résultats de ce travail confirment que le niveau réel des HAPs dans les produits de la pêche dépend de variables telles que le type de combustible utilisé dans le processus de fumage. Ainsi, la concentration relativement élevée de (B(a)P) au-delà de la limite de 5 µg/kg du poisson fumé dans la fumage traditionnelle du poisson-chat utilisant le feu de bois pourrait être attribuée au processus de fumage direct par rapport au charbon, charbon/feu de bois et gaz de poisson fumé dans lequel les benzo(a)pyrenes n'ont pas été détectés.

*Mots clés: HAPs, HAP4, Benzo(a)pyrene, Poisson fumé, C. gariepinus, Four de fumage amélioré, Four de fumage traditionnel, Composition globale*

## 1. INTRODUCTION

Smoking as a way of preserving fish using smoke, typically from hardwood or charcoal is predominant in Nigeria: Certain compounds given off by burning wood have a preservative or antimicrobial effect on the food, and add flavour. Other compounds, such as polycyclic aromatic hydrocarbons (PAHs) may have a detrimental effect on human health at certain levels. Hard curing by salting and smoking permits lengthy preservation by removing moisture, which is essential for bacteriological and enzymatic spoilage. Consumers are rediscovering the good taste of smoked seafood, including smoked catfish. To satisfy the consumer demand, it is necessary to produce good quality and safe smoked seafood products.

Various studies have been carried out on the proximate chemical composition (Exler 1987, Chandrasekhar & Deosthale, 1993, Eun et al., 1994,) and fatty acids profiles of different fish species (Wanasundara & Shahidi, 1998; Uauy & Valenzuela, 2000). There is, however, a dearth of accurate basic chemical composition data for fish species, particularly from African and Asian sources (Schonfeldt, 2002). This constitutes a barrier to development of the safe use of the resources. In Nigeria, our present knowledge of the proximate chemical composition of fish species from Nigerian water is very limited. The African catfish, *Clarias gariepinus*, is easily cultured in Nigeria and of great economic interest. It is generally considered to be one of the most important tropical catfish species for aquaculture. It has an almost Pan African distribution, ranging from the Nile to West Africa and from Algeria to South Africa.

Polycyclic aromatic hydrocarbons (PAHs) are a group of chemicals that are formed during the incomplete burning of coal, oil, gas, wood, garbage or other organic material, such as tobacco and charbroiled meat. There are more than 100 different PAHs. They are a major class of environmentally hazardous organic compounds due to their known or suspected carcinogenicity (Martson et al., 2001; Wynder and Hoffmann, 1959; Simko, 2002; Klein, 1963). Some of them are fat soluble and tend to bioaccumulate in living organisms, especially those higher up in the food chain (Atlas, 1991).

PAHs constitute a large class of organic substances containing two or more fused aromatic rings made of carbon and hydrogen atoms. BaP was the first PAH to be identified as carcinogen. As a consequence it has been the most studied (Simon et al., 2006). Benzo(a)pyrene is no longer the only suitable marker for the occurrence of PAH in food; the EU has now added three more PAHs-benzo(a)anthracene , benzo(b)fluoranthene and chrysene making a total of four substances suitable as markers for polycyclic hydrocarbon (PAH4) in food (OJ L 215, 20.8.2011).

BaP has evoked much interest due to its carcinogenic properties and because it is one of the four regulated PAHs for which there is sufficient toxicological evidence to enable guidelines on allowable levels to be set. The EU has set the minimum acceptance level of B(a)P at 2 µg/kg while acceptance level for the sum total of benzo(*a*)pyrene, benzo(a)anthracene, benzo(b)fluoranthene and chrysene (PAH4) is 30 µg/kg; this will be effective from 2012 (OJ L 215, 20.8.2011). B(a)P has been detected in charcoal-broiled meat, smoked grilled foods, seafood, liquid smokes and beverages (Gomaa et al., 1993).

B(a)P has been considered by the International Agency for Research on Cancer (IARC), which concluded that it is a probable human carcinogen (IARC,1987). A variety of tests have shown BaP to be genotoxic. It has produced stomach tumors, mammary gland tumors, lung and respiratory tumors, as well as hepatic tumors in laboratory animals [ATSDR, 1995].

More than 200 PAH compounds have been identified in smoke, with the concentration depending mainly on the kind of wood and the smouldering temperature. High levels of PAHs are associated with the dark discolouration in intensively heated products. In addition, some (but not all) of these compounds are cancer-causing.
Benzo (a) Pyrene is now a big issue in smoked fish not only for export market but also local markets, especially products that are to be sold in supermarkets in Nigeria. However, attainment of the limit of acceptability of total PAH in general and Benzo (a) Pyrene levels in smoked fish as laid down by EU based on their present method of smoking will be an Herculean task for the small scale fish processors especially in meeting the minimum permissible level of 5μg/kg of smoked fish.

Smoked fish as a source of foreign exchange is gradually losing ground. This is adduced to the fact that exportation of processed fish to developed countries is becoming increasingly stringent because of the emerging set of Food Safety and Agricultural Health Standard, along with buyers changing their requirements (Ito, 2005 and Oyelese, 2006). Nigeria artisanal fisheries could benefit considerably from increased trade to the ethnic markets in Europe and United States through export of smoked fish and small dried shrimp. However, consignments of smoked fish are regularly detained and often destroyed by Port Health Authorities at Gatwick and Heathrow Airports due to mould growth and insect infestation (Ward, 2003).

The present comparative study evaluates the proximate composition and PAH content of *Clarias gariepinus* smoked using different smoking methods.

## 2. MATERIALS AND METHODS

### Sample collection and preparation

Catfish samples smoked using Liquefied Petroleum Gas (LPG) gas; firewood; charcoal; charcoal and firewood as sources of smoke were obtained from Lagos State and Borno State, Western and Northern Nigeria. The samples collected were virtually of the same size and were grouped and coded as follows:

- Sample A: Gas smoked catfish from Lagos State (GL).
- Sample B: Firewood smoked catfish from Borno State (FB).
- Sample C: Charcoal and Firewood smoked catfish from Lagos state (CFL).
- Sample D: Charcoal smoked catfish from Lagos State (CL).
- Sample E: Firewood smoked catfish from Lagos State (FL).

Sample A was procured from a fish processor that uses LPG as the sole source of energy to dry the fish. The procedure adopted was to gut the fish, bend to a horse shoe shape and hold in this position by means of a sharp stick. Then gas dry for 12 hours before packaging in a low density polyethylene bags. The kiln used was a NIOMR improved smoking kiln adapted with a gas burner.

Sample B was procured from a fish market in Lagos that sells only smoked catfish and other fish species originating from Lake Chad in Borno State. The mode of processing from previous studies of the Lake is the use of the Banda oven and processing can take between 2 to 3 days to reach the required moisture level before delivery to Lagos market.

Samples C and D were fresh catfish cropped from the NIOMR experimental fish farm and smoked using improved the NIOMR fish smoking kiln. The catfish samples were cleaned and subjected to pre-smoking process operations such as washing, gutting, salting, shaping (bending to horse shoe shape and holding in this position by means of a sharp stick). The prepared fish were spread as uniformly as possible on the 10 trays with each tray carrying 10 percent of the total quantity of fish to be smoked, i.e. 100kg of the catfish. Three stages of smoking process were employed. The first stage (slow firing) to prevent case hardening, the second stage, in which temperature was raised and the third stage after which the product is allowed to cool to room temperature. Sample C was smoked using a combination of charcoal and wood while sample D was only charcoal.

Sample E was also harvested from NIOMR experimental farm but the fish was given to a local processor who uses a traditional 44-gallon drum for smoking. Hardwood is the sole source of fuel energy and smoking was completed after 24 hours.

## Proximate Composition Analysis

The smoked catfish samples were finely grounded and homogenized for chemical analyses. The percentage proximate composition of the smoked fish samples was determined according to the AOAC (1994) methods. Triplicate determinations were carried out for moisture, ash and lipid while duplicate determination was done for protein on each chemical analysis. The content of total protein was estimated using the semi-automated and modified Kjeldahl method (Kjeltec™ 2100, Sweden). Crude fat content was determined using the Soxhlet method. The moisture content was determined by oven drying samples overnight at 105°C until constant weight was achieved (Gallenkamp Oven, UK) and the ash content was determined by incineration of the samples for 6 hrs at 500°C in a muffle furnace (Carbolite Sheffield® muffle furnace, England).

## Reagents for PAHs Analysis

All chemicals and reagents were of analytical grade and of highest purity possible. GC-grade dichloromethane used for extractions was obtained from Fischer Scientific and silica gel used to clean up the extracts was supplied by BDH Laboratories. A PAH standard mixture (Restek) containing naphthalene, acenaphthylene, acenaphthene, fluorene, phenanthrene, anthracene, fluoranthene, pyrene, benz[a]anthracene, chrysene, benzo[b]fluoranthene, benzo[k]fluoranthene, benzo[a]pyrene, benzo[ghi]perylene dibenz[a,h]anthracene and indeno[1,2,3-cd]pyrene was used in this study.

## Analysis of PAHs

For the determination of the PAH content, 5g of each type of smoked dried fish were weighed into amber glass bottles and extracted sequentially by ultrasonication using 25 ml of n-hexane for 1 h. After ultrasonication the supernatant of the extracts were decanted into a vial and 15 ml of fresh solvent was added for another 1h of ultrasonication. The process was repeated with another 10 ml of fresh solvent for 1h. The combined extracts (50ml) were centrifuged at 2500 rpm for 10 min and the supernatant was decanted (Garcia-Falcon et al., 1996). The supernatant was cleaned-up using a Whatman nylon filter membrane. Further clean-up was done using solid phase extraction (SPE) cartridges. The sorbent of the SPE cartridges was first conditioned with n hexane, after which the filtered extracts were loaded on to the cartridges, the analytes were eluted with dichloromethane. The volume of the dichloromethane was blown down to dryness and extract was reconstituted in 200μl of acetonitrile. After purification of the sample with n-hexane solution on Silica SPE column, it was concentrated and analyzed on an Agilent Model 6890 gas chromatograph equipped with the mass selective detector Model 5973.

## Statistical data analysis

Data were analyzed by descriptive analysis and one-way analysis of variance (ANOVA) to explore the general trend of the experimental data. Duncan's Multiple Range Test of difference between means (DMRT) (Duncan, 1955) was also performed. SPSS (version 16.0) statistical software package (SPSS, Chicago, USA) was employed in the analysis. Differences were considered significant at an alpha level of 0.05. All means were given with ± standard error.

## 3. RESULTS AND DISCUSSION

The result of the proximate composition of catfish smoked using different smoking systems is shown in Table 1. The values represent the mean of triplicate determinations and standard error. Mean values of duplicate determination and standard error were reported for protein. Mean values followed by different letters in the same row were significantly different (P<0.05). The protein content ranged from 64.8 percent to 81.11 percent. The lipid content ranged from 9 to 22.43 percent. The moisture content of the samples smoked with charcoal and firewood (CFL) was 8.33 percent while lipid content range was 9 percent to 22.43 percent. Ash content was least in the sample smoked with gas. There was no significant difference (p<0.5) in the ash content of FB and CFL samples. All the smoking systems produced products with an acceptable level of protein. Smoking results in a loss of weight due mainly to loss of water rather than nutrients; in effect, the protein becomes concentrated. The trend of proximate composition in this study showed increasing concentration of nutrients as moisture level decreases.

**Table 1. Proximate composition of catfish smoked using different smoking systems from different parts of Nigeria**

| | Fish samples | | | | |
|---|---|---|---|---|---|
| Parameters measured | GL | FB | CFL | CL | FL |
| Protein content[#] (%) | 64.80±0.75a | 81.11±0.05b | 78.01±0.13c | 66.51±0.37d | 69.80±0.27e |
| Lipid content[*] (%) | 9.0±1.15a | 10.44±0.29a | 16.75±0.75b | 11.0±0.29a | 22.43±0.43c |
| Moisture content[*] (%) | 25.33±0.59a | 13.07±0.13b | 8.33±0.17c | 17.40±0.4d | 10.07±0.23e |
| Ash content[*] (%) | 3.33±0.13a | 5.53±0.13c | 5.33±0.33c | 4.60±0.12b | 4.44±0.1b |

[*] Mean values of triplicate determinations and standard error.

[#] Mean values of duplicate determinations and standard error.

Mean values followed by different letters in the same row were significantly different ($P<0.05$).

Table 2 shows the PAHs found in *Clarias gariepinus* by different smoking methods. BaP was detected in samples smoked with firewood only (FB and FL at levels of 83µg/kg and 11µg/kg respectively). BaP was not detected in other smoking methods investigated in this study.

**Table 2. PAHs contents found in *Clarias gariepinus* by different smoking methods**

| PAHs | GL(µg/kg) | FB(µg/kg) | CFL(µg/kg) | CL(µg/kg) | FL(µg/kg) |
|---|---|---|---|---|---|
| Naphthalene | 76.0 | 94.0 | 86.0 | 14.0 | 7.0 |
| 2-Methylnapthalene | 81.0 | 96.0 | 96.0 | 9.0 | 7.0 |
| Acenaphthlene | 73.0 | 81.0 | 74.0 | 5.0 | 5.0 |
| Acenaphthene | 70.0 | 70.0 | 69.0 | 5.0 | 5.0 |
| Flourene | 75.0 | 86.0 | 75.0 | 5.0 | 6.0 |
| Phenanthrene | 85.0 | 142.0 | 87.0 | 5.0 | 7.0 |
| Anthracene | 71.0 | 83.0 | 72.0 | 4.0 | 5.0 |
| Flouranthene | 81.0 | 116.0 | 81.0 | 8.0 | 7.0 |
| Pyrene | 76.0 | 110.0 | 76.0 | 6.0 | 5.0 |
| Benzo(a)anthracene | 88.0 | 99.0 | ND[*] | ND | 2.0 |
| Chrysene | ND | 93.0 | ND | ND | 3.0 |
| Benzo(b)fluoranthene | ND | 91.0 | ND | ND | ND |
| Benzo(k)fluoranthene | ND | 78.0 | ND | ND | ND |
| Benzo(a)pyrene | ND | 83.0 | ND | ND | 11.0 |
| Benzo(g,h,i)perylene | 59.0 | 99.0 | 45.0 | ND | ND |
| Dibenz(a,h)anthracene | ND | ND | ND | ND | ND |
| Indeno(1,2,3-cd)pyrene | ND | 23.0 | 23.0 | ND | ND |
| Total PAHs | 835.0 | 1444.0 | 784.0 | 66.0 | 70 |

* ND = Not Detected

The FAO and WHO have set a maximum permissible concentration of B(a)P in food of 10 µg/kg (Joint FAO/WHO Expert Committee on Food Additives, 1987) while a level of 5 µg/kg has been set by the European Commission (2005). In the study, the level of BaP found in catfish smoked with firewood only (FB and FL) had concentrations above the EU maximum level (5µg/ kg). The sample FB is caught from the wild. The influence of the environment might have contributed to the high level of BaP found in sample FB. This agrees with (Yang, et. al., 1998) who observed that anthropogenic sources contribute significant amounts of BaP and other PAHs. Nevertheless, it has been shown that PAH are quickly metabolised in fresh fish and do not accumulate in the muscle meat (EFSA, 2008), hence the contribution of direct firing with minimal control, as found with traditional smoking systems would have influenced the high content of BaP in the samples.

Table 3 shows the PAH4 found in the smoked fish samples with the emerging EU standards. Benzo(a)anthracene (BaA) was detected in GL, FB and FL at levels of 88 µg/kg, 99 µg/kg and 2 µg/kg respectively. Chrysene (CHR) was detected in samples FB and FL at level of 93 µg/kg and 3 µg/kg respectively while Benzo(b)fluoranthene (BbFA) was detected only in sample FB at a high level of 91 µg/kg . The sum total of BaP, BaA, CHR and BbFA in the samples GL, FB and FL are 88µg/kg, 366µg/kg and 16µg/kg respectively. These values exceeded the emerging maximum permissible level of 30.0µg/kg set by EU except for FL which was within the limit. No PAH4 was detected in samples CFL and CL.

**Table 3. The four specific substances (PAH4) in the smoked fish samples and the emerging EU standards**

| PAHs | GL (µg/kg) | FB (µg/kg) | CFL (µg/kg) | CL (µg/kg) | FL (µg/kg) |
|---|---|---|---|---|---|
| Benzo(a)anthracene | 88.0 | 99.0 | ND* | ND | 2.0 |
| Chrysene | ND | 93.0 | ND | ND | 3.0 |
| Benzo(b)fluoranthene | ND | 91.0 | ND | ND | ND |
| Benzo(a)pyrene | ND | 83.0 | ND | ND | 11.0 |
| Sum of benzo(a)pyrene, benz(a)anthracene, benzo(b)fluoranthene and chrysene (PAH4) | 88 | 366 | ND | ND | 16 |
| [†]EU Maximum level for the Sum of benzo(a)pyrene, benz(a)anthracene, benzo(b)fluoranthene and chrysene | 30.0 µg/kg (as from 1/9/2012 until 31/8/2014) 12.0 µg/kg (as from 1/9/2014) | | | | |
| [†]EU Maximum level for B(a)P | 5.0 µg/kg (until 31/8/2014) 2.0 µg/kg (as from 1/9/2014) | | | | |

[*] ND = Not Detected.
[†] SANCO/10616/2009 rev. 7 (11.4.2011).

The high levels of Benzo [a] Pyrene (BaP), Benzo(a)anthracene (BaA), Chrysene (CHR) and Benzo(b)fluoranthene (BbFA) found in traditional smoked fish is a function of many parameters which have to be addressed to ensure compliance with the maximum level of PAH4 acceptable for smoked fish. Direct fired smoking kilns are not good in reducing the PAHs, while at the same time the oil dripping from the fish into the flames also produces PAHs that return to the smoked fish. In the traditional method of smoking there is no way to control the PAHs formed during the combustion of firewood. The high level of BaA in gas smoked catfish (GL) is an indication of the presence of the substance in large amounts in the LPG gas used for smoking. The low level of PAH4 detected in FL compared with FB, both being firewood smoked but at a different location, could be attributed to the dexterity of the operators, the time of exposure to smoke and the type of wood used.

## 4. CONCLUSION

Generally the levels of BaP in two of the samples studied were well over the permissible level of 5 µg/kg as specified by European Union. The sum total of the PAH4 in two of the samples was also well over the emerging permissible level specified by the EU. The heavy deposition of wood smoke and soot during traditional smoking of fish, employing direct firing, must have accounted for the high levels when compared with the improved smoking kiln, with some facility to filter some of the wood smoke soot and also without direct smoke firing. While BaP was not detected in charcoal and gas fired smoked fish, it is assumed that they are present in small quantities, likely to be below the recommended permissible level.

Charcoal smoked products are best in addressing the concerns about PAHs in hot smoking of fish and should therefore be documented for dissemination, for the sake of protecting consumers as well as trade promotion.

The type of wood and whether direct or indirect smokes are determinant in the smoking process with firewood needs investigation. Further research should be encouraged and supported to assess the status of the most frequently used fuelwood in small scale fisheries and the extent to which they impart PAH during smoking (direct and indirect firing).

## 5. REFERENCES

**Association of Official Analytical Chemists**. 1994. *Official Methods of Analysis of the Association of Official Analytical Chemists*, Vols. I & II, Association of Analytical Chemists, Arlington. 1298 pp.
**Atlas, R.M.** 1991. Microbial hydrocarbon degradation: Bioremediation of oil spills. *Biotechnology*. 52: 149–156.
**ATSDR**. 1995. Toxicological profile for polycyclic aromatic hydrocarbons (PAH). Atlanta, GA, US Department of Health and Human Services, Public health service, agency for toxic substances and disease registry.
**Chandrasekhar, K. & Deosthale, Y.G.** 1993. Proximate composition, amino acid, mineral, and trace element content of the edible muscle of 20 Indian fish species. *Journal of Food Composition and Analysis* 6(2): 195–200.
**Duncan, D.B.** 1955. New multiple range and multiple F-tests. *Biometrics*, 11, pp 1–42.

EFSA 2008. "Polycyclic Aromatic Hydrocarbons in food Scientific Opinion of the Panel on Contaminants in the Food Chain" (Question N. EFSA-Q-2007-136).*The EFSA Journal* (2008) 724, 1–114. http://www.efsa.europa.eu/en/efsajournal/pub/724.htm. Accessed on 18/10/2011).

Eun, J.B., Chung, H.J. & Hearnsberger, J.O. 1994. Chemical composition and microflora of channel catfish (*Ictalurus punctatus*) roe and swim bladder. *Journal of Agricultural and Food Chemistry* 42 (3): 714–717.

European Commission. 2005. Directive No. 208/2005 of 4 February 2005, amending Regulation (EC) No. 466/2001 as regards polycyclic aromatic hydrocarbons. *Official Journal of the European Union*, Brussels, Belgium.

European Union. 2011. Official journal of the EU 215, 20.8.2011, p. 4, Legislation, 54:4-6 http://eur-lex.europa.eu/LexUriServ/LexUriServ.do?uri=OJ:L:2011:215:FULL:EN:PDF. Accessed on 18/10/2011).

Exler, J. 1987. Composition of foods: Finfish and shellfish products. US Department of Agriculture, Handbook. pp 9–15.

Garcia-Falcon M.S.G., Amigo G.S., Lma, Y, Villaizan, M.J, & Lozano, S.L. 1996. Enrichment of benzo(a)pyrene in smoked food products and determination by high-performance liquid chromatography fluorescence. *J.Chromatora., 753;207–215*

Gomaa, E.A., Gray, J.I., Rabie, S., Lopez-Bote, C. & Booren, A.M. 1993. Polycyclic aromatic hydrocarbons in smoked food products and commercial liquid smoke flavoring. *Food Additives and Contaminants* 10: 503–521.

International Agency for Research on Cancer (IARC). 1987. Monographs on the evaluation of the carcinogenic risk of chemicals to humans. Overall evaluation of carcinogenicity: An updating of IARC monographs, pages 1–42.

Ito, S. 2005. The distribution effect of compliance with Food Safety and Agricultural Health Standards on small producers in developing countries. Proceedings for the Japan Society for International Development, pp.113–116.

Joint FAO/WHO Expert Committee on Food Additives JECFA. 1987. Evaluation of certain food additives and contamination. *WHO Technical Report Series*, 31, 759 pp.

Klein, M. 1963. Susceptibility of strain B6AF1/J hybrid infant mice to tumorigenesis with 1, 2-benzanthracene, deoxycholic acid, and 3- methylcholanthrene. II. Tumours called forth by painting the skin with dibenzpyrene. *Cancer Res.* 23: 1701–1707.

Martson, C.P., Pareira, C., Ferguson, J., Fischer, K., Olaf, H., Dashwood, W. & Baird, W.M. 2001. Effect of complex environmental mixture from coal tar containing polycyclic aromatic hydrocarbons (PAH) on tumor initiation, PAH-DNA binding and metabolic activation of carcinogenic PAH in mouse epidermis. *Carcinog.* 22(7): 1077.

Oyelese, O.A. 2006. Quality assessment of cold smoked, hot smoked and oven dried *Tilapia nilotica* under cold storage temperature conditions. *Journal of Fisheries International* 1(2–4):92–97.

Schonfeldt, H.C. 2002. Food composition program of AFRO FOODS. *Journal of Food Composition and Analysis* 15: 473–479.

Simko, P. 2002. Determination of polycyclic aromatic hydrocarbons in smoked meat products and smoke flavouring food additives. *J.Chromatogr. B.* 770: 3–18.

Simon, R., Palme, S. & Anklam, E. 2006. Validation (in-house and collaborative of a Method based on Liquid Chromtography for the Quantitation of 15 European-priority Aromatic Hydrocarbon in Smoke Flavourings : HPLC-method Validation for 15 EU Priority PAH in Smoke Condensates. Food Chemistry. Belgium.

Uauy, R. & Valenzuela, A. 2000. Marine oils: the health benefits of n-3 fatty acids. *Nutrition* 6(7/8): 680–684.

Wanasundara, U.N. & Shahidi, F. 1998. Lipase-assisted concentration of n-3 polyunsaturated fatty acids in acylglycerols from marine oils. *Journal of the American Oil Chemist's Society* 75(8): 945–951.

Ward, A. 2003. A study of the trade in smoked-dried fish from West Africa to the United Kingdom. *FAO Fisheries Circular*. No. 981. Rome. 2003, 17p.

Wynder, E.L. & Hoffmann, D.A. 1959. Study of tobacco carcinogenesis VII. The role of higher polycyclic hydrocarbons, Cancer. 12: 1079–1086.

Yang, H.H., Lee, W.J., Chen S.J. &. Lai, S.O. 1998. PAH emission from various industrial stacks. J. *Hazard. Mater*, 60: 159–174.

# NOUVEAU PROCÉDÉ DE DÉTERMINATION DU TAUX D'HISTAMINE DANS LES PRODUITS HALIEUTQUES

## [NEW METHODS FOR MEASURING THE LEVEL OF HISTAMINE IN FISHERIES PRODUCTS]

by/par

Alphonse Tine[1], Stephy Edgard Douabale, Lamine Cisse, Moussa Mbaye,
Atanasse Coly et Mame D. Gaye Seye

**Résumé**

Plusieurs techniques relatives au dosage d'histamine ont été établies. Parmi celles-ci on distingue des méthodes biologiques, enzymatiques et chimiques. Malheureusement toutes ces méthodes présentent des insuffisances notoires au regard des limites de détection trop élevées, des taux de recouvrement trop larges aboutissant à des résultats non reproductibles. Même les méthodes de Lerke et Bell et celle de l'Association Officielle des Chimistes en Analyse (AOCA), qui sont les méthodes officielles en vigueur présentent des difficultés. En réalité, le milieu acide utilisé inhibe la fluorescence. C'est pourquoi l'équipe a développé une méthode spectrofluorimétrique de détermination du taux d'histamine dans les produits halicutiques à partir de la cinétique de formation du complexe histamine-orthophthalaldéhyde en milieu basique. Ainsi, le protocole obtenu permet de réduire le temps du dosage, de diminuer les limites de détection et de quantification. De même le taux de recouvrement obtenu permet d'avoir une belle reproductibilité.

*Mots clés: Produits halieutiques, Dosage, Taux d'histamine, Spectrofluorimétre, Milieu alcalin*

**Abstract**

Several methods of histamine determination have been established to measure the content of histamine, including biological, enzymatic and chemical methods. Unfortunately all these methods present significant shortcomings regarding their too high detection limits, non-reproducibility and the inadequate rate of recovery. Even the official methods of Lerke and Bell and that of the Official Association of Analytical Chemists (AOAC) suffer from these deficiencies. In fact, the acid environment used inhibits fluorescence. For this reason we developed a spectrofluorimetric method of determination for histamine in fish products from the kinetics of formation of an orthophthalaldehyde-histamine complex in an optimized alkaline environment. The subsequent protocol allows a reduction in the time of measurement, lowering in the detection and quantification limits, improvement of the recovery rate and good reproducibility.

*Key words: Fish products, Determination, Histamine rate, Spectrofluorometer, Alkaline environment*

## 1. INTRODUCTION

La pêche occupe une place prépondérante dans l'économie de nombreux pays moins avancés, de surcroit exportateurs de produit de pêche et d'aquaculture. En plus la demande des produits halieutiques sur le marché international est en constante progression. Cette expansion s'accompagne cependant d'une plus grande exigence des consommateurs sur la qualité sanitaire des produits. A cela il faut ajouter le fait que la plupart des pays grands importateurs, ont renforcé leur réglementation sanitaire des aliments et sont de plus en plus exigeants en matière de protection de santé publique. L'histamine constitue l'une des premières causes d'intoxication alimentaire due au poisson dans la majorité des pays.

A faible dose, l'histamine joue un rôle bénéfique dans le système nerveux en qualité de neurotransmetteur (Taylor, 1986). Cependant, l'absorption d'une forte dose d'histamine provoque chez les mammifères des désagréments, en augmentant la dilatation et la perméabilité vasculaire. Certains symptômes apparents tels que les palpitations, les picotements de la peau, l'enflure du visage, l'étourdissement, la nausée, les vomissements, la diarrhée, la sensation de brûlure dans la gorge, les douleurs abdominales interviennent immédiatement ou quelques heures après l'intoxication. Ces symptômes peuvent durer quelques heures, voire quelques jours. Dans

---

[1] Faculté des Sciences et Techniques, Université Cheikh Anta, Diop, BP 16404, Dakar-Fann, Senegal. alphtine@yahoo.fr

les cas extrêmes, une intoxication à l'histamine peut aboutir aux complications respiratoires et cardiaques chez les individus présentant les prédispositions: de tels cas peuvent alors devenir fatals.

La plupart des intoxications répertoriées sont liés à la consommation de certaines espèces de poisson appartenant aux familles suivantes : Scombridés (thon, bonite, maquereau), Scombérésocidés (saurel), Clupéidés (hareng, sardine), Coryphénidés (mahi-mahi), Pomatomidés (tassergal), Carangidés (chinchard), Englaulidés (anchois), Xiphiidés (espadon), Istiophoridés (voilier, marlin) (Antoine *et al.*, 2001 ; Becker *et al.*, 2001 ; Bermejo *et al.*, 2003 ; Chytiri *et al.*, 2004 ; Cinquina *et al.*, 2004 ; Den Brinker *et al.*, 2002 ; Henderson, 1830 ; Kanki *et al.*, 2004 ; Kanny *et al.*, 1996 ; Kim, Field, Chang *et al.*, 2001 ; Kim, Field, Morrissey *et al.*, 2001 ; Lopez-Sabater *et al.*, 1994 ; Lopez-Sabater *et al.*, 1996 ; Periago *et al.*, 2003 ; Taylor *et al.*, 1989 ; Taylor *et al.*, 1979). En effet la plupart de ces espèces de poisson contiennent un fort taux d'histidine endogène qui est de l'acide aminé.

L'histamine est une amine biogène qui dérive du mécanisme enzymatique de transformation par décarboxylation de l'acide aminé appelé histidine en amine (dans notre cas : histamine). Ainsi, les poissons nobles (« sains ») contiennent peu d'histidine et pas de bactéries histamino-productrices.

Aussi est-il impératif de mesurer le taux d'histamine de tous les produits de pêche.

C'est pourquoi, depuis la découverte de la molécule d'histamine au début du siècle passé (1916) par Guéggenheim et Loffler (Lorenz & Neugebauer, 1990), plusieurs techniques relatives à son dosage ont été proposées. Malheureusement divers problèmes y sont associés. Les problèmes rencontrés dans les méthodes sont de plusieurs ordres et dépendent de la méthode utilisée. Parmi celles-ci, la méthode fluorimétrique ou officielle est la plus fiable. Cependant, beaucoup reste à faire. En fait, La molécule d'histamine n'étant pas naturellement fluorescente, on la fait réagir avec un fluorogène tel que l'orthophthalaldéhyde (OPA) pour former un complexe fluorescent (isoindole) avec une stoechiométrie de 1 :1.

Les méthodes en vigueur pour le dosage de l'histamine par voie fluorimétrique sont celles de Lerke et Bell (Lerke & Bell, 1976) et de l'Association Officielle de Chimie Analytique (AOCA) (Rogers & Staruszkiewicz, 1997). Leur différence se situe au niveau des solvants d'extraction : la première utilise une solution d'acide trichloracétique (TCA) à 10 pour cent et la seconde du méthanol à 75 pour cent. Dans tous les cas, l'histamine est éluée d'une colonne chromatographique échangeuse d'ions par de l'acide chlorhydrique, pendant que la condensation de l'histamine avec l'orthophthalaldéhyde (OPA) est réalisée en milieu basique (pH proche 11,5).

Cependant, dans la méthode de Lerke et Bell (comme dans celle de AOCA), le complexe obtenu est très fluorescent mais trop instable en ce milieu basique. C'est pourquoi le dosage par fluorescence a été fait en milieu acide (HCl 0,16N).

La comparaison de la méthode Officielle de l'Association Officielle de Chimie Analytique (OAAC) avec d'autres méthodes donne les résultats ci-après (Tableau I).

**Tableau 1. Comparaison des résultats par différentes méthodes**

| Echantillon | Officielle (ppm) | Neogen Agrimeter II (ppm) | Neogen Veratox (ppm) | Immunotech Histamarine (ppm) | IDR K1-HTM (ppm) |
|---|---|---|---|---|---|
| 1 | 67,9 | 45 | 64 | 71,7 | 60,7 |
| 2 | 58,1 | 75 | 66 | 72,6 | 95,7 |
| 3 | 190 | | | 309 | 323 |
| 4 | 300 | | | 372 | 435 |
| 5 | 8,8 | 20 | 9,4 | 17,1 | 17,2 |
| 6 | 19,7 | 30 | 31 | 43,6 | 35,6 |
| 7 | 300 | | | 366 | 455 |
| 8 | 158 | | | 191 | 323 |

Ces tests faits par Rogers et Staruszkiewic (Rogers & Staruszkiewicz, 2000) montrent les difficultés rencontrées pour la détermination du taux d'histamine. En fait, on ne constate pratiquement pas de valeurs de la méthode officielle qui coïncident avec celles des autres méthodes. En plus, en comparant les valeurs trouvées entre les deux premiers échantillons, la valeur trouvée en 2 est supérieure à celle trouvée en 1 pour toutes les méthodes sauf celle de la méthode officielle.

Ainsi, ces résultats montrent que la méthode officielle pose problème pour le dosage de l'histamine. C'est ce qui a conduit au présent travail dans lequel l'innovation a été de faire des mesures en milieu alcalin après avoir stabilisé le complexe entre l'OPA et l'histamine (Tine & Douabale, 2008).

## 2. ETUDE EXPERIMENTALE

### *Produits utilisés*

Le dichlorhydrate d'histamine (98 pour cent, m/m), l'histamine (96 pour cent), l'orthophthalaldéhyde (OPA, 99 pour cent, m/m) et le méthanol de qualité spectroscopique ont été achetés chez Sigma-Aldrich (Taufkirchen, Allemagne) et utilisés sans modifications supplémentaires. L'hydroxyde de sodium (97 pour cent, m/m), l'acide chlorhydrique (36 pour cent, m/m), l'acétate de sodium et l'acide acétique cristallisable ont été obtenus chez Labosi (Oulchy Le Château, France). L'acide trichloracétique (TCA 99 pour cent, m/m) provenait de chez Janssen Chimica (Belgique) et l'acide phosphorique (85 pour cent) chez FlukaChemica (Suisse). La résine Amberlite GC-50 (H) Type I était livrée par Prolabo (France). L'eau distillée a été utilisée pour préparer toutes les solutions aqueuses nécessaires.

### *Instrumentation*

Toutes les mesures de fluorescence ont été exécutées à la température ambiante à l'aide d'un spectrofluorimètre de marque Kontron, modèle SFM-25, connecté à un micro-ordinateur IBM, modèle Aptiva. Les spectres de fluorescence non corrigés ont été acquis grâce au logiciel K-Wind 25 et traités par le logiciel MicrocalOrigin version 6.0. Un agitateur magnétique VELP ARE Scientifica muni d'un chauffage thermostaté a été utilisé. Pour enregistrer les spectres d'absorption électroniques, un spectrophotomètre UV-visible de marque JASCO modèle 7800 associé à un écran PHILIPS de type Computer monitor 80 et une imprimante JASCO PTL-3965 ont été utilisés. Les solutions tampons ont été préparées grâce à un pH-mètre de marque HANNA Instruments modèle HI 190N. Un broyeur (moulinex) était utile pour homogénéiser les échantillons de poisson. Une balance de précision 0,1mg de marque Sartorius a été utilisée pour la pesée des produits. De même, les cellules en quartz à cinq faces polies étaient utilisées pour la mesure en fluorescence, tandis que deux cellules en quartz à deux faces polies étaient nécessaires pour des mesures en absorption. Toutes les cellules avaient 1 cm de trajet optique. La verrerie était constituée de colonnes chromatographiques, fioles, béchers et pipettes de plusieurs dimensions. Une micropipette de 20µl de marque Pipetman et des cônes étaient nécessaires pour des prélèvements.

### *Procédures expérimentales*

#### Préparation des solutions
Les solutions mères d'histamine ($10^{-3}$ M) ont été directement préparées dans l'eau, pendant que celles d'OPA ($10^{-3}$ M) nécessitaient éventuellement un minimum de méthanol (1/10, v/v) pour faciliter la dissolution. Les dilutions en série ont été effectuées pour obtenir des solutions standard de travail. Toutes les solutions ont été protégées contre la lumière avec du papier aluminium et conservées dans un réfrigérateur. Les solutions mères de HCl (1N) et de NaOH (1N) ont été préparées avec de l'eau distillée et utilisées avec ou sans dilutions préalables selon le cas.

#### Préparation des échantillons
La préparation de l'échantillon ainsi que l'extraction et la purification de l'histamine ont été réalisées selon la méthode de LERKE et BELL décrite plus haut. Des changements ont été apportés au niveau de la condensation et de la mesure de fluorescence. En effet, à ce niveau, la solution tampon de phosphate de pH 11,5 était parfois utilisée.

#### Mesure de la fluorescence
L'allumage du spectrofluorimètre a lieu une trentaine de minutes avant le début des mesures afin d'atteindre la stabilité du rayonnement émis par la source.

Deux types de mesure ont été faites:

- Le premier consistait à mettre une solution d'OPA de titre connu dans la cellule du spectrofluorimètre, ensuite on ajoutait progressivement des volumes d'histamine. Après chaque ajout, on enregistrait le spectre de fluorescence ou la cinétique du complexe OPA-histamine. La variation de l'intensité de fluorescence du complexe en fonction du nombre de moles d'histamine ajoutée nous permettait d'établir la courbe de calibration.
- Le second type de mesure consistait à préparer le complexe OPA-histamine dans un réacteur contenant de l'eau distillée, l'OPA, NaOH et de l'histamine. Parfois une quantité de sel NaCl était introduite dans ce réacteur. Dans tous les cas, tous les soins étaient pris pour d'introduire l'OPA ou l'histamine en dernier lieu avant de déclencher le chronomètre car c'est à cet instant que commençait la réaction. Pour étudier l'effet du pH, on ajoutait des quantités connues de HCl dans le réacteur. Le temps d'agitation et/ou chauffage variait entre 0 et 5 minutes, correspondant au temps de réaction de condensation. Une portion de ce mélange était prélevée dans la cuve et portée au spectrofluorimètre pour l'enregistrement de la cinétique de complexation. Après chaque mesure, le réacteur était lavé et rincé avant de reprendre un autre mélange réactionnel. La longueur d'onde d'excitation était fixée à 350 nm, l'intervalle de balayage du spectre de fluorescence était fixé entre 370 et 550 nm. Pour la cinétique de fluorescence, la longueur d'onde d'excitation était à 350 nm et celle d'émission à 430 nm. La cinétique permettait de mesurer la stabilité du complexe fluorescent, et des mesures successives des cinétiques en fonction de la concentration d'histamine permettaient d'établir la droite de calibration.

Toutes les mesures de fluorescence ont été corrigées par rapport au signal du solvant. Les résultats ont été exprimés comme valeurs moyennes des mesures répétées. Le logiciel MICROCAL ORIGIN, version 6.0, a été utilisé pour le traitement statistique des données.

## 3. PROCEDE DE DETERMINATION DU TAUX D'HISTAMINE

Le nouveau procédé consiste à extraire, à purifier l'histamine et à procéder à la formation du complexe entre l'histamine et l'OPA selon la méthode officielle (OAAC). Le suivi de la cinétique de formation du complexe se fait en milieu alcalin avec optimisation du milieu. Deux (2) mL de l'éluat d'histamine est mélangé avec des solutions de NaOH et d'OPA, dans un récipient contenant un barreau aimanté et soumis à une agitation dans les conditions optimales : OPA ($10^{-5} - 10^{-4}$ M) / NaCl (0 – 0,2 M) / pH 11,48. Au bout de 3 minutes de réaction à 50°C, 2,5 mL du mélange dans les conditions optimales sont introduits dans une cuve en quartz et placés dans le spectrofluorimètre. La courbe de cinétique est lancée pour une durée de 10 minutes à la température ambiante. L'intensité maximale du palier de cinétique correspondant au mélange réactionnel est relevée. A partir des intensités maximales correspondant aux paliers de la cinétique, une courbe de calibration est déterminée. Une extrapolation sur cette courbe permet d'en déduire la concentration en histamine. Le taux d'histamine de l'échantillon est déterminé en remontant par le calcul correctif des diverses dilutions selon le protocole (Figure 1).

**Figure 1. Protocole de détermination du taux d'histamine dans les produits halieutiques**

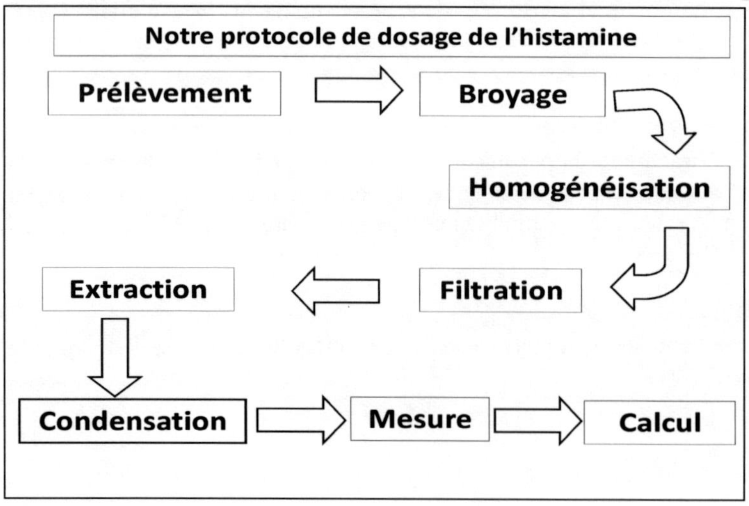

*Optimisation des mesures*

Pour mesurer le taux d'histamine en milieu basique il a été procédé à diverses optimisations:

**Effet de l'acide chlorhydrique (HCl)**

Le nouveau procédé a également permis de montrer que l'acide Chlorhydrique inhibe la fluorescence du complexe (Figure 2).

**Figure 2. Effet de HCl sur la fluorescence du complexe OPA-histamine**

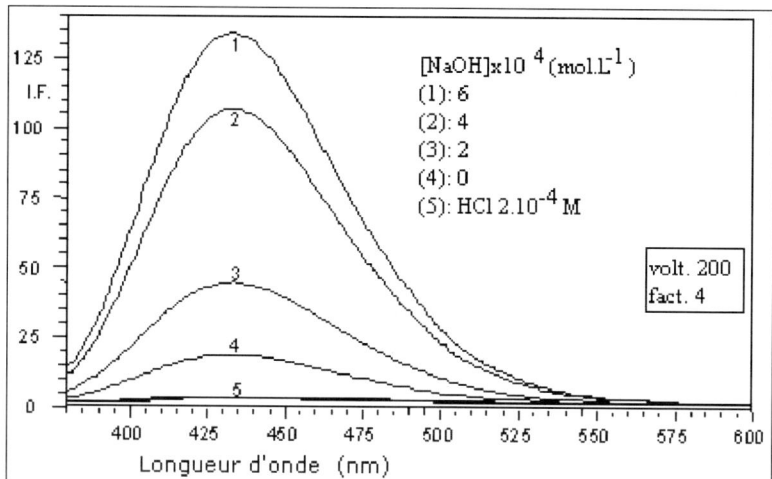

En plus, quand on augmente l'énergie d'excitation (400 V), en milieu acide, on décèle deux types de complexes, l'un émettant aux environs de 430 nm et l'autre émettant aux environs de 510 nm (Figure 3).

**Figure 3 : Mise en évidence de l'existence de deux complexes en milieu acide**

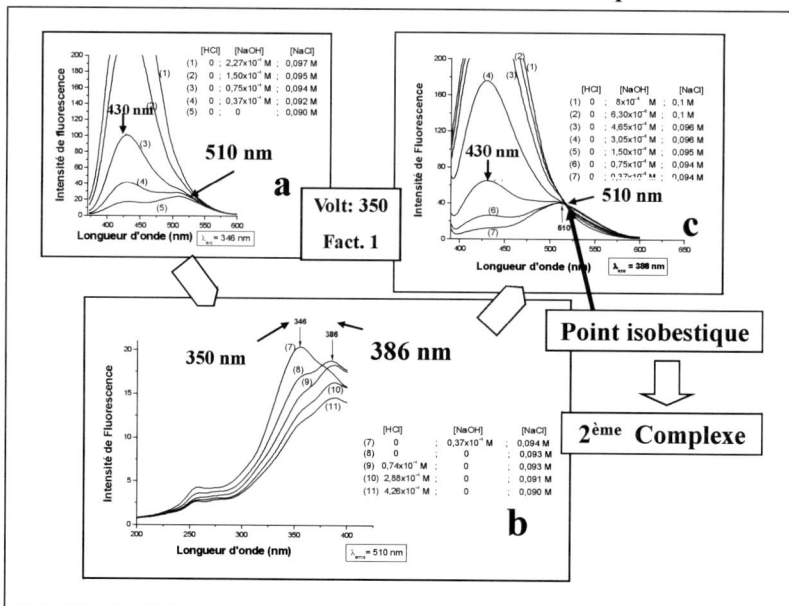

Par conséquent, non seulement l'acide chlorhydrique (HCl) inhibe la fluorescence, mais favorise la formation de deux types de complexe entre l'histamine et l'OPA. Dès lors, une étude quantitative dans ce milieu pour déterminer le taux d'histamine pose problème.

C'est pourquoi il a été envisagé de travailler en milieu basique en observant la cinétique de la formation du complexe OPA-Histamine où l'intensité de la fluorescence est très intense.

## Effet de la soude (NaOH)

L'étude de l'effet de la concentration de NaOH donne une intensité maximale aux environs de $[NaOH] = 10^{-3}$ M (figure 4). Les résultats obtenus sur les courbes de calibration en milieu basique indiquent une extension du domaine de calibration avec l'augmentation de la concentration de NaOH jusqu'aux valeurs proches de $10^{-3}$ M (figure 4). Le maximum du domaine de calibration se situe aux environs du pH 11,5, au-delà, on constate un effet inhibiteur de NaOH.

**Figure 4. Effet de NaOH sur la fluorescence du complexe OPA-histamine**

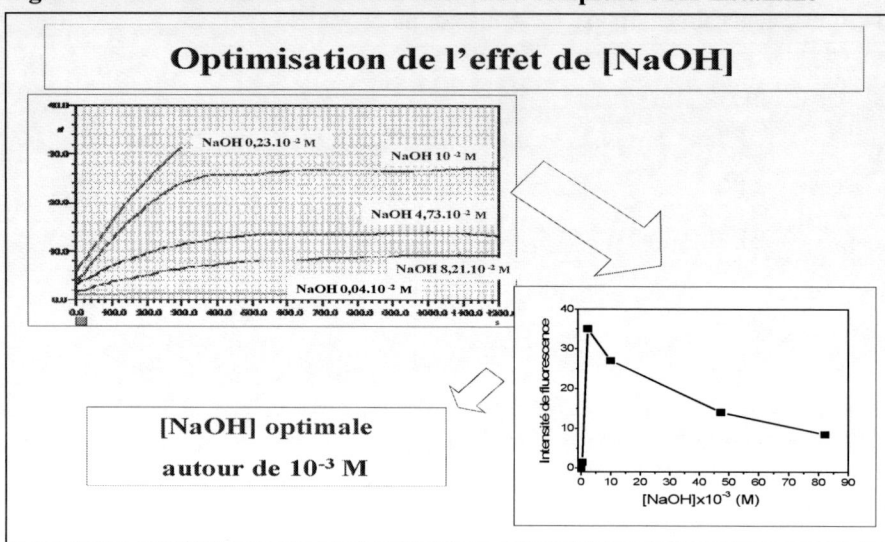

## Effet du pH

En étudiant l'effet du pH sur l'intensité de fluorescence du complexe OPA-histamine, nous avons trouvé un maximum aux environs du pH 11.5 contrairement à la méthode officielle où le maximum se situe aux environs du pH 1 (figure **5**).

**Figure 5. Effet du pH sur la fluorescence du complexe OPA-histamine**

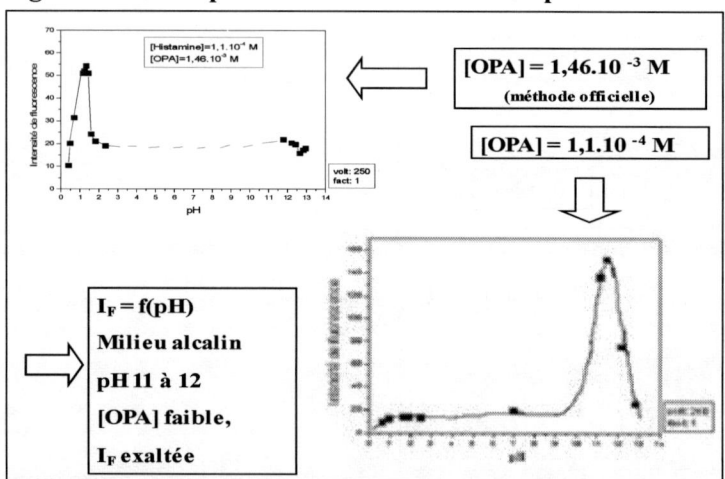

## Effet du sel (NaCl)

L'extraction de l'histamine se fait en milieu très acide, Par conséquent, l'échantillon d'histamine à doser contiendra du NaCl lors de la neutralisation de l'acide chlorhydrique par la soude avant d'être en milieu alcalin. Pour cette raison, nous avons donc déterminé la variation de l'intensité de fluorescence du complexe en fonction de la concentration de NaCl. La figure ci-dessous montre que la concentration de NaCl n'affecte pas d'une manière significative l'intensité de fluorescence du complexe OPA-Histamine (Figure 6).

**Figure 6. Effet de NaCl sur l'intensité de fluorescence du complexe OPA-histamine**

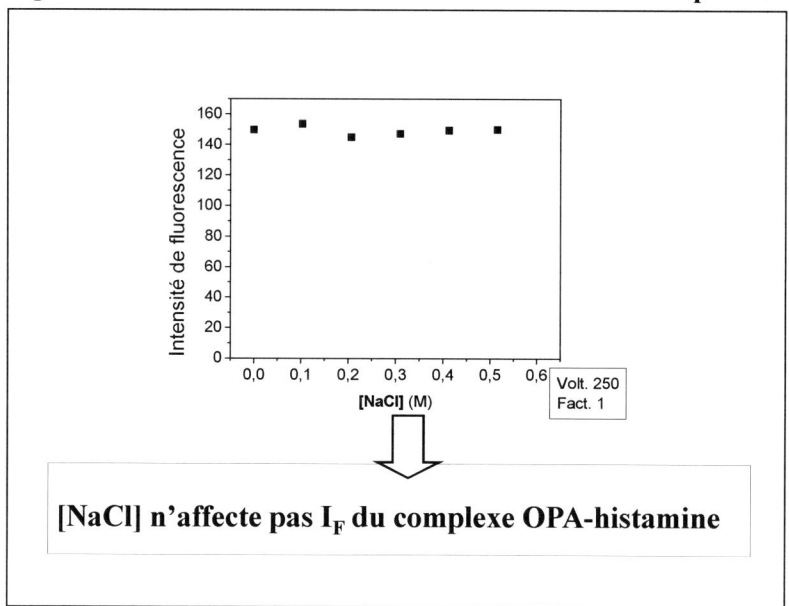

## Effet du chauffage

L'une des principales difficultés rencontrées par les méthodes connues dans le dosage de l'histamine est l'instabilité du complexe qu'elle forme avec l'orthophthalaldéhyde, en milieu basique. Pour contourner cette difficulté, la cinétique de la formation du complexe OPA-Histamine en milieu basique (pH 11,5) a été étudiée. Cette cinétique admet un maximum. Cependant, si le chauffage n'a pas lieu, il est très difficile de déterminer le maximum puisque les courbes présentent des ondulations non négligeables (Figure 7).

**Figure 7. Effet de chauffage sur la fluorescence du complexe OPA-histamine**

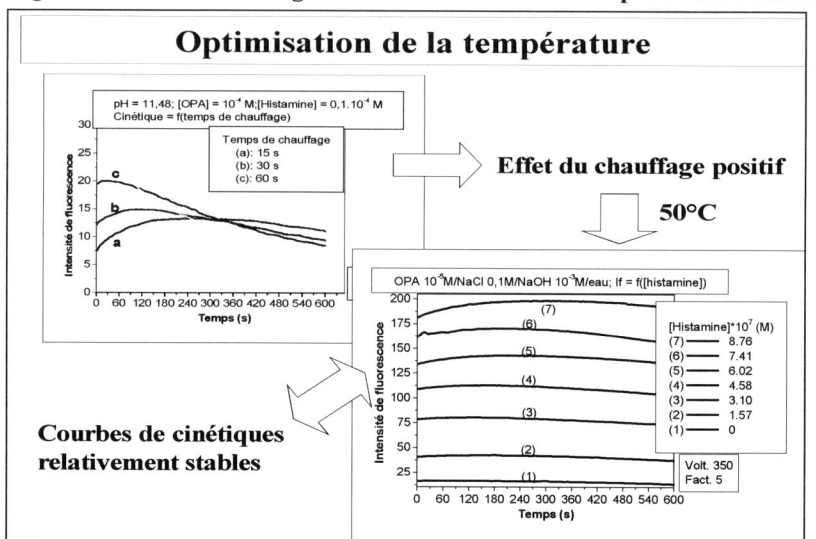

## Courbe de calibration et résultats statistiques en milieu NaOH

Les courbes de calibration sont obtenues à partir des maxima de la cinétique de formation du complexe OPA-Histamine en milieu basique. En milieu acide, les courbes de calibration sont obtenues à partir des intensités de fluorescence du complexe OPA-Histamine. On constate que la pente de la droite de calibration et le domaine de calibration sont dix fois plus élevés en milieu basique qu'en milieu acide (Figure 8). Ce domaine réduit en milieu acide montre le caractère inhibiteur de l'acide chlorhydrique sur l'intensité de fluorescence du complexe OPA-Histamine.

**Figure 8. Comparaison de courbes de calibration en milieu acide et basique**

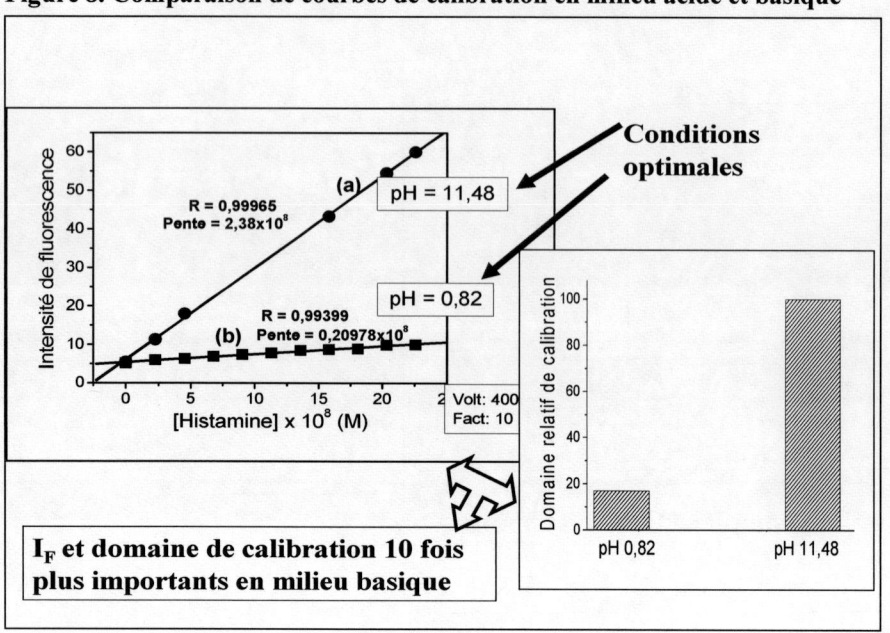

Les résultats statistiques sont résumés dans le Tableau 2. Ces résultats montrent que les coefficients de corrélation sont supérieurs à 0,999 en milieu basique au lieu de 0,993 en milieu acide, ce qui indique la précision des mesures. Par ailleurs, la comparaison des résultats statistiques obtenus dans les deux milieux (basique et acide) montre que le taux de recouvrement est plus étroit en milieu basique et la limite de détection y est plus faible.

**Tableau 2. Comparaison des résultats statiques dans les deux milieux**

| Méthode | Domaine relatif de calibration en If | Limite de détection (ng/mL) | Taux de recouvrement | Coefficients de corrélation |
|---|---|---|---|---|
| **Notre méthode (milieu basique) NaOH** | 0 – 100 | 0,25 | 89% - 114% | 0,9996 |
| **Méthode officielle (milieu acide)** | 0 - 10 | 2,5 | 17% - 217% | 0,9940 |

Meilleure limite de détection et meilleurs taux de recouvrement en mileu basique

***Résultats statistiques en milieu tampon 11,5***

Pour mieux optimiser, le travail a été fait en solution tampon 11,5. A partir des droites de calibration, des taux de recouvrement compris entre 96,87 et 110,8, une limite de détection de 0,55ng/ml et une limite de quantification de 1,50 ng/ml (Figure 9) ont été trouvés. Les limites de détection et de quantification très basses de l'ordre de 1 ng/mL sont parmi les meilleures des méthodes de dosage d'histamine connues à nos jours. Un taux de recouvrement compris entre 17 et 217 pour cent signifie que pour une solution de 100ppm à doser on peut trouver avec la méthode officielle de l'Association Officielle de Chimie Analytique (OAAC) une valeur comprise entre 17 et 217 ppm, méthode non du tout précise (Tine & Douabale, 2008).

**Figure 9. Comparaison des résultats statiques entre les milieux acide et tampon pH 11,5**

**Limites de détection et de quantification**

LD = 0,55 ng.mL$^{-1}$

LQ = 1,50 ng.mL$^{-1}$

**Taux de recouvrement de notre méthode** ⇒ **96,87 à 110,83%**

**Taux de recouvrement par la méthode officielle** ⇒ **17 à 217%**

Ainsi, la nouvelle méthode donne des résultats très fiables et elle est très peu coûteuse. Ce protocole serait donc très utile pour une utilisation industrielle, qui est facile à réaliser avec peu de réactifs. On gagnerait donc à utiliser notre méthode pour diminuer le rejet de produits sur la base des résultats issus d'une méthode non faible.

En réalité, sur la base de ces travaux:

- Le domaine de linéarité est au moins 10 fois plus élevé en milieu basique qu'en milieu acide.
- La limite de détection est 10 fois plus faible par le nouveau procédé d'invention qu'en milieu acide.
- Le nouveau procédé est beaucoup plus précis et plus fiable au regard des taux de recouvrement compris entre 96 et 110,83 pour cent pour le nouveau procédé au lieu de 17 à 217 pour cent. Ce mauvais taux de recouvrement observé en milieu acide est normal car l'acide chlorhydrique inhibe la fluorescence du complexe formé. En plus, le complexe émettant à 430 nm est en compétition avec un autre qui émet à 510 nm en milieu HCl.

Ceci explique pourquoi les valeurs obtenues en amont sont dans la plupart des cas trop distantes de celles trouvées par les pays importateurs et qui expliquerait le rejet trop abusif des produits de pêche.

## 4. APPLICATIONS

Pour illustrer le nouveau protocole de détermination du taux d'histamine, le taux d'histamine contenu dans un échantillon de poisson séché (on aurait pu prendre du poisson frais à la place du poisson séché. Cependant, il est plus difficile de conserver du poisson frais alors que nous voudrions vérifier en même temps la reproductibilité de notre méthode) a été déterminé.

L'expérience a été faite en deux étapes. La première étape consistait à déterminer le taux d'histamine aussitôt après l'achat (échantillon 1), et la seconde étape quelques jours après, en prenant soin de bien garder l'échantillon au congélateur (échantillon 2).

*Détermination du taux d'histamine dans l'échantillon 1*

L'histamine a été extraite de cet échantillon conformément au protocole décrit ci-dessus. Ensuite les spectres d'excitation et d'émission de l'histamine extraite de l'échantillon ont été comparés avec ceux venant de l'étalon dans le même solvant (Figure 10). Une différence de longueurs d'onde nulle a été obtenue pour chaque type de spectre, confirmant alors la présence de l'histamine dans l'échantillon.

**Figure 10. Spectres d'excitation et d'émission du complexe OPA-histamine: histamine authentique (a et b), extrait de l'échantillon (1 et 2); en présence d'OPA $10^{-5}$ M et du tampon phosphate pH 11,5**

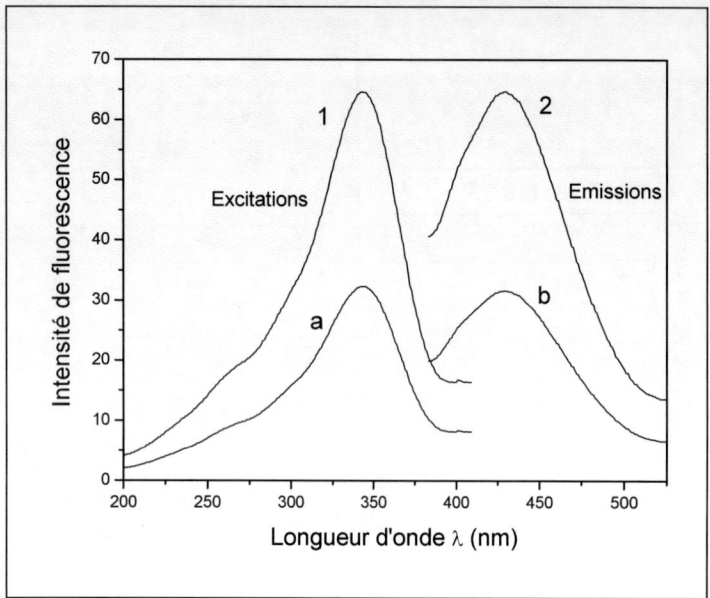

Le taux d'histamine de l'échantillon a été déterminé à partir de la droite d'addition standard et la courbe de calibration de l'histamine (Figure 11). Les coefficients de corrélation correspondant aux deux droites sont de l'ordre de 0,999 proches de l'unité. Ce qui indique la bonne précision des mesures. Une bonne extrapolation de l'ordonnée à l'origine de la droite d'addition standard sur la droite de calibration donne la concentration de l'histamine contenue dans le réacteur. Avec la concentration de $1,39.10^{-7}$M trouvée, le calcul donne un taux d'histamine égale à 54 ppm en utilisant la formulation du protocole.

*Détermination du taux d'histamine dans l'échantillon 2*

Deux semaines après l'échantillon 2 issu du même poisson mis au frais dans un congélateur a été repris. Avec le même procédé, une droite de calibration standard (en fonction de la concentration de l'histamine de l'étalon) et une autre d'addition standard (Figure 11) a été realisée. Il a été trouvé pour les deux droites des coefficients de corrélation de l'ordre de 0,999 indiquant la précision des mesures. L'utilisation de ces deux courbes a permis par extrapolation d'en déduire la concentration de l'histamine égale à $1,4.10^{-7}$ M dans le réacteur. Ce qui correspond toujours à un taux de 54 ppm.

Ce taux d'histamine de 54 ppm est identique à celui obtenu lors de la première mesure de l'échantillon 1.

**Figure 11. Addition standard (dessus) et calibration (dessous) de l'histamine en présence d'OPA 10$^{-5}$ M et du tampon phosphate pH 11,5**

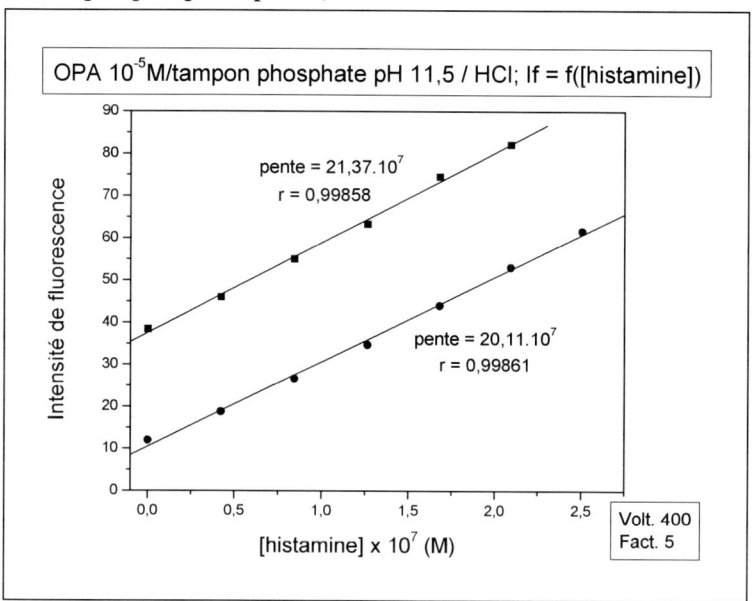

Ce résultat prouve donc qu'il n'y a pas eu d'évolution significative du taux d'histamine quand on a conservé l'échantillon au froid, les bactéries histamino-formatrices étant soit absentes ou détruites dans le poisson séché. Ce résultat prouve également la reproductibilité de notre méthode.

## 5. CONCLUSION

Après avoir cherché les conditions optimales (température optimale, pH optimal 11,48), il a été obtenu une stabilité du complexe OPA-histamine en milieu alcalin à 50 °C.

Ensuite des courbes de calibration linéaires ont été obtenues entre l'intensité de fluorescence correspondant aux maxima de la cinétique de formation du complexe en fonction de la concentration de l'histamine en milieu alcalin. Les coefficients de corrélation correspondant à un grand domaine de calibration sont supérieurs à 0,999, proches de l'unité. Dans ce milieu, des limites de détection aux environs de 0,25 ng/mL ont été obtenues. L'analyse des données spectrales et expérimentales ont montré que la détermination du taux d'histamine était plus probante et plus fiable en milieu basique qu'en milieu acide (conditions de LERKE et BELL ou de l'Association Officielle de Chimie Analytique). En effet, il a été montré l'existence de deux types de complexes en milieu acide et un domaine de calibration très faible.

En solution tampon 11,48 en tenant compte des conditions optimales, il a été également établi des corrélations linéaires entre les intensités de fluorescence correspondant aux maxima de cinétique du complexe et la concentration de l'histamine. Les coefficients de corrélation de 0,999 montrent la précision de nos mesures. La comparaison des résultats statistiques des mesures avec ceux obtenus par la méthode de Lerke et Bell ont montré que la nouvelle méthode (milieu alcalin) est de loin plus fiable, plus précise et plus reproductible que la méthode officielle de Lerke et Bell ou de l'OAAC (milieu acide).

Forts de ces résultats, il a été donc établi un protocole expérimental avec lequel a été déterminé le taux d'histamine de deux échantillons du même poisson à des intervalles de temps différents respectivement. Dans tous les cas, l'on obtient les mêmes résultats, ce qui explique la fiabilité et la reproductibilité de la nouvelle méthode.

Les difficultés rencontrées pour la détermination du taux d'histamine par la méthode officielle sont donc résolues et le nouveau protocole peut servir valablement, d'une manière plus précise et plus nette, pour la détermination du taux d'histamine aussi bien dans les milieux industriels que dans les hôpitaux.

La conséquence des insuffisances de la méthode officielle est que beaucoup de poissons sont rejetés par les marchés a importateurs et d'autres organismes très souvent, non parce que les poissons sont altérés mais surtout en grande partie parce que cette méthode ne fournit pas des résultats fiables. Puisqu'il s'agit d'une affaire de santé publique, l'humanité toute entière gagnerait à coup sûr d'utiliser une méthode fiable et reproductible. Par conséquent nous recommandons sa normalisation et son adoption par les instances habilitées en tant que méthode officielle de détermination du taux d'histamine.

Actuellement une expertise est en train d'être faite pour normaliser la méthode au niveau national et régional. Il est recommandé qu'une évaluation de cette méthode soit conduite par les partenaires (CEE, CODEX Alimentarius, Royaume Unis, USA, Canada, Chine, Japon, Australie, etc.) dans l'optique de promotion d'un commerce international responsable

## 6. REFERENCES

**Antoine, F.R., Wei, C.I., Littell,R.C., Quinn, B.P., Hogle, A.D. & M.R. Marshall**. 2001 Free amino acids in dark- and white-muscle fish as determined by O-phthaldialdehyde pre-column derivatization. Journal of **Food** Science, 66, N° 1,72–77.

**Becker, K., Southwick, K., Reardon & J. MaccormackJ.N**. 2001. Histamine poisoning associated with eating tuna burgers. Journal of the American Medical Association. 285, N° 10, 1327-1330.

**Bermejo, A., Mondaca, M.A., Roeckel, M. & Marti, M.C**. 2003. Growth and characterization of the histamine-forming bacteria of jack mackerel (*Trachurus symmetricus*). Journal of Food Processing and Preservation, 26, N° 6,401-414.

**Chytiri, S., Paleologos, E., Savvaidis, I., & Kontominas M.G**. 2004. Relation of biogenic amines with microbial and sensory changes of whole and filleted freshwater rainbow trout (*Onchorynchus mykiss*) stored on ice. Journal of food protection, 67, N° 5,960–965.

**Cinquina, A.L., Longo, F., Cali, A., De Santis, L., Baccelliere, R., & Cozzani, R**. 2004. Validation and comparison of analytical methods for the determination of histamine in tuna fish samples. Journal of Chromatography A, N° 1-2 , 79-85.

**Den Brinker, C., Rayher, C. & Kerr, M**. 2002. Investigation of Biogenic Amines in Fermented Fish and Fish Products; Ed. 1, Public Health Division, Victorian Government Department of Human Services.

**Henderson, P.B.**1830. Case of poisoning from the bonito (*Scomber pelamis*); Edinburgh medical journal, 34, 317–318.

**Kanki, M., Yoda, T., Ishibashi, M. & Tsukamoto, T**. 2004. *Photobacterium phosphoreum* caused a histamine fish poisoning incident. International Journal of Food Microbiology, 92, N° 1, 9-87.

**Kanny, G.I., Moneret-Vautrind, A. (dir.) & Grignon, G. (dir.)**. 1996. Role of food histamine in false food allergy. Studies of histamine metabolism in food histamine intake models. Thèse Université de Nancy 1, 1.

**Kim, S.H., Field, K.G., Chang, D.S., Wei, C-I. & An, H**. 2001. Identification of bacteria crucial to histamine accumulation in Pacific mackerel during storage. Journal of food protection 64, N° 10, 556–1564.

**Kim, S.H., Field, K.G., Morrissey, M.T., Price, R.J., Wie, C.I. & Haejung, A.N**. 2001. Source and identification of Histamine-producing bacteria from fresh and temperature-abused albacore. Journal of food protection, 64, No 7, 1035-1044.

**Lerke, P.A.D. & Bell, L**. 1976. A rapid fluorometric method for the determination of histamine in canned tuna. Journal of Food Science, 41,1282.

**Lopez-Sabater, E.I, Rodriguez-Jerez, J.J., Hernandez-Herrero, M. & Mora-Ventura, M.T**. 1996. Incidence of histamine-forming bacteria and histamine content in scombroid fish species from retail markets in Barcelona area. International Journal of Food Microbiology, 28, N° 3, 411–418.

**Lopez-Sabater, E.I., Rodriguez-Jerez, J.J., Roig-Sagues, A.X. & Mora-Ventura, M.A.T**. 1994. Bacteriological quality of tuna fish Thunnus thynnus) destined for canning: Effect of tuna handling on presence of histidine decarboxylase bacteria and histamine level. Journal of food protection, 57,318.

**Lorenz, W. & Neugebauer, E**. 1990. Fluorometric assays, Curent techniques of histamine determination, 2, 9-30.

**Periago, M.J., Rodrigo, R.J.G., Rodriguez-Jerez, J.J. & Hernandez-Herrero, M**. 2003. Monitoring volatile and non-volatile amines in dried and salted roes of tuna (*Thunnus thynnus* L.) during manufacture and storage. Journal of food protection, 66, N° 2,335-340.

**Rogers, P.L. & Staruszkiewicz, W.J**. 1997. AOAC International, 80, N° 3,591-602.

**Rogers, P.L. & Staruszkiewicz, W.J**. 2000. Histamine test kit comparison; Journal of Aquatic Food Product Technology,9, N° 2,5-17.

**Taylor, S**. 1986. Histamine food poisoning: toxicity and clinical aspects. Critical Reviews in Toxicology, 91-128.

**Taylor, S.L., Stration, J.E. & Nordlee, J.A**. 1989. Histamine poisoning (Scombroid fish poisoning): an allergy-like intoxication. Journal of Toxicology,27, 225.

**Tine, A. & Douabale, S.E**. 2008. New method for determining histamine rate in halieutic products. Reviews in fluorescence, 5, 195-218.

# THE LEVEL OF TOTAL MERCURY IN SWORDFISH (*XIPHIAS GLADIUS*) CAUGHT IN THE WESTERN INDIAN OCEAN

## *[LE NIVEAU DU MERCURE DANS L'ESPADON (XIPHIAS GLADIUS) CAPTURÉ DANS L'OUEST DE L'OCÉAN INDIEN]*

by/par

Christopher G. Hoareau[1]

### Abstract

The relationship of body weight to levels of total mercury in the muscle tissue of swordfish, *Xiphias gladius*, caught in the Seychelles water was studied during the period 2002 and 2003. The samples were collected from fish landed by semi-industrial long liners operating in the Seychelles EEZ between latitude $00^0 29'$ – $12^0 47'$ South and longitude $43^0 10'$ – $59^0 30'$ Fish were categorized into six different sizes and samples were collected from each category. Analysis was carried out for total mercury by the national reference laboratory for Fish and Fishery Products (Seychelles Bureau of Standard). The level of total mercury ranged from 0.13 mg/kg in a 5 kg fish to 4.302 mg/kg in a 240 kg fish. This observation is in agreement with several publications on the subject demonstrating the bio-accumulative nature of contaminants such as mercury in large predatory marine organisms The first five weight categories demonstrated a clear increase in the average concentrations showing that the larger the fish, the higher the level of mercury. It was observed that 61 percent of the samples from individual fish tested had levels above the maximum allowable limit set by the European Union. The data obtained will assist the Competent Authority in making informed decisions to ensure that swordfish exported by the Seychelles to the international market, mainly to the EU, are not a health hazard to consumers and the swordfish fishery remains viable and a continued source of foreign exchange for the country.

***Keywords: Swordfish, Total mercury, Methyl mercury, Bio-accumulative, Seychelles water***

### Résumé

La relation entre poids du corps et niveaux de mercure dans le tissu musculaire de l'espadon, *Xiphias gladius*, capturé dans les eaux des Seychelles a été étudiée durant la période de 2002 et 2003. Les échantillons ont été prélevés de poissons débarqués par les palangriers semi-industriels opérant dans les ZEE des Seychelles entre la latitude $00^0 29'$ – $12^0 47'$ Sud et la longitude de $43^0 10'$ – $59^0 30'$. Les poissons ont été classés en six tailles différentes et les échantillons de chaque catégorie ont été prélevés. L'analyse a été effectuée pour le mercure par le laboratoire national de référence pour les poissons et produits de la pêche (Bureau de Normalisation des Seychelles). Le niveau de mercure total varie entre 0.13 mg/kg chez un poisson de 5 kg à 4.302 mg/kg chez un poisson de 240 kg. Cette observation est en accord avec plusieurs publications sur le sujet démontrant la nature bio-accumulable des contaminants tels que le mercure dans les grands organismes de prédateurs marins. Les categories des cinq premiers poids ont démontré une nette augmentation dans les concentrations moyennes montrant que plus le poisson est gros, plus le niveau du mercure est élevé. Il a été vérifié que 61 pour cent des échantillons di individus de poissons examinés avaient des niveaux au-dessus de la limite maximale autorisée par l'Union européenne et l'Administration des aliments et drogues des Etats-Unis. Les données obtenues aideront l'autorité compétente à prendre des décisions pertinentes afin d'assurer que l'espadon exporté par les Seychelles sur le marché international, essentiellement vers l'Union européenne, ne soit pas un danger pour la santé des consommateurs et que la pêche à l'espadon reste viable et une source continue de devises étrangères pour le pays.

***Mots clés: Espadon, Mercure total, Méthyle de mercure, Bio-accumulable, Eau des Seychelles***

---

[1] Fish Inspection and Quality Control Unit, Seychelles Bureau of Standards, PO Box 953, Victoria, Mahe, Seychelles. vetfiqcu@seychelles.net

# 1. INTRODUCTION

Mercury occurs naturally in the environment. Inorganic mercury is present in the earth's crust and when it enters the aquatic environment, some of it gets converted by microorganisms into the more toxic form: methyl mercury. Consequently all fish contains some methyl mercury usually in small amount (FDA, 1994). According to FDA toxicologist, Bolger, approximately 2700 to 6000 tonnes of mercury is released annually into the atmosphere naturally by degassing from the earth's crust and oceans. Another 2000 to 3000 tonnes are released annually into the atmosphere by human activities primarily from the burning of household and industrial wastes, especially fossil fuel such as coal (FDA, 1994).

The total mercury level for most fish species ranges between 0.01 to 0.5 mg/kg. It is only in large predatory species such as sharks, swordfish, tuna and marlin that levels often exceed the 1.0 mg/kg (FDA, 1994).

Several authors have reported high levels of mercury in swordfish *(Xiphias gladius)* taken from different oceanic waters (Montiero and Lopes, 1990) High concentrations of total mercury are probably related to mercury biomagnification through the marine food web and the high trophic level of swordfish (Stillwell & Kohler, 1985). Bio-concentration of mercury by branchial extraction through the gills may be important in species like swordfish, which have high metabolic rates and are extremely mobile, filtering thousands of liters of water over a short period of time (Montiero & Lopes, 1990).

Maximum acceptable levels of mercury in fish have been published by several Food Safety Agencies in order to protect the consumer against high exposure.

- The Food and Drugs Administration of the US (FDA) has set a limit of 1.0 mg/kg of methyl mercury for all fish species.
- The European Union (EU) set a limit of 1.0 mg/kg of total mercury in predatory fish such as sharks, swordfish, tuna, marlin, etc, with a limit of 0.5 mg/kg for other species of fish (Commission Regulations 1881/2006).
- The Codex Committees on Food Additives and Contaminants and on Fish and Fisheries Product has a guideline level of 0.5 mg/kg methyl mercury in all fish except predatory species and 1.0 mg/kg in predatory fish (sharks, swordfish, tuna, pike, and others)

Commercial swordfish fisheries in the Seychelles started in the mid 1990s with a small fleet of semi industrial long liners fishing in the Seychelles Exclusive Economic Zone. The catch is landed fresh, in chilled sea water (CSW). Most is exported as loins and fillets to the European Union, mainly the UK and France.

*Objectives*

This study was therefore conducted to:

- evaluate whether swordfish caught in Seychelles water exceeds the maximum level for total mercury set by the EU Commission, FDA and the Codex Alimentarius Commission;
- provide reliable data upon which informed decisions can be made by the Competent Authority on issues related to the export of swordfish;
- compare the level of total mercury in swordfish caught in Seychelles EEZ with other parts of the world;
- investigate the relationship between size of swordfish and the concentration of total mercury; and
- collect and compile reliable statistical data for future research.

## 2. MATERIALS AND METHODS

### Sampling

Samples of swordfish were collected from catches of semi-industrial long liners landing at the export establishments during the period of 2002 and 2003. These vessels operate within the Seychelles Exclusive Economic zone between latitude $00^0 29'$ – $12^0 47'$ South and longitude $43^0 10'$- $59^0 30'$ East. Swordfish landed ranged from juvenile fish, as low as 5 kg to adults of over 200 kg. The fish were therefore divided into the following weight categories:

- Less 20 kg;
- 21–40 kg;
- 41–60 kg;
- 61–80 kg;
- 81–100 kg;
- Above 100 kg.

Total mercury is uniformly distributed in swordfish edible tissues (Monteiro & Lopes, 1990). Samples of at least 500g each were collected from the anterior portion of the carcass by transverse dissection near the dorsal fin. Twenty fish were sampled for each weight category out of which ten were carefully selected. The selection was based on a fair distribution below and above the mid-point in each weight category. This allowed the average weight of each category to be close enough to the midpoint. However for fish above the 100 kg category, the weight distribution did not follow the mentioned order. This was due to the fact that the fish weights ranged from 130 to 245 kg.

**Table 1. Example of selection of samples for category less than 20 kg**

| Sample No | Category < 20 kg |
|-----------|------------------|
|  | 5 kg |
|  | 6 kg |
|  | 7 kg |
|  | 8 kg |
|  | 9 kg |
|  | 12 kg |
|  | 14 kg |
|  | 15 kg |
|  | 16 kg |
|  | 17 kg |
|  | **mean weight - 10.9 kg** |

*Source: Fish Inspection and Quality Control Unit.*

Samples were wrapped in foil paper, packed in polystyrene bags and frozen before submission to the National Reference Laboratory for Fish and Fisheries Products, the Seychelles Bureau of Standard (SBS).

An aggregate sample was also prepared for each weight category by combining an equal weight of fish taken from each of the ten individual samples.

### Estimation of fish weight

The weight of each carcass was recorded in order that the weight of the whole fish can be properly estimated. In swordfish, the head and offal account to approximately 23 percent of the weight of the whole fish (Seychelles Fishing Authority, 2002). The weight of the whole fish is therefore estimated by multiplying the carcass weight with a factor of 1.3 that has been derived on the basis of 23 percent of the fish weight being the head and offal.

**Example of calculation of whole fish weight**

- Carcass weight - 75 kg
- Factor used - 1.3
- Approximate weight of whole fish: 1.3 x 75 kg = 97.5 kg
- Approximate weight of head, gut and offal: 97.5 kg - 75.0 kg = 22.5 kg

NB: 22.5 kg = 23% of 97.5 kg

*Analytical determination of total mercury*

**Sample preparation**

Fish tissue was homogenized using a 'Wareing Blender'. Three replicates of 2g each were put into CEM Teflon Microwave vessels and 3ml concentrated nitric acid, Aristar Grade', was added. Three replicates of procedural blanks were prepared in parallel with the fish tissue.

The vessels were placed into the CEM Microwave oven and digested for 50 minutes as per the CEM Microwave oven program of the manufacturer. The vessels were allowed to cool down to room temperature and the digests transferred to polyethylene vials. Then 2ml of 10 percent potassium dichromate was added to the vessels and diluted to 20 ml with double distilled water.

*Sample measurement*

The Varian Atomic Absorption Spectrometer coupled with the Varian Cold Vapour Accessories was used to measure the digested fish samples. The instrument was set as per the manufacturer's specifications. The instrument was calibrated using aqueous inorganic mercury standards within the established working range of the spectrometer.

The procedural blanks and samples were read and the final concentration of total mercury in the fish tissues was calculated taking into consideration the procedural blank value, final volume of digests and the sample weight.

## 3. RESULTS AND DISCUSSIONS

*Fish weight and level of total mercury*

The level of total mercury in swordfish caught in western Indian Ocean clearly demonstrates that in general, the larger the fish, the higher the level of total mercury. This is illustrated in Figure 1 which shows a clear trend of the increase from the lowest level of 0.13mg/kg in a 5 kg fish to 4.302 mg/kg in a 240 kg fish. This reflects observations made in previous studies in different parts of the world. In a study made in 1990 in the Azores Exclusive Economic Zone, the levels ranged from 0.06 mg/kg to 4.91 mg/kg (Monteiro and Lopes, 1990). It is known that the western Indian Ocean region around the Seychelles and East Africa are areas of low industrial activities and therefore pollution with mercury from land based activities is believed to be low. It is therefore believed that the main source of the contaminant could be through natural occurrences.

The higher level of mercury in the larger sized fish is due to bioaccumulation and bio-magnification in predatory species such as swordfish that have a high trophic level amongst marine organisms. Swordfish is known to feed on a wide variety of other fish, all of which are known to contain small amounts of mercury in their tissues. These include mackerel, barracuda, redfish, herring, hake, tigerfish, viperfish, crustaceans and molluscs. Based on tagging data from 1961 to 1976, Beardsley, (1978) reported that swordfish live to a maximum age of at least nine years, assuming it takes a period of two years to grow large enough to tag. (Chalabi *et al.,* 1996).

**Figure 1. Total mercury concentration in aggregate samples of swordfish tissues by weight categories**

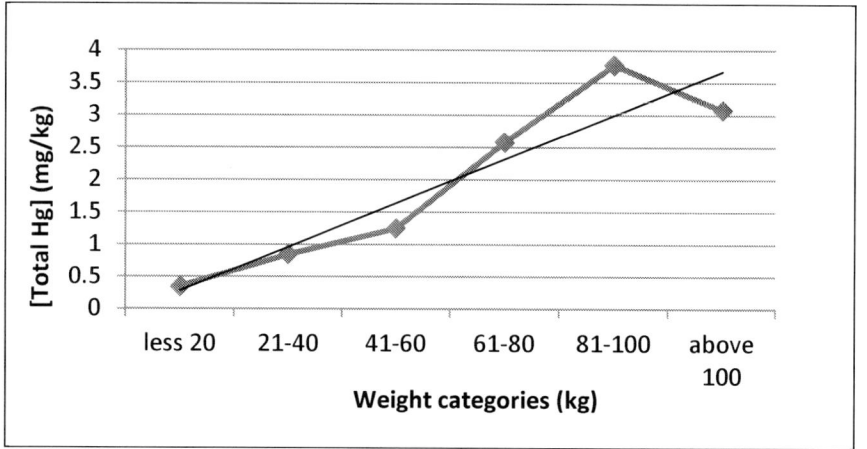

Figure 1 shows that the aggregate sample above 100 kg category does not reflect the trend or the increase in level as observed in the lower weights. Although seven samples only were analyzed in this category as compared to 10 in all other categories, it is not believed that the limited number of samples could have contributed to such different observation. Based on the trends observed and literature consulted, it is possible that if the weight categorization above 100 kg had increased by the same interval as the lower weights (Example 101 to120) a smoother linear relationship would have been observed. Other factors that need to be considered include the sex and age of individual. Female swordfish grow faster, live longer and are proportionately heavier than their male counterparts It is known that during the shedding of gametes, contaminants, such as heavy metals that are more concentrated in reproductive organs, are also released in the process. Consequently this has a reducing effect on the total mercury content in the muscle tissues. Swordfish spawn for the first time at 5–6 years old and at this age they are reported to be over 110 kg (Chalabi *et al.*, 1996).

Another important observation shows that there could be big variation in the level of mercury between individuals of more or less the same weight as illustrated in Figure 2. The most significant difference was observed in the 41–60 kg category. The mercury level in a 60 kg fish (sample No 6) individual was 0.364 mg/kg and that of a 58 kg (sample No 7) was 3.992 mg/kg, a difference of 3.628 mg/kg.

**Figure 2. Total mercury concentration in muscles tissue of swordfish, 41–60 kg**

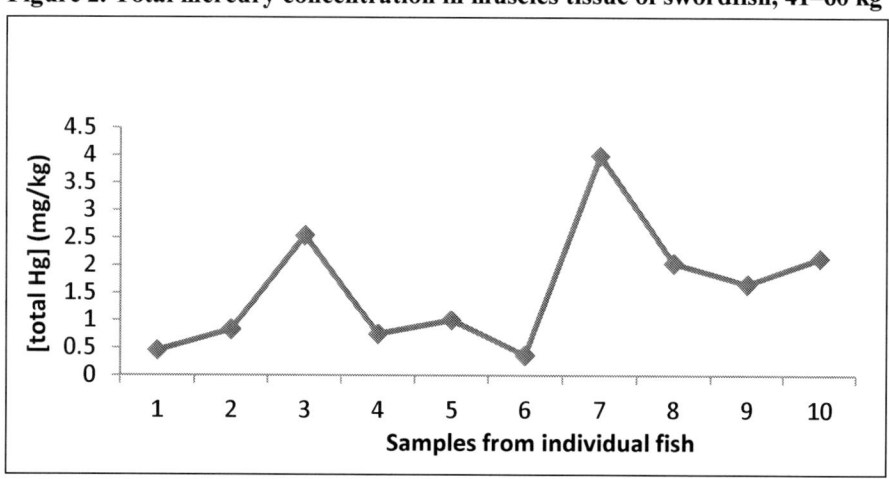

Preliminary evidence of the sex effect in the net accumulation rate of mercury implies that for the same size or age, mercury concentration in muscles tissues can differ significantly between sexes. Differential growth rate of males and females is not enough to explain this and further research is needed to clarify the implication of physiological differences in mercury bioaccumulation kinetic between sexes (Monteiro and Lopes, 1990).

Figure 3 illustrates the levels of mercury for category 81–100 kg and shows a much narrower variation in levels between the individual samples compared to the 41–60 kg category. In this case there is a difference of 0.925 mg/kg between the highest weight, an estimated 91 kg fish and the lowest, an 81 kg. It is worth noting that in this category, the average weight was 85.1 kg, 5 kg lower than the ideal average weight. This inconsistency had been particularly due to difficulty in obtaining sufficient samples in that weight category. Had there been more samples above 90 kg, possibly the variation of the lowest and highest observations would have been greater hence closer to observation made for other weight categories.

**Figure 3. Total mercury concentration in white muscles tissue of swordfish 81–100 kg**

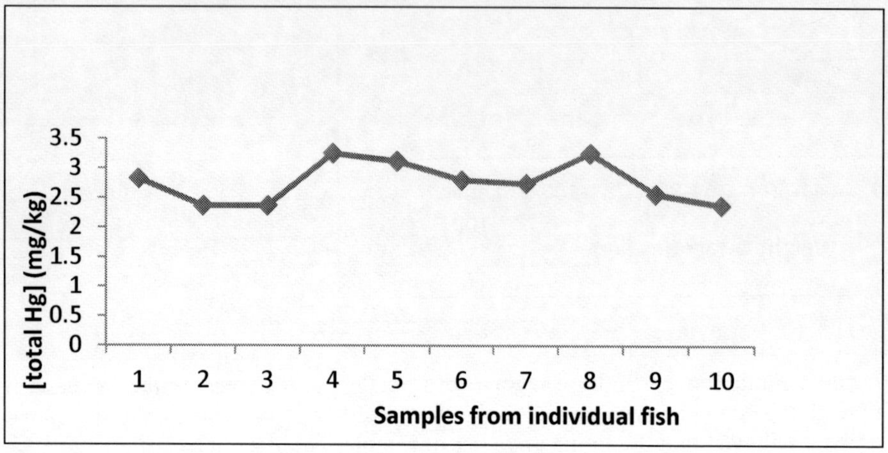

*Levels Detected versus Maximum Allowable Levels*

It was observed that 35 out of the 57 individuals tested, representing 61 percent, have recorded result above the 1.0 mg/kg, the maximum level allowed by the EU Commission. Most of the fish making the 61 percent were above 60 kg although few individuals below 60 kg did record results higher than 1.0mg/kg. The same applies to some individuals above 60 kg having levels below the 1.0mg/kg. It is interesting to note that around 80 percent of all swordfish landed in Seychelles are below 60 kg and probably around 60 percent are between 30 to 60 kg. (Oceana Fisheries and Sea Harvest, 2003).

However, the interpretation of the official maximum limit is only valid when the sampling protocol is followed as per Council Regulation 836/2011. It states that an aggregate sample that is effectively a combination of incremental samples collected to represent a lot of products should be tested. A lot is defined as an identifiable quantity of food (fish) delivered at one time and determined by the official to have common characteristic, such as origin, variety, types, packing, packer, consignor or marking. In the case of fish, also the size of fish shall be comparable (Commission Regulations 836/2011).

**Table 2. Minimum number of incremental samples to be collected from a lot**

| Weight of lot (kg) | Minimum number of Incremental samples |
|---|---|
| Less 50 | 3 |
| 50 to 500 | 5 |
| Above 500 | 10 |

*Source:* Commission Regulation 368/2011.

It is noted that of swordfish landed by the semi industrial long liners in the Seychelles, very few exceed 150 kg. Compliance of a lot with the legislative requirements will depend mainly on the average size of the individuals in the lot assuming that sampling is always done in a representative manner. Council Regulation 836/2011 states that the size of individuals forming the lot must be taken into consideration. Before sampling, it may be necessary to divide fish into sub-lots if there are significant variations in size. This will reduce the chances of obtaining biased results in situations where there are wide differences in weight amongst the individual fish in the lot.

## 4. CONCLUSION

It is clear that the total mercury concentration in swordfish caught in Seychelles waters demonstrates the bioaccumulative effect of this contaminant. Despite the difficulty in obtaining sufficient samples for the higher weight categories, the results generally show a clear relationship between the weight and the level of mercury in this species.

The concentration of total mercury in swordfish caught in Seychelles waters does reflect the levels detected in studies carried out in some other parts of the world for example in the study carried out by Monteiro and Lopes in the Azores water in 1990.

Most individual swordfish above 60 kg caught in Seychelles waters exceed 1.0 mg/kg of total mercury. However an aggregate sample taken representatively from a lot where the majority of fish are lower than 60 kg may easily comply with the maximum allowable level.

## 5. RECOMMENDATIONS

It is strongly recommended that the study be continued to adequately represent the larger fish above 100 kg, applying the same weight interval for categorization.

In future a similar study should be planned with the difference between each weight category reduced to 10 kg as compared to 20 in this particular study. It may ensure a more linear relationship between the weight of the fish and the levels of heavy metals. The relationship between sex and total mercury should also be considered.

The Competent Authority should always ensure that samples collected for each lot for official monitoring is a true representation of the lot. This will provide information whether it is necessary to introduce a maximum weight limit for the export market.

Re-evaluation of acceptable levels of mercury in marine fish, especially predatory species, in light of the new scientific knowledge on the relationship between selenium and mercury absorption in consumers, as well as the size relationship bearing in mind sustainability concerns in terms of minimum size allowed by the relevant management body.

## 6. REFERENCES

**Beardsley, G.L. 1978.** Report of the swordfish workshop held at the Miami Laboratory, Southeast Fisheries Center, Miami, Florida. 7-9 June 1977. Int. Comm. Conservation of Atlantic Tunas. Coll. Vol. Sci . Pap. 7 (1): 149–158.

**Chalabi A., Ifrene F., Kouadri A. & Merazka, N.** 1993. Note sur la présence de divers parasites de l'espadon *Xiphias gladius* L., pêché près des côtes algériennes. Identification d'un *Digenea Didymozidae* Poche, 1907, signalé pour la première fois en Mediterranée. Col.Vol.Sci.Pap. ICCAT, 40(1): 112–112.

**Codex Alimentarius Commission.** 1991. Guideline Levels for Methyl Mercury in Fish (CAC/GL 7-1991) http://www.who.int/fsf/Codex/methylmercury.htm

**European Union.** 2001. Council Directive 22/2001/EC.

Sampling methods and the method of analysis for the official control of the level of lead, cadmium, mercury and 3-MCPD in foodstuff.

**FDA.** 1994. Mercury in Fish: Cause for Concern. http://www.hruschka.com/hg-net/members/july/ fda_consumer_mercury_in_fish_cause_for_concern.htm

**Monteiro, L.R. & Lopes, H.D.** 1990. Mercury Content in Swordfish in Relation to Length, Weight, Age and Sex. *Marine Pollution Bulletin*. 21. 9. Pp. 293–296.

**Oceana Fisheries and Co. Ltd.** 2003. P.O Box 71, Victoria, Mahe, Republic of Seychelles , 2003 & **Sea Harvest Pty Ltd.** Fishing Port, Victoria, Mahe, Republic of Seychelles

**Stillwell, C.E. & Kohler, N.E.** 1985. Food and feeding ecology of the swordfish *Xiphias gladius* in the western North Atlantic ocean with estimates of daily ration. Mar. Ecol. Prog. Ser. 22, 239–247.

**Seychelles Fishing Authority.** 2002. P.O Box 449, Fishing Port Victoria, Mahe, Republic of Seychelles.

# CHALLENGES TO SUB-SAHARAN AFRICAN FISH EXPORTS

## *[DEFIS POUR LES EXPORTATIONS DE POISSONS DE L'AFRIQUE SUBSAHARIENNE]*

by/par

Helga Josupeit[1]
(presented by John Ryder)

**Abstract**

The paper covers some aspects of fish trade in Africa and its developments over the past decade. The main challenges with regard to fish exports from the region are highlighted. The eco-labelling scheme for small-scale fisheries in the Lake Victoria is shown as a successful private-public partnership that could be replicated in other fisheries of the continent.

*Key words: Trade, Africa, Eco-labelling*

**Résumé**

Le document traite des aspects du commerce du poisson en Afrique et son évolution au cours de la précédente décennie. Les principaux défis concernant les exportations de poisson de la région sont soulignés. Le schéma d'éco-labellisation pour les petites pêches dans le Lac Victoria est montré comme un exemple de partenariat publique-privé réussi qui pourrait être répliqué dans d'autres pêcheries du continent.

*Mots clés: Commerce, Afrique, Eco-labellisation*

## 1. INTRODUCTION

Exports of fish[2] from Sub-Saharan Africa[3] are relatively limited, with about 1 million tonnes per year. Thus the Sub-continent represents not more than 3 percent of total world exports in quantity terms (product weight terms). Namibia is by far the main exporting country in the region, accounting for more than one third of total exports. Declines in exports from this country in recent years (lower sardine catches) let to the overall decline in Sub-Saharan fish exports. Value of exports from the region increased during the period under review, both as a result of higher unit value and concentration on higher value fish, but also due to the decline in the value of the USD. In 2008, total fish exports from the region reached USD 2 650 million, which compares to USD 1 565 million eight years earlier. Despite this strong increase, the share of Sub-Saharan Africa in overall world fish trade stayed marginal, and even declined from 2.9 percent in 1976 to 2.6 percent in 2008.

Canned tuna is by far the main commodity exported by the region, with total export value reaching USD 550 million in 2008, accounting for about one quarter of total export earnings. The tuna canning industries of Sub Saharan Africa are mainly based on French investment, in countries such as Senegal and Côte d'Ivoire, some 25 years ago, when canning in France started to be uneconomically viable.

Sub-Saharan African countries had benefited for many years from zero duties on fishery products exports to the EU, as part of ACP agreements. In recent years, these agreements have been transformed into EPAs (Economic Partnership Agreements), under which most African countries still have zero duty access to the EU market. This duty exemption is very important for canned tuna, as the normal tariff is 24 percent. New canneries have been built on the Indian Ocean side of Africa, following the move of the Spanish and French tuna fishing fleet to this part of the world. This change is well reflected in the trade statistics: in 1986, Senegal was the main canned tuna exporting country in the region, with USD 66 million, closely followed by Côte d'Ivoire. In 2008, the main canned tuna exporting country in Sub-Saharan Africa was Mauritius (USD 207 million), followed by Côte d'Ivoire and Seychelles. The Senegalese canned tuna industry practically disappeared.

---

[1] FAO, Products Trade and Marketing Service (FIPM), Viale delle Terme di Caracalla, 00153 Rome, Italy. helga.josupeit@fao.org
[2] In the following text, fish includes all forms of finfish, molluscs, crustaceans, and any other seafood, including fishmeal and fish oil. Source of all tables and statistics cited in the text is FAO FISHSTAT 2011.
[3] In the following text, Sub-Saharian Africa will be nominated as "region".

**Table 1. Exports from Sub-Saharian Africa (in tonnes)**

| Country | 1976 | 1986 | 1996 | 2000 | 2008 |
|---|---|---|---|---|---|
| Namibia | . | . | 266 783 | 411 198 | 373 572 |
| South Africa | 125 050 | 74 332 | 85 420 | 162 359 | 136 051 |
| Senegal | 45 917 | 93 975 | 107 033 | 88 033 | 96 903 |
| Tanzania | 269 | 650 | 22 596 | 49 843 | 71 883 |
| Mauritius | 1 219 | 3 492 | 14 006 | 18 160 | 49 731 |
| Côte d'Ivoire | 10 911 | 25 756 | 58 299 | 64 870 | 47 640 |
| Uganda | . | . | 16 396 | 14 911 | 27 286 |
| Ghana | 2 993 | 33 020 | 31 568 | 54 742 | 23 664 |
| Madagascar | 3 301 | 4 531 | 26 743 | 14 233 | 22 802 |
| Kenya | 550 | 1 297 | 16 186 | 16 896 | 22 638 |
| Other | 31 256 | 24 323 | 65 246 | 99 278 | 82 288 |
| TOTAL | 221 466 | 261 376 | 710 276 | 994 523 | 954 458 |

*Source:* FAO Fishstat, 2011.

**Table 2. Exports from Sub-Saharian Africa (in tonnes)**

| Country | 1976 | 1986 | 1996 | 2000 | 2008 |
|---|---|---|---|---|---|
| Namibia | . | . | 198 906 | 283 931 | 576 978 |
| South Africa | 121 500 | 115 530 | 201 620 | 272 550 | 521 013 |
| Senegal | 52 480 | 259 110 | 310 541 | 260 354 | 222 967 |
| Mauritius | 2 827 | 10 055 | 42 190 | 36 659 | 214 987 |
| Côte d'Ivoire | 14 526 | 59 616 | 222 871 | 128 876 | 198 535 |
| Tanzania | 386 | 3 176 | 41 344 | 99 012 | 188 218 |
| Madagascar | 12 935 | 25 940 | 100 682 | 38 075 | 160 537 |
| Uganda | . | . | 39 780 | 30 986 | 134 554 |
| Seychelles | 506 | 1 045 | 39 715 | 113 465 | 97 155 |
| Mozambique | 12 434 | 34 423 | 86 343 | 99 889 | 76 913 |
| Others | 19 920 | 64 654 | 219 903 | 201 909 | 255 926 |
| TOTAL | 237 514 | 573 549 | 1 503 895 | 1 565 706 | 2 647 783 |

*Source:* FAO Fishstat, 2011.

Frozen fish fillets are the second most important fish product exported from the region. (USD 307 million) but exports concentrate on two countries only: South Africa and Namibia, which are the main exporters of this commodity with USD 100 million each. Hake is the main species exported from these two countries, both in fillets and in whole form. Frozen shrimp is another important commodity exported from the region, with foreign exchange earnings of some USD 260 million. Also for this product, exports concentrate in very few countries, namely Madagascar, Nigeria and Mozambique. The former country very successfully changed from shrimp fishing to shrimp aquaculture, receiving premium prices in the French market. One producer was even granted the "label rouge", the prestigious quality mark of the French food industry. With these positive investments into a high quality product, Madagascar managed to expand its exports from USD 11 million in the year 2000 to USD 109 million in 2008.

**Table 3. Exports from Sub-Saharian Africa (in 1000 USD) by commodity**

| Commodity | 1976 | 1986 | 1996 | 2000 | 2008 |
|---|---|---|---|---|---|
| Canned tuna | 32 915 | 125 677 | 415 557 | 360 050 | 554 028 |
| Fish fillets frozen | 3 459 | 14 705 | 141 438 | 248 183 | 306 538 |
| Shrimps frozen | 45 305 | 150 500 | 261 295 | 142 680 | 272 031 |
| Hake frozen | 8 978 | 43 000 | 92 085 | 87 704 | 251 221 |
| Fresh Fish fillets | . | 11 | 16 598 | 65 511 | 216 653 |
| Miscellaneous fishes frozen | 9 713 | 29 043 | 113 263 | 76 492 | 109 099 |
| Squids and cuttlefishes frozen | 974 | 24 218 | 85 005 | 78 802 | 93 656 |
| Miscellaneous fishes fresh | 1 836 | 964 | 57 309 | 70 956 | 84 117 |
| Mackerels frozen | . | . | 1 687 | 3 203 | 76 035 |
| Fish meat | . | . | 1 062 | 7 639 | 74 554 |
| Other | 134 334 | 185 431 | 318 573 | 424 481 | 609 819 |
| TOTAL | 237 514 | 573 549 | 1 503 895 | 1 565 706 | 2 647 783 |

*Source:* FAO Fishstat, 2011.

Intra-regional trade is important, but tends to escape official statistics. Comtrade, the United Nations Commodity Trade Statistics Database, reports the value of intra-regional trade of fish among Sub-Saharan countries at USD 190 million, which is less than 10 percent of total export value. It can be estimated that intra-regional trade exceeds this figure by 3-4 times.

About 70 percent of all fish exports from Sub Saharan Africa are going to the EU market. This dependency on just one major market creates a potential risk for exporters, in case this market closes down for some reason. This happened to products from Eastern African countries in the late 90s, when they were found not to be compliant with EU legislation. These countries were banned from exporting to the EU for about 18 months, and during this period, the companies were successful in opening up other markets. However, once the embargo was lifted, all the exports from the Eastern African countries shifted back to the EU market, considered far more convenient.

## 2. SMALL-SCALE FISHERIES IN AFRICA AND EXPORT TRADE

Small-scale fisheries are very common in both developed and developing countries. However, the two main fish exporting nations from the region, Namibia and South Africa, rely on industrial fisheries for their fish export earnings. The number three country, Senegal, has strong small scale fisheries, which is important both for domestic market supplies and for export earnings. It is estimated that about 60 percent of the Senegalese exports originate from small-scale fisheries. An important share of the Senegalese fish exports go to neighbouring African countries (about 40 percent), and the bulk of these exports are from the artisanal fisheries, processed and marketed by women.

Tanzania, the fourth major fish exporter from the region, also relies heavily on small scale fisheries for its exports, mainly Nile perch from Lake Victoria. Overall, taking together all countries from the region, it can be estimated that 160 000 tonnes, or one sixth of total exports, originate from the small scale fisheries.

Stringent quality and sanitary measures in importing countries have had an important impact on small scale fisheries, as all the exports to the EU have to come from certified establishments. This has had a detrimental impact on the performance of small scale fisheries. Another important factor impacting the economic performance of the small scale fisheries is the increased penetration of supermarkets in the retail sector, also in African countries. Supermarkets prefer to buy from large suppliers, as this guarantees a constant supply of a standardized product, while catch from small scale fisheries vary in quantity, size, composition, and quality. The only way out from this dilemma is that small scale fisheries producers get themselves organized into sales associations, thus guaranteeing a more standardized supply. In addition, when fishermen get organized this puts them into a better marketing position versus supermarkets and other purchasers. They are enabled to discuss prices on a stronger position than individually.

**Table 4. Exports of products from SSF in Sub Saharan African countries and main markets[4]**

| Species type/group and product form | Countries exporting from small-scale fisheries | Markets sensitized to environmental issues |
|---|---|---|
| Large tuna | West Africa | Mainly to EU |
| Demersal (marine) (mainly fresh whole/gutted) | From Senegal to Gabon, Djibouti | Focus on EU markets for West Africa, for East Africa mainly Near East markets |
| Shellfish e.g. clams, lobster, crabs, scallops | South Africa, some West African countries (e.g. Cape Verde) lobster | EU and US markets |
| Cephalopods e.g. cuttlefish, squid (fresh & frozen) | Senegal | EU and Japanese markets |
| Shrimp | Senegal, Gambia, | EU markets |
| Freshwater - Nile perch | Tanzania, Kenya, Uganda | Strong concentration on EU, but some USA sales |

Another stumbling block for export earnings from small scale fisheries is the recently imposed EU regulation on IUU fisheries. All catch, even when originating from small scale fishing boats has to receive certification – in the format indicated by the EU, if this catch is envisaged for export to the EU. The specific situation of small scale fisheries in export trade is taken into account in the catch certification scheme. The certification requirements have been adapted in order to facilitate the request for validation, which will be done by the exporter following certain criteria based upon this specific situation. A simplified catch certification scheme can apply to catches obtained by fishing vessels:

- with an overall length of less than 12 metres without towed gear;
- or a boat with an overall length of less than 8 metres with towed gear;
- or without a superstructure;
- or of less than 20 GT.

## 3. SMALL-SCALE FISHERIES AND ECOLABELLING

The primary difficulty for small-scale fisheries in relation to ecolabelling is also the primary difficulty with their management more generally – that the cost of meeting the requirements for management (i.e. monitoring, assessment, decision-making and implementation of management measures) can be out of proportion to the value of the fishery. This is particularly true if management is required to provide detailed and formal scientific proof to justify stock status conclusions and management actions.

A critical issue when considering whether certification schemes may have disadvantaged small-scale fisheries is whether not being certified (as opposed to not being able to demonstrate sustainability through other ways) is itself sufficient to lose markets. This is a different question from whether buyers encourage certification, whether being certified may bring about market advantages, and whether not being able to demonstrate environmental sustainability through other means may result in disadvantages.

These problems have been recognised by the Marine Stewardship Council (MSC). They have developed new methods to enable certifiers to assess small and data-poor fisheries against the MSC standard (MSC Risk-Based Framework), in recognition that many developing country fisheries do not have the detailed scientific data needed to demonstrate a conclusive case for their sustainability, but may nonetheless be able to demonstrate they are operating sustainably and can make the case for certification. MSC is facing the challenge on how to deal with SSF and is striving to serve regionally traditional and small scale fisheries to get certified, even though there are only a few cases of MSC certification acquired by the traditional fisheries in developing countries and by small or medium sized fisheries. In fact, the results of this initiative still have to be seen, especially in Sub Saharan Africa, as the only fisheries in the region certified by MSC is an industrial fishery.

FoS seems to be more accessible to the SSF, as the cost of certification is relatively lower (around USD 3 000 per fishery) than those of the MSC which range from a few thousand dollars to USD 20 000 for a pre-assessment and USD 10 000-USD 500 000 for full assessment, which makes FoS attractive to small-scale fisheries. Rather than MSC's strict chain of custody (CoC) approach (whereby all those in the supply chain must

---

[4] Table adapted from Graeme Macfadyen and Tim Huntington, Review of potentially negative impacts on small scale fisheries of market-based environmental sustainability requirements, 2009.

be audited against the MSC CoC standard), FoS has onsite monitors to ensure the origin of certified products. Otherwise FoS does not distinguish between small-scale and larger fisheries.

Fair trade is an organized social movement and market-based approach that aims to help producers in developing countries make better trading conditions and promote sustainability. The movement advocates the payment of a higher price to producers as well as higher social and environmental standards. It focuses in particular on exports from developing countries to developed countries, most notably handicrafts, coffee, cocoa, sugar, tea, bananas, honey, cotton, wine, fresh fruit, chocolate, flowers and gold. Some years ago, the German development agency sponsored some investigation into the potential of fair trade certification for fishery products. This experience was carried out in Senegal and India. However, it was found, that despite the strong interest of buyers in Germany, the small scale producers were unable to deliver the quantity requested by the importers in time. The experiment was abandoned and no further investigations are known. Small scale aquaculture producers seem to be in a better position to comply with the expectations of buyers rather than small scale fishermen.

## CASE STUDY: NATURLAND CERTIFICATION OF SMALL SCALE FISHERIES IN TANZANIA

Naturland, in general a certifier of organic products, including aquaculture products, started wild fish certification in 2007. The Naturland Standards for Sustainable Capture Fishery[5] defines "Sustainability" as a holistic concept, including the ecological, the social, and the economic dimension. Sustainability in the ecological sense means that the fishery is performed in such a way that integrity of the ecosystem is maintained long-term, concerning both the stocks of the economically relevant species as well as the other components of the ecosystem. Sustainability in fisheries in the social sense means that those employed in this industry enjoy fair working conditions, and that the living conditions of other members of the same community are not adversely affected. Sustainability in fisheries in the economic sense means that the marketing of fish encourages stable business relationships distinguished by the mutual sense of responsibility of all members of the value chain. Naturland's sustainability guidelines focus on environmentally friendly use of fish stocks and the entire ecosystem, avoidance of critical and environmentally harmful fishing methods ecologically sound processing without artificial additives or genetic engineering and a publicly open, transparent approval process for all parts of the value chain. This approach makes the Naturland label for wildfish very interesting for small scale fisheries.

In a nutshell, the certification standards are based on:

- monitoring of fish stocks;
- fishing techniques;
- social institutions; and
- fair working and income conditions.

In 2007, Naturland, co-financed by the German development agency, started the certification of small scale fisheries in Tanzania. At present, three small scale Nile perch fisheries are certified, namely in Bukoba, Mwanza and Musoma. At present, some 4 000 tonnes of Nile perch fillets exported from Tanzania to Europe have the Naturland certification.

Activities started in Bukoba, in the western region of Lake Victoria, in July 2007. Private partners are Vicfish as processor, Anova as importer, and Naturland as certifier, while GTZ financed the project start up. The certification area in Bukoba spreads over an area of 1 500 km², with eight landing sites and involves approximately 1 000 fishermen. The fishery is small scale in nature with 2 to 3 fishermen per boat using gillnets. The Naturland certification obliges the fishermen to wear lifejackets on their trips.

The Beach Management Units (BMUs) in the 8 certified landing sites guarantee that only fish from fishermen covered by the Naturland scheme land the fish for processing in the certified factory. In Bukoba, this plant is VICFISH. The plant has to assure that the minimum size of 50 cm, as imposed by the national legislation, is respected for all fish that is processed in the plant. The plant employs 2 000 workers, representing an important source of employment in a very marginalized area of Tanzania. The employment standard of these workers, about 50 percent women and 50 percent men, is also being scrutinized by Naturland.

---

[5] www.naturland.org

At present most of the fillets are airflown in fresh form to the European market, however, the idea is to reduce the number of air-flown products. This is probably because of carbon foot prints that are associated with airtransport. However, it should be assessed how much carbon foot prints are created through freezing and the keeping of the frozen chain, even though the transport (by road to Dar Es Salam and then on a container ship to Europe) is considered less impacting than air transport.

Processing waste is hygienically processed into food for humans (salted fish heads and frames), in different processing companies. The products are exported to neighbouring countries, where tasty soups are prepared by using the fish heads, this is playing an important role for food security in some of the neighbouring countries. A waste water purification plant is also part of the Vicfish company. UgoCert, an inspection body based in Uganda, carries out third party inspections in the plants and landing sites, to ensure that all Naturland criteria are met. Mesh sizes of fishing nets are controlled, and catch documentation and traceability is checked accurately. The inspection reports are the basis for the certification by Naturland, and the reports are published on the Naturland home page. The Naturland criteria also include that fishermen have access to medical treatment, and through the project some hospitals have been opened in the areas.

On the other hand, the importer "Deutsche See" aims at a partnership which is "helping people to help themselves". In the medium term fishermen will have greater profits if they can negotiate directly with a single vendor such as "Deutsche See". The Bremerhaven fish factory pays fair prices and supports the development initiative. In addition, the importer is carrying out some social projects in the island, such as the construction of a kindergarten for 80 children.

The project in Mwanza started in April 2011. The processing plant collaborating with the project is. Tanzania Fish Processors Limited (Alpha Group), and the certifier is Naturland. This project involves a total of 952 fishermen operating from nine landing site. The certification and inspection service follows the same principles spelt out under the Bukoba project. Each fisher community participating must be able to have access to health services within 60 minutes (240 minutes for islands on good weather conditions) from landing site. Mobile health services are made available at least once a month to the island landing sites.

The Lake Victoria project in Musoma involves the processors (Alpha Group: Musoma Fish Processors Limited), and various customers in Europe. The project was started in the May 2011 and involves 573 fishers operating from four landing site. The processing company commits to train fish supplier, fishers and workers to ensure increased and sustainable fish production by managing capture of Nile perch adopting legal gears for fishing. In addition, the processor commits to train her workers for maximum utilization of captured fish by enhancing yield of products; promoting value added and restructured foods, utilizing waste and reducing post harvest losses.

## 4. CONCLUSIONS AND RECOMMENDATIONS

Sub Saharan Africa is a relatively small player in global fish trade. Its position has even declined over the years. One of the reasons for the declining importance of the region as supplier to the international market is the limited development of the aquaculture industry, which is the driving force behind increased fish trade in recent years. Many countries in Asia and Latin America are culturing shrimp, tilapia and pangasius mainly for the export market. More stringent quality control and sanitary measures have put a halt to exports of fishery products from Sub Saharan countries. Fewer companies than before the introduction of the HACCP regulations in the EU are at presently authorized to export to this major market. About 70 percent of all fish exports from Sub Saharan Africa are going to the EU market. This dependency on just one major market creates a potential of danger in case this market closes down, as happened to products from Eastern African countries in the late 90s.

Intra-regional trade plays an important role in fish trade in Sub-Saharan Africa, especially from the Gulf of Guinea to the neighbouring land-locked countries, from the Lake Victoria countries to Central African countries, from Angola and Namibia to Southern African countries, however, very little of this trade enters official statistics. From the official trade matrix it looks as if less than 10 percent of all export earnings are coming from intra-regional trade, the reality is completely different. Sub-Saharan African countries should find ways of better recording the trade flows among regions and countries. This would give an enhanced picture of the importance of intra-regional trade, and also prompt development agencies to address the issue in a better way. From studies carried out in the past, it becomes apparent that despite low tariffs among economic groupings, informal trade barriers are working against a transparent intra-regional fish trade system. SSF and small scale fish traders play an important part of the informal intra-regional trade.

In some countries of the region, SSF plays an important role as provider of animal protein to the population, but also as supplier of a product for foreign exchange earnings. Exporters of SSF fishery product have encountered problems with HACCP certification, traceability and in recent years, with IUU regulations by the EU. These import regulations are easier to implement for industrial fisheries and large scale processing plants rather than for small scale operators and small scale exporters. Somehow, during the more than 15 years of implementation, the producers, even the smaller sized ones have learned to cope with the HACCP certification. Traceability also seems not to be a huge problem anymore. The more recent EU regulation with regard to IUU certification still presents a problem, but the EU has promoted various training courses and capacity building activities to explain the core of this regulation, and it can be expected that slowly but steadily SSF operators will learn how to cope with them.

Another important factor impacting on the economic performance of the small scale fisheries is the increased penetration of supermarkets in the retail sector, also in African countries. Supermarkets prefer to buy from large suppliers, as this guarantees a constant supply of a standardized product, while small scale fishery catches varies in quantity, size, composition, and quality.

Yet an additional problem for small scale fisheries is the recent tendency of supermarket chains in developed countries to request that the fish products they are selling be eco-certified. However, not being certified does not necessarily mean that the product will not find a market. At present, with a strong economic crisis dominating the market, ecological concerns give ground to economic considerations. Buyers in the developed world are more attentive to lower prices, and not willing to pay a price premium for a certified fisheries product. Buyers are under enormous pressure and competition to ensure supplies, so even for those that would ideally like producers to demonstrate sustainability, if producers are unable to do so it may not result in the cessation of sales.

In principle, a SSF, with a community based management scheme in place, should be an ideal candidate for eco- or social labelling. Even though initiatives in recent years have resulted in failures, more donor money should be mobilized to promote this type of eco-labelling. An excellent example, which highlights the importance of involving the private sector and the importers, is the Naturland/GTZ initiative in the small scale Nile perch fisheries in Tanzania. This example could be replicated in many other areas worldwide. More work in this field should be carried out, including investigation into the possibility of fair trade certification. FAO should make some funds available to follow up this issue.

## 5. REFERENCES

**Blaha, F.** 2011, EU Market Access & Eco-Labelling, SIPPO, Zürich. 46p.

**FAO.** 2003. Report of the Expert Consultation on the Development of International Guidelines for Ecolabelling of Fish and Fishery Products from Marine Capture Fisheries. Rome, Italy, 14–17 October 2003. *FAO Fisheries Report*. No. 726. Rome, FAO. 2003. 36p.

**Macfadyen, G. & Huntington, T.** 2009. Review of potentially negative impacts on small scale fisheries of market-based environmental sustainability requirements. Final Report, September. www.consult-poseidon.com

**Wessells, C.R., Cochrane, K., Deere, C., Wallis, P. & Willmann, R.** 2001. Product certification and ecolabelling for fisheries sustainability. *FAO Fisheries Technical Paper*. No. 422. Rome, FAO. 83p.

## EXISTING ECOLABELS AND CERTIFICATION SCHEMES

Ecolabelling is a market-based economic instrument that seeks to direct consumers' purchasing behaviour so that they take account of product attributes other than price. Such attributes can relate to economic and social objectives (fair trade; support to small-scale fishers; discouragement of child labour) in addition to environmental and ecological ones. The label helps consumers to distinguish a product according to desirable attributes without requiring them to have the detailed technical knowledge.

Northern European and North American consumers with good incomes and a high level of education have a moderate, and sometimes, strong, tendency to choose an ecolabelled product over a non-labelled one, even when the former costs slightly - but not much - more. There is evidence that ecolabels covering product attributes that relate not only to lower environmental impacts, but also to assumed higher product quality in terms of nutritional and/or health benefits can realize significant price premiums and show strong growth in market shares, although such products are still operating from a small base. This applies to organic food products, for example.

The feasibility of achieving fisheries management objectives through ecolabelling schemes depends on certain requirements being met. The economic incentive created by the labelling scheme needs to be sufficiently high to encourage the fishery management authority and participants in the fishery to seek certification and cover the related fisheries management and labelling costs.

There is no guarantee that the widespread adoption of ecolabelling programmes for marine fisheries would result in the better management of global fisheries in toto. At present, only a small fraction of global fish consumers (most of them living in Europe and North America) are likely to be responsive to ecolabels. Most of the future growth in global fish demand, however, will be in Asia, Latin America and Africa.

The first "green stamps" for fishery products were launched in the early nineties in the USA. They focused on a specific by-catch issue, such as the dolphin by-catch by tuna seiners (Dolphin safe), or the turtle by-catch of shrimpers (Turtle safe). In 1997 a public awareness campaign in the USA "give swordfish a break" turned out to be the first wide scale campaign asking consumers to help impact fishing practices.

In 1997 the Marine Stewardship Council (MSC), today's world leading ecolabel for fishery products, was created. Since 2000, three other major ecolabels are on the market: Krav, Friend of the Sea, and Naturland. Existing schemes use different certification criteria, and in some cases are liable to cause market access problems. There prove to be reserves on the part of developing countries, which fear their products may be excluded from the markets of developed countries. Furthermore, it is not always easy to ascertain the credibility of environmental claims displayed on the labels and the criteria used for issuing the label.

### MSC

The MSC is now perhaps the best known of the environmental schemes for capture fisheries, and arguably the most widely regarded in terms of its standards and certification processes. There are 126 certified fisheries in the world, which amount to close to 9 million tonnes. It incorporates a process of third party certification of fisheries and supply chains, and the use of labels. The MSC is an independent, global, non-profit organization whose role is to recognize well-managed fisheries and to harness consumer preference for seafood products bearing the MSC label of approval.

The map of MSC certified fisheries shows that the vast majority of these fisheries are in Europe, Canada, and the USA. Africa is really a blank spot on the map, with just one certified fishery: the South Africa hake trawl fishery, which was certified as sustainable in April 2004 and re-certified in March 2010. The species covered are *Merluccius paradoxus* and *Merluccius capensis*.

### Friend of the Sea

The Friend of the Sea scheme works closer to the point of sale than production, by approving products if (a) target stocks are not overexploited; (b) fisheries use fishing methods which do not impact the seabed and (c) they generate less than 8 percent discards (the global average as per recent FAO publications). Products/fisheries are audited and certified against published information/data, following application by fisheries using a standard application form. Fisheries are assessed against: FAO data on stock status in different fisheries areas; the IUCN red list of endangered species; fishing gear types felt to be harmful to the seabed; IUU and Flags of Convenience; and compliance with TACs, use of the precautionary principle, and national legislation.

Friend of the Sea (FoS) has certified some Sub-Saharan African fisheries: Senegal - Asociacion Atuneros Caneros Dakar Tuna fleet - Pole and line - *Thunnus obesus, Katsuwonus pelamis, Thunnus albacares* and South Africa - AKA Global Fish fleet - Pole and line - *Katsuwonus pelamis, Thunnus albacores*. In addition to these fisheries, FoS certified some purse seine fisheries in Morocco, namely the anchovy, mackerel and sardine fisheries. Only the Senegalese pole and line fisheries can be considered as small-scale fisheries.

Label Rouge is an official sign of quality, granted by the French Ministry of Agriculture and Fisheries. It certifies the superior taste and flavour of the shrimp, validated by a panel of experts and consumers. It identifies shrimp that have followed production and preservation methods set to exacting standards. This prestigious label was granted to the shrimp products from one producer, the UNIMA group. This group is a pioneer in the Malagasy shrimp industry. It has developed ecological fishing and farming models by applying the best practices, which respect and preserve the environment. In addition to being a pioneer in shrimp quality, the company has also a socio-economic side. Located in landlocked regions, the company's sites provide employment for 4 000 people and thus real opportunities for development to these often deprived areas. The company has several community development activities, including schools, public health-care centres, including maternity facilities, sewage infrastructures, roads, ecological, solid earthen brick houses and market gardens, providing commerce and income for families. It has also monitored installations to supply drinking water.

## NATURLAND CHECKLIST FOR INSPECTION

### FISHERIES

#### *LVFO stock assessment on the whole Lake Victoria*
There must not be any indication (by LVFO) that the Nile Perch stock in Lake Victoria is critically overfished (acute danger of not recovering). Remarks: The overall assessment of the Nile perch stocks in Lake Victoria will be done by the LVFO. The project should take into consideration (will implement) the recommendations concerning fishing activities given by (LVFO and) competent authorities based on the results of the overall assessment and up-date the Standards/ criteria accordingly.

#### *Prohibited gear*
The following gear types and techniques shall not be used by the fishermen participating in the pilot project: monofilament gillnets; gillnet with a mesh size of less than 6 inches; drift nets; beach seines; trawl nets; hooks below size no.9 or above size no.11; weirs; Explosives and chemicals; splashing (katuli); harpooning (spear guns)

#### *Monitoring duties*
A thorough monitoring of the catch development shall be performed regarding Nile Perch and the other fish species. The monitoring shall consist of: 1. Reporting total weight of catch per boat per trip of all participating BMU landing sites and Fishers; 2. Length frequency measurements of fish landed of all (certified) suppliers at the factory once a month (100 sample size); 3. Catch composition and length frequency measurements (catch assessment survey according LVFO) of six eco labelled boats per supplier at all landings at one day (randomly selected) once in every month of the year in all participating BMUs. 4. Unannounced random sample of catch composition and length frequency measurements (done by Tanzania Fish Processors Ltd.) of ten eco labelled fishing boats per participating BMUS on the lake once a year.

#### *Results of Monitoring*
Monitoring results must show that: 1. Fishing activities of BMUS involved do not contribute to illegal fishing; 2. Size of 100 percent of fish going into processing are within the agreed minimum limits of 50 cm and above; 3. Level of undersized fish (less than 50 cm) should be within acceptable limits and will be monitored.

#### *Obligatory fishery practices*
To use gillnets with a mesh size of more than 6 inches To observe closed seasons and closed areas (sanctuaries gazette by Dept. Fish)

#### *Health services*
Each fisher community participating must be able to have access to health services within 60 minutes (240 minutes for islands on good weather conditions) from landing site. Mobile health services are made available at least once a month to the island landing sites.

#### *HIV/AIDS information and care*
Each fisher community participating must have access to HIV/AIDS (including PMTCT)/STD/TB information, education, testing, care and treatment.

#### *Access to school for children*
Each child (age 7–14) must be able to attend primary education during the complete cycle. In order to attend the school, children may not be exposed to hazardous situations or may not have to put in an effort which is not suitable for a child of that age. In populations larger than 1500 people, a primary school with trained teachers has to be present. Walking distance to the next primary school shall not be more than 5 km.

#### *Housing*
The fisher community shall live in houses that are weather proof and well ventilated. Settlements must allow basic hygiene.

#### *Potable Water*
Population must be sensitized on the issue of drinking water quality; good quality potable water must be available to the fisher community.

*Monitoring of the fisher-folk*
Specific standards are: 1. Records shall be available concerning the assets (fishing boats, fishing gears, engines) owned by these players; 2. All Fishers working for the boat owners shall be registered and these registers shall be updated at least once a year; 3. All suppliers, collectors and boat owners participating in the project shall be known by name, family situation (voluntary: number of dependents at landing site /person to be contacted in case of accident), place of origin and physical address; 4. Staff working for the suppliers and the collectors shall be registered and follow the working conditions as spelled out in the Naturland Standards (if formally employed). 5. Immigrant staff working for suppliers and collector should produce introductory letter from relevant authorities where they come from to BMUs.

*Life jackets*
Compulsory life jackets must be carried on the boats for all crew members; by 30th September 2010. All crew members participating in eco label project should be trained on safety measures during their service period.

*Toilets/Latrines*
Gender separated toilets/latrines environmental friendly eco (as minimum 1 pit per 25 persons) must be in good hygienic condition and accessible for all.

*Correct weighing of fish*
Measures shall be in place to ensure the participating fishing community of correct weighing of fish. Participating BMUs can supply the processor directly.

*Financial Services*
Participating fishing communities shall have access to financial services.

*Transparency in price development*
Fish price developments along the value chain should be transparent (current telephone number procurement manager will be made available to all BMUs).

*Contracts*
The participating fishermen must be directly or indirectly linked by contract to Naturland certification.

*Documentation of Project Management*
The management of the fishery project must be able to prove that the requirements laid down in the standards and the project-specific management conditions are implemented systematically, effectively and promptly at every level. This proof includes: consistent records and analysis of the catch data; feedback between the current catch data and the fishing practice in place; knowledge of current national and international regulations and fulfilment of the duties arising there from - establishment of mechanisms guaranteeing regular communication between the project and the fishermen with regard to social matters; existence of and compliance with a development plan (e.g. for deficient issues).

*Training*
All participants (landing site monitors, fishers, suppliers, BMU members, members and factory staff) are familiar with the requirements of the certification and trained. Training records (participants, trainers, dates and topics) and materials should be kept and available within the project (training based on the need).

*Monitoring of environmental contaminants*
The quality of the fish should be checked by the factory lab or an external lab or by the competent authority. The system must be able to trace back contaminated fish. Moreover, the processor and Naturland together determine: 1) a list of the contaminants and noxious substances (from both anthropogenic and natural sources) that are relevant to the region and the type of production will be submitted by TFP to Naturland; 2) alert values of max. 50 percent of the critical German legal level, at which Naturland must be notified; 3) The frequency of and processes used in the analysis of these pollutants (with reference to the water, sediment and products) will be submitted by TFP; 4) Maximum permitted values must be according to the EU legal level and if above Naturland must be notified; 5) Threshold values leading to the exclusion of the product from marketing.

## PROCESSING COMPANY

Last year's instructions and conditions made by Naturland have been fulfilled. (If not, list deviations in Naturland list of deviations.) The instructions and deviations are mentioned in the Naturland Certification Letter which is sent together with the certificate.

### Subcontracted Companies for Naturland Products
Please add a list of subcontractors which are involved in the handling of Naturland products and specify the kind of handling. If there are any new subcontractors please give also information on the certification of the company and the Naturland products which are handled. Please also note if there are no subcontractors at all.(list of suppliers).

### End Consumer Labels/packaging with reference to Naturland
End consumer labels/packaging and promotion material referring to Naturland have been approved by Naturland. A written confirmation for the logo use of the Naturland Zeichen GmbH is available. (Please note that this confirmation is different from the product application approval which is issued by the Naturland Certification Committee.) (TFP labels).

### Particular labelling rules
The use of ethanol, iodized salt and gelatine is explicitly mentioned on the end consumer label/package.

### Multiple certifications
The operator/company is not certified by any organic certifiers other than Naturland. Where other organic certifiers do certify the operator, those are listed below. There are currently no serious infringements with any other organic certifier.

### Separation of products in space or time
Products to be certified as Naturland products are produced separately in space or time from other products. Special attention is to be paid here on keeping the products separate from conventional goods. Where other goods are produced for recognised certifiers, it is sufficient for the containers to be emptied by mechanical means. Otherwise production lines have to be cleaned prior to processing of Naturland goods.

### GMO in conventional production (only if company also produces conventionally)
No genetically modified ingredients, additives or processing additives are used in the conventional production of the company. Otherwise please describe where and list additional quality assurance measures taken by the company.

### Identical products in the product range
If the product range contains products that are physically identical but produced in different qualities besides Naturland such as EU-Organic or conventional, the products are capable of being differentiated by means of labelling, packaging and/or shape.

### Approval of new products with the Naturland logo
All Naturland products are approved by Naturland. A product is approved if a written confirmation by the Naturland certification committee for a product is available or if the product is already listed on last year's certificate.

### List of substitutes for Naturland raw goods
The raw goods used corresponds to Naturland list of priorities/substitutes. Uppermost priority is the use of Naturland certified raw goods. Other raw goods have to be applied for. If not available in Naturland quality, raw goods of other recognized certifier may be used (demeter, Bioland, Naturland); then recertified goods, then raw goods in EU-organic quality and finally conventional goods (max. 5 percent).

### Aromatic substances, Flavours
Aromatic substances and flavours are not used. If they are used, please attach a product specification sheet of the substance and specify in which Naturland product the aromatic substance is used.

### Water and salt
Water used water corresponds to the drinking water regulation. Salt and iodized salt are free of anti-caking agents (except for calcium carbonate E170 which is permitted).

*Minerals, trace elements and vitamins*
Only those minerals, trace elements or vitamins are used that are mentioned in the specific Naturland standards for certain products.

*New processing procedure*
Since the last inspection no new processing procedures have been introduced. When a new procedure is introduced, the list of ingredients and a description of the new procedure must be included with the inspection report.

*Prohibited procedures*
There is no direct (through raw materials, ingredients) or indirect (via half finished products) use of micro-waves, ionizing rays or genetically modified goods.

*Clear labelling during processing and storage within the company*
The labels and raw goods are labelled correctly, clearly and unambiguously with reference to organic certification (Naturland or other).

*Cleaning of operating rooms, machinery and equipment*
Cleaning agents and cleaning methods are documented. Cleaning protocols are kept.

*Warehouse pesticides*
The only warehouse pesticides used are those listed under Appendix 3 of the Naturland standards. In the case of pesticides used which are not listed in Appendix 3, please provide the following information: name of substance, ingredients, application field, and frequency of application.

*Gassing*
Before using any gassing methods, an application has been filed by the operator and was approved by Naturland.

*Pest Control Company informed (if applicable)*
Enterprises dealing with pest control have been informed about the Naturland standards in written (abstract of the Naturland standard has been sent).

*Residue Analysis*
Naturland has been informed about all considerable findings of residues in Naturland products.

*Complaint Register*
Complaints are recorded in a register and followed up (documents are made available to Naturland). Please make a copy of last year's complaints related to Naturland or Naturland products and attach it to the Naturland inspection report, if any.

*Training of employees*
Employees are regularly trained on organic topics and Naturland.

*Complete documentation*
Production records are complete (see Naturland Standards Part 3.C.9).

*Current description of operation*
The current description of the operation is up to date. In case of first Naturland inspection it is enclosed with the report (project description).

*Current list of products (please attach to Naturland inspection report)*
A current list of products is available (Naturland products are marked as such). Please attach the product list to the Naturland inspection report (list of products).

*Current list of suppliers (please attach to Naturland inspection report)*
A current list of suppliers is available, listing all the raw goods supplied with their certification details. Please attach the suppliers list to the Naturland inspection report (list of suppliers).

***Calculation of flow of goods (please attach a sample calculation to the Naturland inspection report)***
Calculation of flow of goods has been performed (record of incoming and outgoing goods). Calculation is attached with the inspection report (Flow of products).

***Traceability of goods (please attach a sample calculation to the Naturland inspection report)***
A trace back audit has been performed. Calculation example is attached with the inspection report.

***Reference to Naturland on Invoices/delivery notes***
For all commercialized certified goods, the licensee is indicating the Naturland certification of these goods on all business documents (invoices, bill of sale, shipping order etc.). If not, please attach an example of such a document to the inspection report and send it for evaluation to Naturland.

***Social responsibility***
The Naturland Checklist on Social Responsibility has been completed in 2009. No notifications/conditions were issued in Naturland's latest certification letter. (Naturland demands inspection according to the standards in the chapter Social Responsibility at least every 3rd year. In 2012 all operators will have to undergo a complete social audit. Conditions/ notifications resulting from previous audits must be checked for compliance in the following year.)

# OCTOPUS VALUE CHAIN AND IMPLEMENTATION OF AN UPGRADING STRATEGY: KEY FINDINGS IN THE CASE STUDY OF NIANING AND POINTE SARENE (MBOUR SENEGAL)

## *[LA CHAINE DE VALEUR DE LA PIEUVRE ET MISE EN OEUVRE D'UNE STRATEGIE DE MODERNISATION: PRINCIPALES CONCLUSIONS DE L'ETUDE DE CAS DE NIANING ET DE POINTE SARENE (MBOUR SENEGAL)]*

by/par

Papa Gora Ndiaye[1], Moustapha Kebe, Betty Lette Diouf and Daba Ndione

### Abstract

This paper provides a case study of potential implications of implementation of upgrading strategy in a value chain approach within a developing country context by drawing on a project that was been undertaken in Senegal (in the department of Mbour) by ENDA/REPAO with support from IDRC (International Development Research Centre) and ODI (Overseas Development Institute). The project is working with the octopus value chain, and the objectives of the project are to i) support pro-poor and gender sensitive 'upgrading' strategies to improve the value chain and ii) support an approach to certification that addresses social and economic issues within the value chain. The first step of this project has therefore been to understand the current value chain and the inequalities within it, assess implications of certification and determine how the value-chain can be improved through upgrading. Upgrading refers to improving the value and efficiency of a supply chain and can refer to i) vertical upgrading: doing better with the same product throughout the chain; or ii) Horizontal upgrading where improvements are made at one level e.g. management capacities at the production level.

*Key words: Value chain, Senegal, Octopus, Small-scale fisheries, Certification, Procurement, Upgrading*

### Résumé

Ce document présente une étude de cas d'implications potentielles de la mise en oeuvre de stratégie de la modernisation dans une démarche chaîne de valeur dans un contexte de pays en développement, en s'appuyant sur un projet qui a été entrepris au Sénégal (dans le département du Mbour) par ENDA/REPAO avec le soutien du CRDI (Centre de Recherche pour le Développement International) et l'IDO (Institut du Développement Outre-mer). Le projet travaille avec la chaîne de valeur des pieuvres, et les objectifs de projet sont i) soutien en faveur des pauvres et aux stratégies de modernisation genre-sensibles pour améliorer la chaîne des valeurs et ii) soutien à une approche de certification qui aborde les questions sociales et économiques dans la chaîne de valeurs. La première étape de ce projet a donc été de comprendre la chaîne de valeur actuelle et les inégalités internes, évaluer les conséquences de la certification et déterminer comment la chaîne de valeur peut être améliorée à travers la modernisation. La modernisation fait référence à l'amélioration de la valeur et à l'efficacité d'une chaîne d'approvisionnement et peut faire référence à i) la modernisation verticale: en faisant mieux avec le même produit sur toute la chaîne; ou ii) la modernisation horizontale où les améliorations sont apportées à un niveau par ex.: la capacité de gestion au niveau de la production.

*Mots clés: Chaîne de valeur, Sénégal, Pieuvre, Pêche artisanale, Certification, Achat, Modernisation*

## 1. INTRODUCTION

Almost all the octopus (*Octopus vulgaris*) caught in Senegal are flash frozen and exported to Italy (60 percent by weight), Japan (15 percent), Spain (9 percent) and Greece (6 percent) by large processing companies based in-country. Octopus is a high value global product with an average price of US$ 5/kg in importing countries (with the price generally increasing with the size of the fish, up to US$ 19/kg). The aim of this study was to promote the sustainable management of the octopus resource, improve quality management and increase returns for fishermen. The research was undertaken by ENDA/REPAO (a Dakar-based fishery research organisation) and MRAG (a UK-based marine research and fisheries consultancy).

---

[1] Réseau sur les Politiques de Pêche en Afrique de l'Ouest (REPAO), PO Box 47076, Liberte 4, No 5000, Dakar, Senegal. gndiaye@gmail.com

Senegal has the third largest octopus catch in Africa (after Mauritania and Morocco). The most recent statistics (2006) indicate that the Senegal octopus catch was 8,800 tonnes and the 6,030 tonnes exported had a commercial value of US$ 21m in 2006. The octopus catch varies significantly each year and there is general perception that the resource is being overfished.

The octopus project was implemented in two locations (Nianing and Pointe Sarene) in the Department of Mbour, about 150 kms south of the Dakar – the Capital of Senegal in West Africa. These choices of location were justified on the basis that these two areas had well-organised local fishery organisations, having benefitted from significant support from the Japanese bilateral aid programme. Almost all the octopus caught in Senegal are from Dakar and Thies Region and the two regions account for 86 percent of the motorised pirogues (fishing canoe with an outboard engine) in Senegal. The study areas are not, therefore, necessarily representative of rural fishing communities in Senegal. The study areas are characterised by artisanal fishers, who fish from open pirogues and jig for octopus. The fresh molluscs are landed and sold to fishmongers, who then take the catch to Ikagel Ltd.'s processing factory. Some 40 percent to 50 percent of fisher incomes are derived from octopus, the remainder being from pelagic species.

## 2. METHODOLOGY

The octopus project has two main components, which are critical for its overall success:

- First, the proposed objectives and activities are defined in such manner that they do adequately address and outmost take into account local realities of the stakeholders and targeted interest groups of the Cephalopod fishery in the Nianing and Pointe Sarrene area.
- Second, the programme is strongly oriented towards creation of market-based incentives to facilitate and support changes in economic responsibilities, dependencies and performance of job for daily subsistence.

The proposed linking of eco-labelling, as a market based incentive and practical tool to guide producers towards sustainability, with the methodology of value chain promotion along the products' supply chain has been proven successful in a number of other initiatives in other sectors such as small-scale aquaculture and organic farming.

The basic underlying question is: How to break the vicious circle of vulnerability and dependency of small scale primary producers towards traders and processing companies that are having complete control on pricing and therefore directly exert pressure on natural resource exploitation? Taking into account the needs of smallholders for daily subsistence such a combination often leads to resource over-exploitation. Especially, as it is the case in Senegal, if there are no fishery management and enforcement procedures in place.

## 3. VALUE CHAIN DESCRIPTION AND TARGET GROUP

In both study areas the fishers are organised into local a CGRH (local marine resource management committees). There is some evidence of conservation measures being taken by the local committees. These committees have organised the donor-financed earthenware vases (like flower pots) to be placed on the sea-bed to encourage octopus to lay eggs and stricter enforcement of the ban on fishing activity during the biological rest period during the October breeding season.

The motorised pirogue owners split the value of the daily catch in one-third shares between the boat, the engine and the crew – a division of catch revenue which is seen in many fishing communities. The average daily net revenue for the total crew (of normally six members) in 2009 is estimated to be US$ 52 (FCFA 26,291). The monitoring data suggest that the daily net revenue earned by crews increased significantly, 64 percent - from 2008 to 2009 (from about US$ 31 to US$ 52). The reason for this is not clear because the physical catch did not increase and neither did the unit prices received by the fishers. Assuming the crew revenue figures are correct (and the price data and catch totals), a rapid reduction in fishing trips in 2009 (perhaps caused by the reduction in the number of fishers) could explain this finding.

Men dominate the octopus value chain as pirogue owners, fishers and the larger-scale fish mongers. The role of women is confined to processing pelagic fish and some micro-processing of octopus.

**Table 1. Average selling prices for octopus in the Nianing and Pointe sarène (US$/kg), 2009**

| Grade | Fishers (US$/kg) | Fish mongers (US$/kg) |
|---|---|---|
| PP: (< 0.5 kg) | 0.75 | 1.13 |
| P: (between 0.5 and 1.0 kg) | 1.92 | 2.24 |
| M: (between 1.0 and 2.0 kg) | 2.63 | 3.03 |
| G: (between 2.0 and 3.0 kg) | 3.47 | 4.54 |
| GG: (> 3.0 kg) | 4.64 | 5.98 |

*Source:* REPAO, 2010.

The prices achieved by fishers for octopus sold in 2009 is described in the final report: figures given on: PP[2] = FCFA 300; P[3] = FCFA 1,700; M[4]= FCFA 2,300; G[5]= FCFA 2,600 and GG[6] = FCFA 2,800 (the current exchange rate in June 2009 was FCFA471=US$ 1), which are significantly higher than those presented in the able above.

Applying the lowest estimate of unit prices for octopus (i.e. those from Table 1) to our estimates of the catch in 2009, suggests that the gross revenue for the artisanal fishers was some US$ 60,000. The average sized octopus caught in 2009 was 1.5 kg and was worth just under US$ 4 to the fishers. Applying the highest estimate of unit prices for octopus, suggests gross revenue for the artisanal fishers was some US$ 93,000 and the average 1.5 kg octopus was worth just under US$ 6 to the fishers – a significant difference.

However, even if we accept the higher price figures and divide the estimated total fisher revenue of US$ 93,000 by 1,360 fishers – this equates to only US$ 68 per fisher per year (or US$ 116 if there are 800 fishers) which suggest that the daily catch earnings of US$ 52 are incorrect. If fishers receive half their income from octopus (as reported in post-project monitoring) this implies an average total income of US$ 137 for fishers (or US$ 232 if there are 800 fishers). Assuming the figures reported are correct, this suggests extremely low per capita income levels amongst fisher households.

**Table 2. Octopus production by grades, 2007–2009**

| Year | Grade | Nianing Quantity (kg) | % | Pointe saréne Quantity (kg) | % | Total Quantity (kg) | % |
|---|---|---|---|---|---|---|---|
| 2007 | PP | 12,578 | 59.6 | 10,310 | 37.9 | 22,888 | 47.4 |
| | P | 5,173 | 24.5 | 9,163 | 33.7 | 14,336 | 29.7 |
| | M | 3,041 | 14.4 | 6,588 | 24.2 | 9,629 | 19.9 |
| | G | 292 | 1.4 | 745 | 2.7 | 1,037 | 2.1 |
| | GG | 20 | 0.1 | 421 | 1.5 | 441 | 0.9 |
| | Total | 21,104 | 100.0 | 27,227 | 100.0 | 48,331 | 100.0 |
| 2008 | PP | 21 | 7.2 | 5,872 | 23.9 | 5,893 | 23.8 |
| | P | 133 | 45.9 | 7,996 | 32.6 | 8,129 | 32.8 |
| | M | 84 | 29.0 | 5,157 | 21.0 | 5,241 | 21.1 |
| | G | 52 | 17.9 | 5,148 | 21.0 | 5,200 | 21.0 |
| | GG | 0 | 0.0 | 349 | 1.4 | 349 | 1.4 |
| | Total | 290 | 100.0 | 24,522 | 100.0 | 24,812 | 100.0 |
| 2009 | PP | 41 | 1.1 | 3,005 | 15.0 | 3,046 | 12.8 |
| | P | 1,839 | 48.9 | 4,005 | 20.0 | 5,844 | 24.6 |
| | M | 1,129 | 30.0 | 8,010 | 40.0 | 9,139 | 38.4 |
| | G | 376 | 10.0 | 3,003 | 15.0 | 3,379 | 14.2 |
| | GG | 377 | 10.0 | 2,002 | 10.0 | 2,379 | 10.0 |
| | Total | 3,762 | 100.0 | 20,025 | 100.0 | 23,787 | 100.0 |

*Source:* REPAO, 2010.

---

[2] PP in French "Plus Petit" is the smallest size of octopus with a weight less than 500 gr.

[3] P in French « Petit » is the small size after PP with a weight between 500 gr and 1 kg.

[4] M in French « Moyen » is the medium size of octopus with a weight between 1 kg and 2 kg.

[5] G in « French « Grand » is the big size of octopus with a weight between 2 kg and 3 kg.

[6] GG in French « Grand Grand » is the biggest size of octopus with a weight more than 3 kg.

Octopus are sold by the fishers, to the fishmongers. There are two categories of fishmongers; micro fishmongers without factory quotas) or large traders (wholesalers with factory quotas). Octopus sold to micro fishmongers is then resold to larger-scale traders who then sell the product to the Ikagel fish processing factory. In the study area some 99 percent of octopus is sold to Ikagel for export. Ikagel, clean, sort, freeze, package, market and export the octopus to the overseas markets.

What is striking about the value chain is that, after operating costs have been deducted, the fishers and boat owners operate at high gross margins, about 45 percent (REPAO, 2010). Despite the fact that Ikagel have a captive supplier base with several thousand artisanal fishers supplying octopus and presence of one factory with sophisticated processing machinery and a highly perishable product, their gross margins are estimated to be lower, at about 31 percent. The processing factory is the key link between artisanal fishers and a very sophisticated global market.

**Table 3. Octopus prices at different nodes of the value chain in Senegal**

| Grade: | Fishing | | Fishmongers | | Processing company | |
|---|---|---|---|---|---|---|
| | US$/kg | FCFA/kg | US$/kg | FCFA/kg | US$/kg | FCFA/kg |
| P | 3.34 | 1,700 | 3.93 | 2,000 | 4.92 | 2,500 |
| M | 4.52 | 2,300 | 4.72 | 2,400 | 5.91 | 3,000 |
| G | 5.11 | 2,600 | 5.90 | 3,000 | 7.28 | 3,700 |
| GG | 5.51 | 2,800 | 6.89 | 3,500 | 7.87 | 4,000 |

*Source:* REPAO, 2010.

*Critical issues*

The objective of the project is to promote sustainable management of the octopus resource, improve quality management and increase returns for fishers in the study area. The critical issues required to achieve objective were: the over-exploitation of octopus; the need to locate and support access to more lucrative octopus supply chains; and, low quality of octopus caused by inadequate handling of the catch. Table 4 overleaf outlines the horizontal and process and product upgrading strategies which the study intended to follow (as the final report outlines, a significant proportion of these activities were not implemented during the project period).

**Figure 1. Elements of Upgrading for fishers and the horizontal impacts (Environmental benefits and Poverty alleviation)**

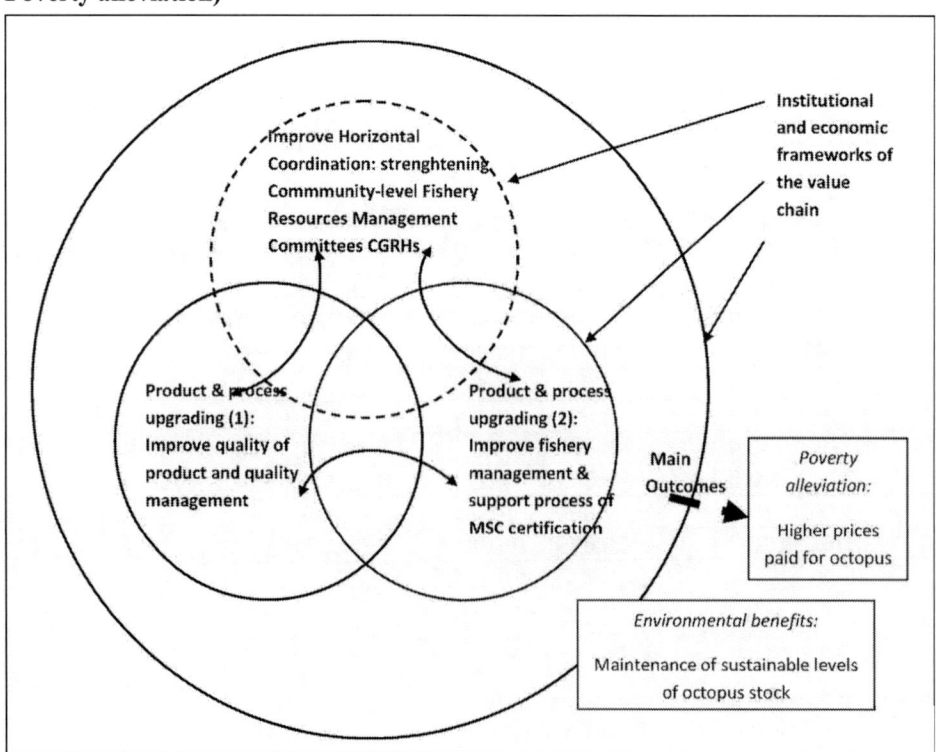

**Table 4. Summary of the intervention**

| Critical issues | Upgrading strategy (2) | Action points (3) | Activities (4) | Outputs (5) |
|---|---|---|---|---|
| Over-exploitation of octopus resource. | Through horizontal coordination, improve the conservation of octopus and increase the size (and value) of the octopus landed. Process upgrading through preparation for certification against Marine Stewerdship Council ecolabel. | CGRH (Fisheries Resource Management Commitee) | Workshops to disseminate 'best practice' on sustainable management of the resource (rest periods, repopulation vases, etc). Disseminate the objectives, process and possible outcomes of eco-label. Establish regulatory framework for environmental management in Department of Mbour. | Monitor Implementation of rest period and 2,000 repopulation vases in study area. Collaboration with fishers and processors to progress towards MSC eco-label certification. Establish and implement a participatory resource management plan including appropriate mobilisation, regulations, surveillance and enforcement. |
| Locate and support actors in Senegal to access a more lucrative Octopus value chain. | Horizontal cooperation to improve bargaining power of fishers and market analysis to identify new markets. | CGRH (Fisheries Resource Management Commitee) | Strengthen CGRHs, through remobilization of members in accountable management structures. Organize training on financial management, commercialization, IT and marketing to support the transformation of the CGRH to a commercial entity. Identify higher value export markets for traceable, sustainable, 'fairer' and high quality Octopus. Distribute up-to-date information on Octopus prices at all nodes of the value chain. Support CGRH in setting up of a revolving fund for the purchase of octopus and collective sales. | More representative CGRHs with strengthened fishery resource management and internal managerial and financial competences. CGRH buys octopus from its members and organises transactions with processing plants.Higher value markets for their Octopus from the case study areas identified and price information disseminated. |
| Low quality of octopus and inadequate hygienic handling. | Product and process upgrading: improve quality management of octopus from sea to sale to factory and quality of octopus sold to processor factories. | CGRH (Fisheries Resource Management Commitee) | Fishers receive training on hygienic handling, quality management and traceability of octopus (including writing of a manual).Insulated tanks installed in each site and independent ice delivery organized so that fishers can store their octopus if necessary. | More hygienic and better quality of octopus sold by fishers resulting in higher prices for octopus. Manual of standard procedures for hygienic handling available to, and implemented by, fishers. |

*Key Findings*

1. This study illustrates how large numbers of low-income fishers can access a high-value globally-traded product with comparatively low barriers to entry. The revenue generated from octopus sales constitutes an important livelihood for fishers[1].

2. The octopus is a common-pool resource which attracts a high price (some US$ 4 to US$ 6 per octopus, depending upon which unit revenue data are used) and for which there is virtually no government restraint on exploitation in practice. In this context, one would expect the resource to be over-exploited to the point that the scarcity of octopus made fishing unfeasible. There is some evidence of over-exploitation of octopus. For instance, the decline in the aggregate catch in recent years, the capture of small and immature octopus and the low catch rate per unit of fishing effort are evidence of over-fishing. However, the quite rapid increase in the size of octopus landed in recent years, together with the sharp increase in fishers' incomes and the evidence that some conservation measures are being implemented by local community structures should at least question the assumption that the octopus population is exploited beyond the sustainable rate of extraction (unfortunately we do not have data on the maximum sustainable yield for octopus). What is clearly evidenced from Morocco is that effective government regulation of octopus extraction can quickly and effectively bring a natural resource back from the edge of collapse to one of sustainable extraction.

3. This study highlights the importance of having accurate baseline information. We do not know with any confidence, the number of fishers, their income, the price they receive for octopus, or their catch. With data of this quality, it is not possible to assess the impact of any upgrading strategy.

4. A related issue is the importance that the analysis is led by the empirical data rather than assumptions. For example, this project was formulated on the assumption that fishers are exploited by the monopoly buyer (the processing factory) and that the quality of octopus was a critical issue. Notwithstanding all the caveats about data, it appears that the fishers and boat owners are capturing a reasonable price for octopus landed (US$ 4 to US$ 6 per octopus on average). The finding that only 0.5 percent of the octopus catch was rejected on quality groups in 2008 strongly suggests that the post-capture handling of octopus is not regarded as a problem by the market. This suggests that a strategy based on extracting higher octopus prices for fishers from better quality stock, is unlikely to succeed. In addition to serious project implementation issues, the limited impact of this project could reflect the fact that several of the three critical issues do not, in fact, exist.

5. The project illustrates well the weakness of focusing the analysis and value chain development on the upstream (production end) of the value chain to the exclusion of the market. The consequence of this traditional, supply-side preoccupation with producers is that, fishers invest time and resources in upgrading themselves, but with no benefit in terms of access to new, more advantageous markets. To the extent that upgrading absorbs resources, such an approach risks increasing poverty amongst producers (by increasing cost without increasing revenue) – not reducing it.

6. The development sector often confuses the participation of women with their gain from value chain upgrading. The focus of this project on getting women to participate in the steering committee structures of the fishery resource management committee, to the exclusion of understanding how or whether women have any control over the resources earned by men from the chain, illustrates this weakness well. If women can control the resources earned by men from octopus, then the project can have positive gender impacts – irrespective of who participates in local management structures. However, if men control the resources generated by octopus and spend the proceeds in a way that does not benefit the household, the potentially adverse gender impacts of developing this chain are not reversed by the presence of women co-opted onto management structures.

---

[1] See table 4 and figure 1.

**Figure 2. Octopus value chain map**

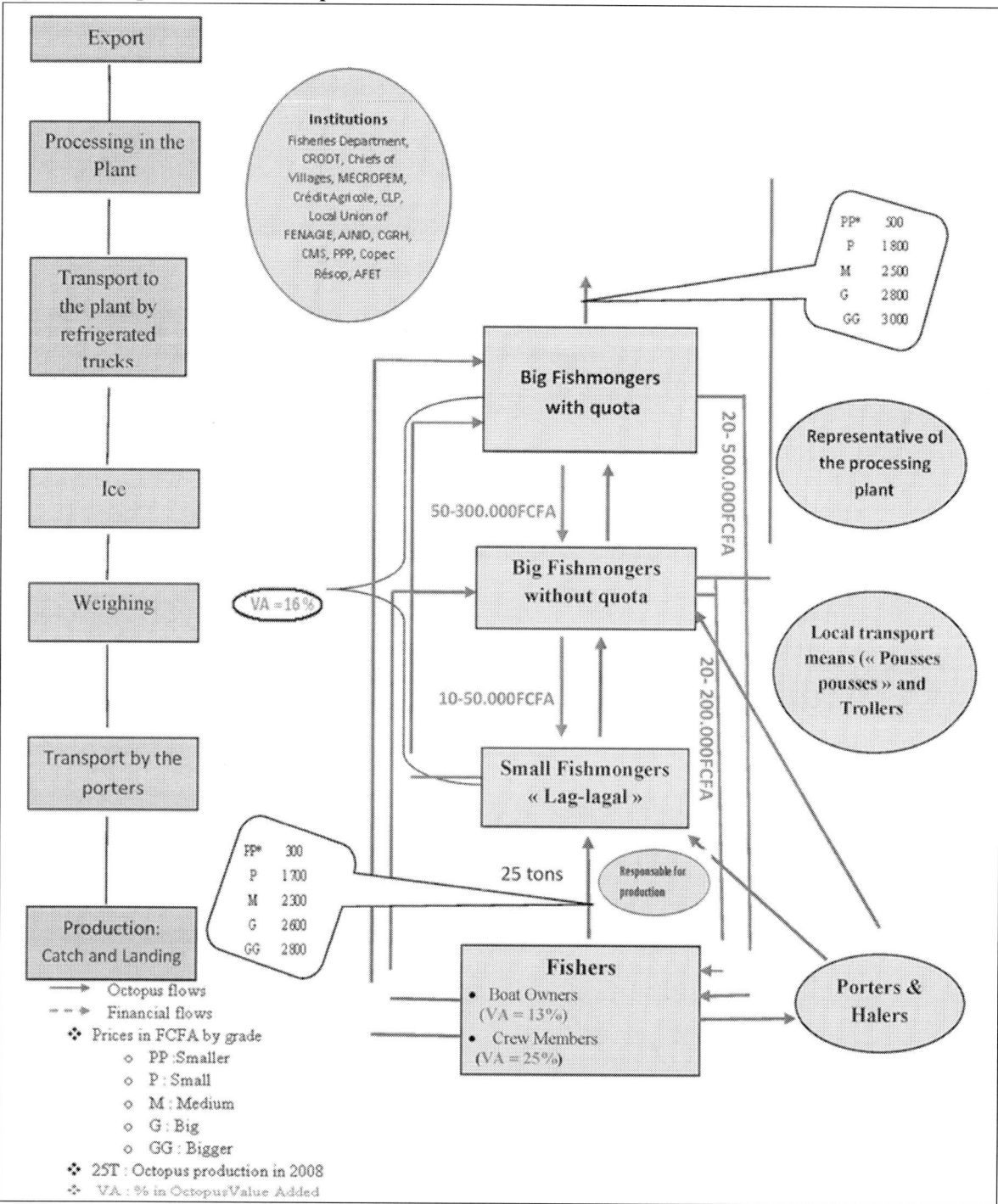

## 3. CONCLUSION

This project on octopus value chain in Senegal shows the importance of considering the markets factors for both achieving sustainable management of fisheries resources and poverty alleviation with combining better prices for products from sustainable management systems. Knowledge of the evolution of the selling price of octopus in local and international markets permits the fishermen to know in advance the revenues they can except to earn after and before the landing. The lack of conservation material and information on the prices system often forced them to sell their products to the factory and wholesalers lower at the last time. Better control of the means of conservation products and formulation of selling prices would strengthen the position of local fishing communities' effort in sustainable management of the resource and in poverty alleviation.

## 4. LIST OF FURTHER READING

**Blue You, ENDA/Repao & WWF**. 2007. Funding Proposal: Eco-Labelling and Value-Chain Promotion of a Senegalese Cephalopod Fishery.

**Coulibaly, M., Dème, M., Diop, N. & Kane, A.** 2003. Etude du profil de pauvreté des communautés de pêche des départements de Mbour et de Foundiougne, Sénégal. Document de travail. Programme pour des moyens d'existence durables dans la pêche en Afrique (PMEDP). 41 p.

**Dème, M.** 2002, Rentabilité économique et financière des pêcheries céphalopodières artisanales sénégalaises. In Caveriere.

**DPM (Direction des Pêches Maritimes).** 2007, Résultats Généraux des Pêches Maritimes 2004

**ENDA/REPAO.** 2008. Etude sur la faisabilité de l'éco-labellisation dans les sites pilotes du projet GIRMAC

**FAO.** 2003, Report of the Expert Consultation on International Fish Trade and Food Security Casablanca, Morocco, 27-30 January 2003. FAO Fisheries Report. No. 708. Rome, FAO. 213p.

**FAO.** 2005. FAO Fishery Subcommittee for the Eastern Central Atlantic: Reports from the 17th session in Dakar, Senegal 2004 & 4th session of the scientific sub-committee in Accra, Ghana 2005.

**Josupeit, H.** 2008. World Octopus Market. Globefish Research Programme, Vol. 94 Rome, FAO. 80p.

**Kurien, J.** 2005. Responsible fish trade and food security. FAO Fisheries Technical Paper, No. 456. Rome, FAO. 2005 102p.

**MRAG, Soil Association & IIED.** 2000. Relevance of certification to Fisheries in Developing Countries, DFID Contract: 7381 CA

**Ponte, S.** 2008. Greener than thou: The Political Economy of Fish Ecolabeling and its Local Manifestations in South Africa, World Development, Vol. 36, No. 1, pp. 159–175.

**Samudra.** 1996. Samudra: Brussels and Chennai: International Collective in Support of Fishworkers (ICSF) Issue: 15 (July 1996)

**Thiam, A.M. & Jouffre, D.** (éd. Scient.). Le poulpe Octopus vulgaris: Sénégal et cotes nord-ouest africaines. Colloques et séminaires, IRD Collection: pp. 169–18

**Washington, S.** 2008. Ecolabels and marine capture fisheries: current practice and emerging issues. Globefish Research Programme, Vol. 91, Rome, FAO. 52 p.

# SUN-DRIED MUKENE (*RASTRINEOBOLA ARGENTEA*) VALUE-CHAIN ANALYSIS IN UGANDA

## [*ANALYSE DE LA CHAINE DE VALEUR DU MUKENE* (RASTRINEOBOLA ARGENTEA) *SECHE*

by/par

Margaret Masette[1]

**Abstract**

With improved security in the neighbouring countries, marketing of mukene (*Rastrineobola argentea*) has become a lucrative business in Uganda after decades of underutilization. However, little was known about the value-chain from capture to market. A 2-week study was undertaken at two selected landing sites in the Ugandan portion of Lake Victoria and several Kampala markets known for a significant sun-dried mukene business. A total of 200 fisher-folk were interviewed, using a structured questionnaire, to identity key stakeholders, linkages between them and economic variables along the mukene value-chain. Preliminary results indicated that boat-owners incurred the highest input per 100 kg-bag of dried at a cost of UGX[2] 60,000 followed by regional traders and local traders at UGX 10,000. However, the profit margins increased from the boat-owners to the regional traders who earned 2 and 4 times the cost of input respectively. This was expected in some instances where some traders were known to offer advance payment to fishers cum processors that trap the latter in perpetual indebtedness and inadvertently compromise final product quality. Although fishers and processors determined mukene quality which ultimately determined the price of the final product, they benefited least from their efforts with profit margins 10 percent and 12 percent respectively. The profit margin for the boat-owners cum traders selling mukene for human consumption, varied between UGX 180,000 and UGX 240,000 per 100 kg-bag depending on whether the market was regional or local. A similar weight of mukene designated for animal feed earned the manufacturer between 44 percent to 52 percent profit, depending on the mixing ratio with other feed ingredients. The market retailers in the local markets as well as in supermarkets earned substantial profit margins, varying between 50–80 percent depending on type of product and rental charges. Since the mukene business seemed to be so lucrative most dealers used it as a transitional stint for only 1–2 years and then moved on to less risky enterprises like hardware shops and commercial houses. It was therefore concluded that key-players along the mukene value-chain played indispensable roles regardless of the profit margins. However, for comprehensive analysis of the whole mukene fishery and respective interactions between different actors, it was recommended that further studies should be undertaken to include different products, cover regional markets and consider seasonal variations.

*Key words: Muken, Value-chain, Key-players, Lake sardine*

**Résumé**

Avec l'amélioration de la sécurité dans les pays voisins, la commercialisation du mukene *(Rastrineobola argentea)* est devenue une activité lucrative en Ouganda après des décennies de sous-utilisation. Toutefois, on sait peu de choses sur la chaîne de valeur de la capture au marché. Une étude de 2 semaines a été entreprise sur 2 sites de débarquement et sélectionnés dans la partie ougandaise du Lac Victoria et de plusieurs marchés de Kampala connus pour l'importance de l'activité du mukene séché au soleil. Un total de 200 pêcheurs ont été interrogés, à l'aide d'un questionnaire structuré, pour 'identifier les principaux acteurs, les liens entre eux et les variables économiques le long de la chaîne de valeur du mukene. Les résultats préliminaires ont indiqué que les propriétaires de bateau engrangent les plus hauts intrants par 100 kg de sacs de séché au coût de UGX[2] 60.000 suivi par les commerçants régionaux et les commerçants locaux à UGX 10.000. Cependant, les marges bénéficiaires ont augmenté des propriétaires de bateaux aux commerçants régionaux qui ont gagné respectivement 2 et 4 fois le coût de l'intrant. Cela était attendu dans des cas où certains commerçants étaient connus pour offrir un payement à l'avance aux pêcheurs et transformateurs qui enferme ce dernier dans l'endettement perpétuel et par inadvertance compromet la qualité du produit fini. Bien que les pêcheurs et les transformateurs déterminent la qualité du mukene qui en fin de compte détermine le prix du produit fini, ils ont bénéficié moins de leurs efforts avec une marge respective de 10 pour cent et de 12 pour cent.

---

[1] Food Biosciences Research Centre (FBRC), National Agricultural Research Laboratories (NARL), PO Box 7852, 17 Bombo Road, Kampala, Uganda. mmasette@yahoo.com
[2] Conversion rate US$ 1 equivalent to UGX 2,800.

La marge bénéficiaire pour les propriétaires de bateaux avec commerçants vendant du mukene pour la consommation humaine, a varié de UGX 180.000 à UGX 240.000 par sac de 100 kg selon que le marché était régional ou local. Un poids semblable de mukene destiné à l'alimentation animale rapporte au fabricant un profit de 44 à 52 pour cent, selon la proportion du mélange avec d'autres ingrédients. Les détaillants des marchés locaux ainsi que dans les supermarchés ont des marges importantes variant de 50 à 80 pour cent de profit selon le type de produit et de frais de location. Étant donné que l'entreprise du mukene semblait tellement lucrative la plupart des marchands s'en servait comme un passage transitoire pendant 1 ou 2 ans vers des entreprises moins risquées comme des magasins et des maisons de commerce. Il a été donc conclu que les acteurs clé le long de la chaîne de valeur du mukene jouent des rôles indispensables quelque soit les marges bénéficiaires. Cependant, pour une analyse approfondie de l'ensemble de la pecherie du mukene et les interactions respectives entre les différents acteurs, il a été recommandé que plusieurs études soient conduites afin d'inclure différents produits, couvrir les marchés régionaux et tenir compte des variations saisonnières.

**Mots clés: Mukene, Chaîne de valeur, Acteurs clé, Sardine du lac**

## 1. INTRODUCTION

Value-chain analysis is a method or techniques used to examine supply of food product from producer to a consumer (Taylor, 2005). It demonstrates relationships between different key players, allows management of challenges like provision of market information, coalition and optimization of activities along the value-chain (Dekker, 2003). In addition, Kaplinsky (2000) noted that value-chain analysis provided insight for policy formulation and implementation.

In the sun-dried mukene (*Rastrineobola argentea*.) value-chain, the relationship between different actors is dictated by gender roles, capital input and type of market outlet.

Mukene is a silvery sardine-like fish with an average length and weight of 5 cm and 15 g respectively. It is the third highest commercially exploited fish species in Uganda after Nile perch (*Lates niloticus*) and Nile tilapia *Oreochromis niloticus*. Although the recent total mukene catches in L. Victoria seem to be decreasing, the value has tended to stabilize at about US$ 1million, which is a significant contribution to the national economy (DFR, 2010). A large quantity of mukene is processed from a horde of landing sites scattered around the numerous islands of Lake Victoria. Traditionally, most mukene in Uganda was processed into animal feed probably about 80 percent was usually processed for animal feed, while only 20 percent was marketed for human consumption. Indeed there are three large-scale and numerous small-scale mukene processing plants for animal feed around Kampala. However, since 2009, there has been a slight increase in mukene for human consumption as the price of other sources of animal protein has sky-rocketed. Some local consumers, who had previously attached a social stigma to mukene, have reverted to its consumption as source of animal protein. Besides, as security in the neighbouring countries has also improved slightly, a substantial undocumented shift has occurred as evidenced by increased trade across the porous borders with the Republic of South Sudan (RSS), Rwanda and the Democratic Republic of Congo (DRC) where most mukene is used for human consumption. The market across the borders is so insatiable and lucrative that mukene of questionable quality is marketed expensively probably to the middle-class. Without doubt, there is market for sun-dried mukene locally as well as regionally for both uses i.e. human consumption and animal feed manufacture.

The value added to mukene by the respective key players as it moves along the distribution chain varies with the level of contact and ability to cause change. Key-players in direct contact with the product like boat-owners, fishers, processors and traders add value by either extending the shelf-life through preservation or packaging. However, value-addition in the mukene fishery may depend on other variables like quality, nutrient content, intended use of the product, level of competition dictated by supply and demand, handling practices and available facilities, stakeholders' behavior regarding quality patterns and the level of income for the respective player in the value-chain. Whereas the policy-makers may not have direct contact with the mukene product, the enforcement of certain policies may affect the mukene value chain. Admittedly, value-chain analysis in the Uganda mukene fishery is quite complex but this study was designed to understand a small component of the interactions between the different players and the value added to sun-dried mukene. The overall objective was to try and analyze the economic market variables along the mukene value chain in Uganda

## 2. SPECIFIC OBJECTIVES

- To identify key-players in the mukene fishery;
- To determine linkages between key-players;
- To establish inputs and profits at various segments along the mukene value-chain; and
- To generate data for policy formulation thereby streamlining the mukene sector.

## 3. METHODOLOGY

Information on economic market variables along the mukene value chain was collected from Kasekulo and Kiyindi landing sites in the Districts of Kalangala and Buikwe respectively using a structured questionnaire. The criteria for selection of the landing sites was based on quantities of mukene landed per day in comparison with other landing sites along the Ugandan portion of L. Victoria and the cosmopolitan nature of the two landing sites. They were the most economically vibrant sites in their respective districts with most of the key mukene value-chain actors. The choice of Kampala was based on its status as a capital city with several fish markets playing a pivotal role in the mukene value-chain. About 132 randomly selected people directly engaged in the mukene business were engaged in the survey. They included boat owners, processors and traders. Their responses were coded and analysed using a statistical package. The cost of inputs involved at each segment of the chain were determined and used to compute the final input cost per 100 kg-bag of sun-dried mukene. The profit margin was calculated from the final selling price of the bag at the retail segment.

## 4. RESULTS AND DISCUSSION

### Key players in the mukene fishery

At the landing site, the key players in the mukene fishery include local authorities at Sub-county level in charge of revenue collection, fishers who conduct the actual fishing operations, and processors. A monthly landing fee is charged for every mukene fishing boat, which the boat-owners pay. About 35 percent of the collected revenue is left at the Sub-county level to finance other government services like education and health. The other key players include boat owners, fishers, processors, Beach Management units (BMUs), traders, fish inspectors and policy makers. The boat owners are both female and male. Fishers are always male with a cross-section of nationalities including Rwandese and Congolese. Processors on the other hand are principally female and young males. Beach Management Units (BMUs) consist of members of the fishing community at respective landing sites, are mandated by act of parliament to co-manage the natural fisheries resources. BMU committees (females and males) are responsible, among other tasks, for ensuring fish quality at landing sites, although quite often it does not feature in their list of priorities. The disregard of quality by consumers may be attributed to lack of basic knowledge about factors that influence fish quality and the high cost of living, which forces most consumers to live within their means. Most consumers would rather consider quantity as opposed to quality when purchasing fish regardless of species or type of product. The traders appear to be the main drivers of the mukene fishery because in their absence, the upstream segment of the value-chain becomes ineffective and totally cut-off. This group consists of both genders and different nationalities depending on landing site. At the study landing sites, Rwandese, Congolese and Burundians were the most common. Since the implementation of the decentralization policy in Uganda, recruitment of fish inspectors is the sole responsibility of districts with a mandate to ensure fish quality among other duties. However, the assurance of mukene quality has eluded their scrutiny with negative consequences to the sector. Finally, policy-makers at the pinnacle of the fisheries sub-Sector formulate policies but it is the District authorities that ensure implementation. However, from the perspective of mukene quality, the implementation segment has been left to the market forces.

### Inputs at various segments along the mukene value-chain

The value added to the product along the distribution chain depends on several variables. However, in the mukene fishery, the variables include quality, nutrient content, intended use, level of competition (dictated by supply and demand), handling practices, available drying facilities, attitude of stakeholders to quality parameters and the level of income for the respective player in the value-chain. At each segment in the chain, the inputs and outputs (Table 1) vary depending on weather conditions, geographical location and accessibility to markets, infrastructure (road network and communication) and source of funding.

**Table 1. The inputs, costs and average investment incurred by various actors along the mukene value-chain (consumables are charged for 100 kg bag of mukene product**

| Value-chain actor | Basic Required Inputs | Production costs (UGX'000) | Average Investment (UGX'000) |
|---|---|---|---|
| Boat owner | Fishing vessel (1)<br>Net (1x7rolls x100m) + floats+ sinkers<br>Boat maintenance<br>Engine<br>Kerosene<br>Net repair<br>2–4 crew members | 3,000–5,000<br>1,000<br>350 per month<br>6,500<br>10 per trip<br>5 per day<br>20–30 per fishing trip | 10,000–15,000 |
| Fisher | Food<br>Landing fee<br>Offloading charge<br>BMU charge | 5 per day<br>1–2 per day<br>0.2–0.3 per day<br>0.3 per day | 10–15 |
| Primary processor | Labour | Equivalent 1 basin of fresh Mukene | |
| Local trader | Vehicle hire<br>Dried mukene | 100 per trip<br>120–150 per bag | 2,000–1000 |
| Regional trader | Vehicle hire<br>Dried Mukene | 600 per trip<br>120–150 per bag | 20,000 |
| Secondary processor | High quality raw materials (mukene + composite flour from cereal or tuber)<br>Hammer mill<br>Workforce (3–4) | 120 per bag<br><br>25,000<br>300–600 per month | 100,000–150,000 |
| Mukene retailer for human consumption | High quality mukene<br>Market dues<br>Monthly rentals | 135 per 100 kg–bag<br>1–15 per month<br>5 per month | 200–250 |
| Supermarkets | High quality and packaged<br>Monthly rent for shelf | 150,000<br>10–15 | 500 (on Mukene alone) |
| Primary consumer | Transport to market<br>Cost mukene (1 kg) | 1–2 per occasion<br>1–2 | 5 |
| Feed manufacturer | Raw materials (mukene + maize or rice bran/husks)<br>Hammer mill<br>Workforce (1–2) | 100–120 per bag<br><br>7,000<br>250–500 per month | 10,000 |
| Local feed trader | Vehicle hire<br>Mukene based feed | 5–10 per bag<br>2–3 per kg | 5000–7000 |
| Regional feed trader | Vehicle hire<br>Dried mukene | 600 per trip<br>90–120 per bag | 100,000+ |
| Retailer for animal feeds | Mukene-based feed<br>Monthly rentals | 2 per kg<br>50–100 | 10,000 |
| Secondary consumer | Transport to feed shop<br>Charge for mukene (10 kg) | 1–2 per day<br>20–30 | 20–50 |

As may be appreciated, the inputs, production costs and average investments highlighted in Table 1 varied significantly among various actors depending on the type of service being offered, terms of payment and whether the business enterprise was solely mukene or mixed with other merchandise. The case in point was the *supermarket*. Most supermarkets in Kampala deal with other products unrelated to mukene which complicates calculation of investment in the entire business.

However, for purposes of clarity only the principal areas of operation along the value-chain will be discussed. They include fishing ground, landing site (together with drying areas) and markets encompassing local as well as regional. For purposes of this paper, key value-points along the mukene distribution chain and respective key player operating within the jurisdiction of the value-point will be discussed singly although it is a well-known fact that interactions occur between different players (Figure 1).

**Figure 1. Interactions between different players along the mukene value-chain**

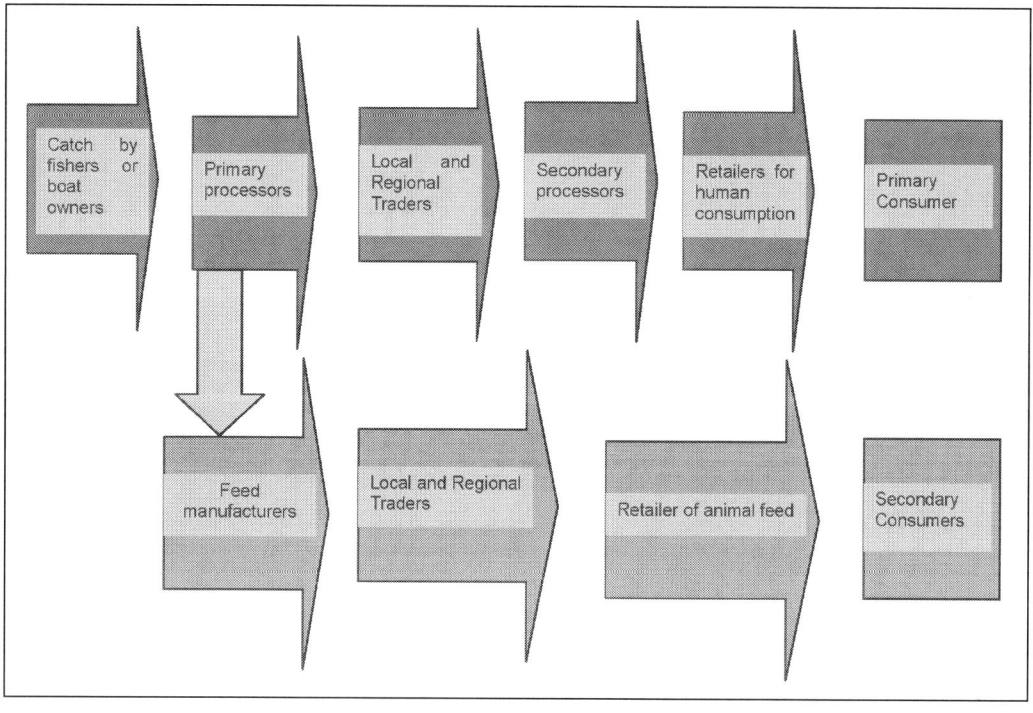

*Landing site segment of value-chain*

Almost all key players were represented at landing sites at any one time and their input in the value-chain varied with the chain segment. They include fishers, boat owners, processors, transporters and traders. The gender disparity and nationalities among the key-players was observed at all segments along the value chain as shown in Table 1. Most of the key-players were Ugandans but there were some players from neighbouring countries. It appeared each player had a designated role at specific segment along the value-chain. However, African Arrow Aquaculture (AAA) a firm based near Kiyindi landing site and considered large-scale by Ugandan standards, was involved at every segment of the value-chain for mukene intended for human consumption. The firm had eight rigs for commercial fishing with a capacity of 400 kg each per day, about 15 raised drying racks with a capacity of 3 tonnes per day, an air conditioned room for packaging and storage. Products were traded at local as well as regional level. In comparison with the local fishers or processors, the firm was significantly ahead of other key-players at Kiyindi landing site in terms of quality and quantity.

**Table 1. Gender disparity among key-players in the mukene value-chain: Case study of Kiyindi landing site**

| Key-players | Percentage disparity | | Percentage disparity in terms of Nationality | | | |
|---|---|---|---|---|---|---|
| | Males | Females | Ugandan | Rwandese | Congolese | S. Sudanese |
| Boat owners | 80 | 20 | 98 | 2 | - | - |
| Fishers | 100 | 0 | 95 | 5 | - | - |
| Processors | 10 | 90 | 85 | 15 | - | - |
| Transporters | 100 | 0 | 85 | 15 | - | - |
| Traders (Whole sale) | 85 | 15 | 75 | 20 | 5 | - |
| Market retailers | 95 | 5 | 100 | - | - | - |

**Boat owners**

Boat owners in the mukene fishery were usually medium-to-high income earners who may have been civil servants, business entrepreneurs or fishers with enough capital to invest in large scale investments. It appeared that most boat owners used the mukene businesses as a transitional occupation for 1 to 2 years (Figure 2) and then moved on to other less risky enterprises like hardware shops and commercial houses. According to 70 percent of the respondents interviewed, they regarded the mukene business as risky because of its seasonality (Figure 3), inability on their part to have collateral in case of credit access, rampant theft of fishing gears and an unstable market. Boat-owners normally resided in urban centres or at landing site depending on their other business interests. Nevertheless, they dealt directly with fishers, processors, traders, BMUs and local government officials. Apart from the boat landing fee, boat owners also incurred the initial cost of the boat and operational costs that entailed fishing gear and boat maintenance at an estimated cost of UGX 3–5 million, depending on size of boat and a net made from 7 rolls of mesh (each roll is 100m). It was reported that the boat could be hired out together with corresponding gear to a fishing crew of 2–4 depending on the capacity of the fishing boat. Alternatively the boat owner recruited the crew and paid them on a daily basis.

**Figure 2. Duration of boat-owners in fish business**

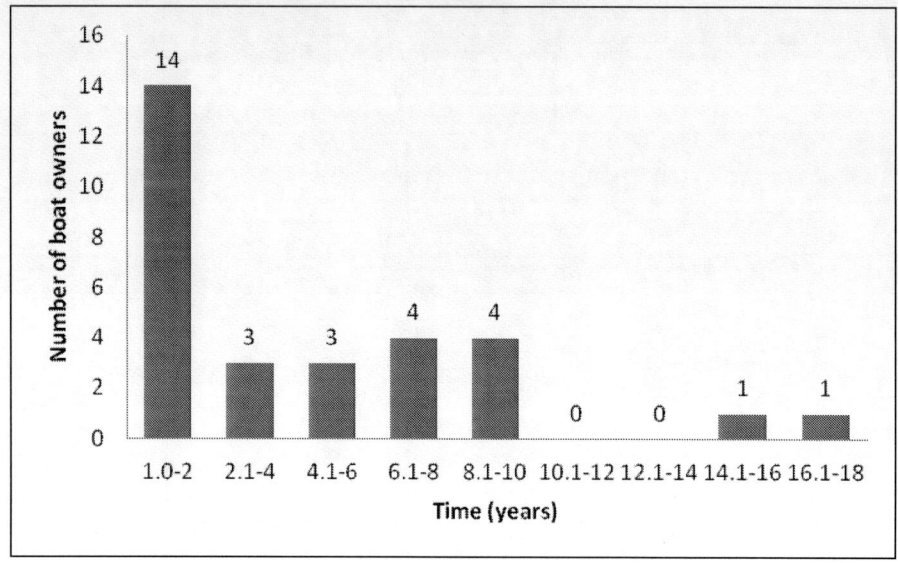

The terms and conditions of boat and crew hire, varied with landing sites. They also paid for hire of drying surface, drying operation and packaging, which was always performed by processors at a cost of UGX 5,000–6,000 per day. Based on the production of a 100 kg bag of sun-dried mukene, the boat-owner was known to incur UGX 60,000 as cost of inputs and sold it between UGX 180,000–240,000 depending on the market and seasonality. During glut seasons, the prices were normally lower than seasons of scarcity and regional markets were normally more lucrative than local markets. The profit margin varied between 26.67 percent and 35 percent, a rare occurrence in other business enterprises. However, the apparent risk involved did not allow most of them to stay long in the mukene business for fear of total loss as a result of any of the risk factors.

**Fishers**

Mukene fisheries are seasonal with distinct monthly variations but in the present study, most respondents indicated that December to February was more lucrative than April to August (Figure 3). Owing to the open access policy in Uganda, fishermen did not pay a fee for accessing mukene. However, some landing sites charged a minimal landing fee ranging from UGX 10,000 to 20,000 per month, which also varied with districts. Whereas, Kalangala District where Kasekulo landing site is located charged UGX 10,000, Kiyindi landing site under Buikwe District charged UGX 300 per basin of fresh fish (30 kg) or 500 per bag (50 kg) of dried mukene. In addition, BMU charged UGX 250 per bag in Kiyindi and Kalangala charged nothing. Essentially, Kiyindi fishers paid more charges on every quantity of mukene landed than their counterparts in Kasekulo although the charge was transferred to the boat owner. The difference has been attributed to by-laws enacted by different districts. Buikwe District where Kiyindi is located charged higher taxes than Kasekulo by virtue of being on the mainland and therefore accessible by road. It had enacted by-laws ostensibly to reduce the number of exploiters and thereby optimizing the natural resource base. On the contrary, Kalangala District where Kasekulo is located could not afford the luxury of prohibitive charges because it is only accessible by water and many traders interviewed feared crossing large expanses of water.

**Figure 3. Annual mukene fishing periodicity and availability in Lake Victoria**

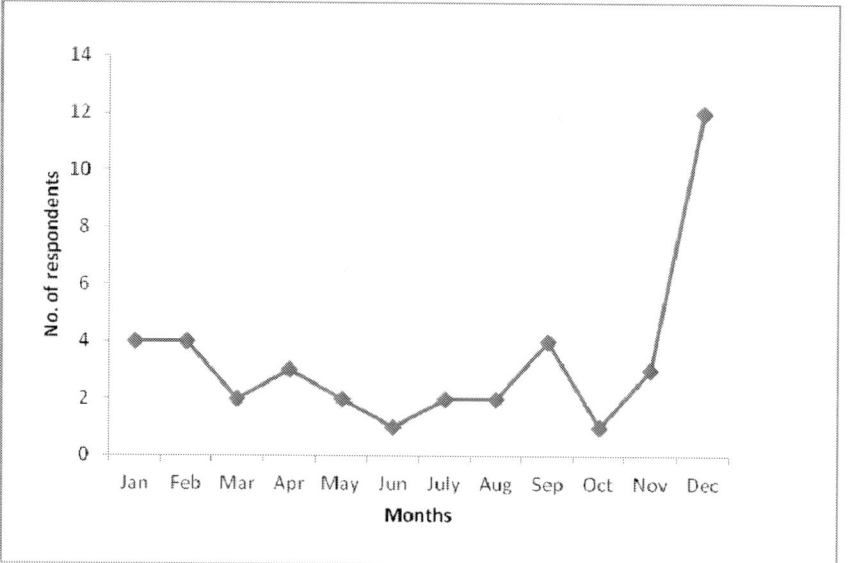

Generally, when fishers went fishing, they bought inputs like fuel, paraffin and food. Upon their return, they subtracted cost of inputs and boat maintenance from cost of the catch and then divided the remaining catch between the crew and boat owner. At some landing sites, fishers were paid UGX 5000 per day or 1 basin of mukene valued at UGX 4,000. In both scenarios, unscrupulous fishers who sold some of the catch whilst at the fishing grounds earned more from daily catches than boat owners. Based on production of 100 kg bag of sun-dried mukene, a self-employed fisher incurred UGX 64,000 on paraffin and petrol per day to catch an average of 12–15 tins (each tin weighing 35–40 kg) of fresh mukene which may be sold at UGX 10,000–12,000. If the fisher decided to sun-dry the mukene, he would realize the same number of tins which he sold at UGX 12,000–14,000. That was an extra UGX 2,000 for every dried tin of mukene. Since 12 tins made up 1 bag, it implied the profit margin was 56.9 percent on fresh mukene and 63.53 percent on dried product.

**Transporters**

There were different types of transportation along the mukene distribution chain. They included boats, head-loads, trucks, pick-ups, trucks and occasionally trains. Fresh mukene was usually carried from the fishing boat upon people's heads (women or youth) that were normally hired by the boat owners. The charge for transportation varied with landing sites and distance to final destination. At some landing sites, labourers were paid one basin for every 30 basins carried from the fishing boat to the drying ground. At other landing sites, UGX 5,000 was charged for the service. The already sun-dried- mukene from the islands was transported to the mainland by water and offloaded by young men at a fee of UGX 500–1,000 per 100–120 kg bag. Since the youth and womenfolk did not have any financial input in the service rendered, they earned revenue that varied with the number of times they were hired. On a busy day during the glut season, it was possible to realize UGX 10,000. Transportation of the same quantity of mukene to distant markets within Uganda using trucks or pick-ups or trucks was charged UGX 10,000 per bag of dried mukene and transportation across borders to regional markets cost UGX 13,000 per bag.

**Processors**

The processors who were usually women or youth were either employed by traders or boat owners. They were normally paid according to the quantity of mukene processed. For example when 1 basin of fresh mukene was sun-dried, the processor was paid UGX 500–1000 depending on landing site. In some cases, the women who carried mukene from the boat continued with the sun-drying operation and charged 1 basin of raw mukene. Part of the drying operation involved turning over mukene every so often to facilitate the drying process. Some processors were paid in kind such that for every 30 basins of mukene collected from the fishing boat to the drying ground, they got a basinful of mukene that they dried separately from the trader's or boat owner's lot. After drying, the processor accumulated their product until it reached saleable quantities (a basin or a bag). The quantities stored were related to the capital input. Small scale processors like labourers operating with less than UGX 100,000 capital base kept their dried mukene in their shacks, while large scale traders cum processors with a capital base of more than UGX 5,000,000 kept their products in commercial stores. Kasekulo and similar landing sites with large stores acted as primary collection centres, while small temporary landing sites supplied large ones. Small-scale processors sold their products to local or regional traders who in turn hired 7 tonne capacity boats to transport their consignments to the mainland. Often processors sold their products to retailers at UGX 15,000–18,000 per tin or to wholesale traders at UGX 13,000–15,000. Each tin of adequately sun-dried mukene weighed 4–5 kg equivalent to 42 cups and each cup weighed 100g and cost UGX 800–1000 at retail local market.

### Traders (local and regional)

Apart from the study landing sites, traders habitually purchased sun-dried mukene from other landing sites along the shores of L. Victoria like Masese, Dimo and Lambu. From these lucrative landing sites, mukene might have been sold to other traders or the same traders might have continued with their consignment to major urban centres like Masaka, Jinja and Kampala. Most of the traders at these other major urban centres were nationals of the neighbouring countries and therefore involved in regional trade. Interestingly enough most traders from DR-Congo were women while traders from Rwanda were men and in most cases got involved in the actual fishing. Very few Sudanese were involved in the mukene fish trade. Usually mukene from L. Victoria and Kyoga was marketed to various markets including local urban centres. All the sun-dried mukene from Uganda traded regionally was invariably used for human consumption regardless of the quality. The quantities of mukene exported across the porous borders of Uganda were not known but information from the study landing sites, indicated that the daily quantities varied from 30 tonnes at Kiyindi to 20 tonnes at Kasekulo per week when in season. At the time, Rwandan traders took over 60 percent of all the mukene processed from these two landing sites, followed by DR Congo (20 percent) and then S. Sudan (10–15 percent). Insignificant quantities were exported to Burundi and Central African Republic. It was reliably learnt that Rwanda re-exported large quantities to Burundi while S. Sudan also re-exported to Central African Republic. Recently a local firm (AAA) started exporting salted and sun-dried mukene to Zimbabwe, S. Africa and Zambia.

Usually traders bought sun-dried mukene from either boat owners or processors but on several occasions, some traders especially from Rwanda bought mukene from boat owners before it even left the water. In that particular instance, they employed both the fishers and processors to fish and dry the mukene respectively. The practice of advance payment to fishers or boat owners that was rampart with Rwandese traders and estimated at UGX 2 billion per month, had a bearing on subsequent handling and processing practices. Such contractors were subjected to immense pressure to meet the conditions of the transaction that ensuring quality was disregarded, which resulted in quality-compromised products. From the socio-economic view point the processors were in a poverty trap that ensured eternal indebtedness since it was kind of credit/loan they had to pay back. The system weakened the bargaining position of fishers and therefore they could not expect better returns than those imposed by the financiers. The trap was so inescapable that even when AAA offered to buy their raw mukene at the market price of the end (dried) product, have it sun-dried hygienically for a better market price of end product/dried mukene, they did not accept the offer because they had an advanced commitment to meet.

Given the stiff competition prevailing at most landing sites, most traders kept their merchandize in the available stores at landing sites or the nearest urban centre until they accumulated enough tonnage for the available vehicle capacity. The storage charges varied with locality. Whereas it was UGX 500 per bag per day at Kiyindi, it was UGX 200 at Kasekulo. There was another group of traders who purchased mukene in relatively large quantities for distribution to local retailers at UGX 120,000–150,000 per 100 kg-bag. The quantities purchased varied from 1 tonne to14 tonnes of sun-dried mukene depending on final destination. While traders involved in distant markets like South Sudan or DR- Congo purchased large quantities to ensure business profitability, local traders purchased small amounts like 1 tonne. In the absence of a large quantity of mukene and traders at a landing site, mukene was packed in gunny or hessian bags and stored in individual houses or BMU stores until substantial quantities were accumulated. Whereas an average house at landing site could store 2–5 bags (100–120 kg each), the BMU stores and similar stores could handle 100–500 bags at a cost of UGX 500 per bag per day. At the final destination, each 100 kg-bag was sold at UGX 300,000 in Kampala, while the same bag cost the equivalent of UGX 600,000–900,000 on regional markets; with S. Sudan offering the highest price. Apparently, Rwanda purchased mukene from Uganda, re-packaged it and exported it S. Sudan at 20–30 percent profit. There was also unsubstantiated information that some of the Ugandan mukene was marketed in the Central African Republic. Owing the long distance involved, the retail price probably exceeded the Sudan price of US $ 5.00 per 100g sachet.

*Market segment of value-chain*

Mukene markets may be at landing sites or urban centres and the key-players included traders of two categories; wholesalers and retailers for human consumption, animal feed manufacture, and Supermarkets. Mukene quantities required by each key-player varied with demand and supplies from landing sites. During the month of December (Figure 3) when there was a glut, some local government authorities designated special stores for mukene storage. In the absence of storage space, it was the responsibility of the trader to find alternative storage facilities, invariably costs are charged. At Kiyindi for example a public store with the capacity to keep 20–30 bags of dried mukene was rented at UGX 50,000 per month. In both cases, a fee was charged on a daily basis on every bag of mukene stored, which varied from UGX 200–500 per bag, depending on landing site or local markets in urban centres. At Kiyindi landing site the storage fee was free for the Fisheries Departmental store but UGX 300 for a private store while at Kasekulo it was UGX 200. Most stores in Kampala charge UGX 500 per bag. There were also some medium-scale feed processing enterprises scattered around major urban centres. They include, Samba Youth, Formula Feeds and Kagodo Feeds. Most of these plants produced about 5 tonnes per day. Then there were numerous small-scale feed production units in the slums of Kampala producing 1 tonne of mukene based feed per day. However, the quality of feed from these slum areas was highly adulterated (about 40 percent) with wood dust and sand (Masette, 2008).

## Supermarkets

There are about 5 large-scale supermarket chains in Uganda with the highest concentration in major urban centres. Small-to-Medium scale "supermarkets" are numerous and could be found along almost every street in urban centres. Until a few years ago, mukene for human consumption was not sold in supermarkets like Shoprite and Capital Shoppers but since 2009, 500g sachets packaged in 5–20 kg packs are freely available on the shelves. The 500 g sachets were sold at UGX 2,500–3,500 depending on supermarket. Mukene is also available in other forms like spiced-deep fried and packaged in 50 g sachets with a price tag of UGX 2,500. Mukene powder was also available in 100g packages costing UGX 2,500. The various products were purchased from processors with enough capital to sustain regular supply at a pitifully low price of UGX 500 for the 500 g sachets and 50 g deep-fried versions which allowed them to make colossal profits. Small-to-Medium scale (SMS) "supermarkets" sold similar mukene products but at 20 percent reduced price with corresponding profits. It appears these SMS were supplied by small-scale processors or small-scale traders from landing sites. The demand for high quality mukene product surpassed supplies especially during periods of scarcity.

## Feed manufacturers

There are several feed manufacturing plants using mukene as source of protein and minerals like calcium and phosphorous. They vary in capacity from small to large and are located around major urban centres. The large-scale plants include Ugachick located 27Km North East of Kampala, Novita in Jinja town and Biyinzika in a suburb of Kampala. Each of these plants require substantial quantities of raw material to produce 5-10 tonnes per day on average but last year (2010), Ugachick switched from mukene to soybeans as source of cheap plant protein on account of low mukene quality. However, according to Arrow African Aquaculture (AAA), a local firm which was supplying them, Ugachick was offering UGX 3,500 per kg for high quality sun-dried mukene when some market outlets in Southern Africa were offering the equivalent of UGX 20,000 per kg of the same product. Undoubtedly, the switch from mukene to soy was partly due to stiff regional competition with corresponding high mukene cost. Besides the low availability of good quality ("clean") mukene, compelled feed manufacturers to use highly adulterated (sand and other extraneous material) mukene, which often destroyed their extruder/feed processing equipment. The profit margin for the feed manufacturer was not only narrowed but also the secondary consumer was affected negatively. For example, Masette (2008) reported low performance in the poultry sector.

Despite the switch from predominantly mukene-based feed to soy, Ugachick was still the sole supplier of high quality floating fish feed in the whole region using small quantities of mukene in the formulation and breaking even. The selling price for mukene intended for animal feed manufacturer varies between UGX 1,900–2,400 per kg depending on quality of mukene. The powder with off-flavours sold cheaply, while whole mukene and inadequately sundried characterized by off-flavours sold expensively. The mixing ratio of mukene with other ingredients varies with intended use of the final feed product. In the poultry enterprise, the ratio varied from 12–15 percent depending on age whether used for layers or broilers, and yet the final product was sold at UGX 1,500. However, in the formulation of any animal enterprise, it did not exceed 20 percent which meant that for every 100 kg bag of pure mukene, the manufacturer mixed 5 bags of feed which were sold at UGX 150,000 giving a profit margin 44 percent but it increased with a low mixing ratio. For example with 12 percent mixing ratio, the profit margin would be as high as 52.3 percent. The small-scale plants with capacities between 100 to 1000 kg were still using mukene in their feed formulation but the quality of their feed was highly compromised probably due to inclusion of sawdust, sand and other extraneous matter as a strategy to improve profit margins.

### Retailers

Retailers basically purchased sun-dried mukene from traders who were invariably wholesalers, either at the landing site or local markets, at the average price of UGX 135,000 per 50 kg bag, equivalent to about 12 tins (1 tin = 4 kg). At the sales price of UGX 700 per 100g of freshly sundried mukene in the local market with revenue of UGX 336,000. Generally, the retailer paid market dues or taxes on a monthly basis, which varied with localities. Whereas the monthly market dues did not exceed UGX 1,000 at the landing site, most fish markets in Kampala charged UGX 15,000. In most cases, the storage charge was prohibitively high for the majority of retailers. As such, they kept their small quantities in individual market lockers that were charged at UGX 5000 per month by the market tax revenue authorities. There were two principal factors that the retailer seemed to consider whilst engaged in mukene trading: the profit margin and the intended use of the mukene. The quality status of mukene rarely featured as a point of concern especially when the product had been branded as feed. However, if a product intended for human consumption did not meet consumer expectations with regard to quality, the retailer sold at a reduced price that inevitably reduced profit margins.

The retailer also incurred some expenses when the store was damp and the retailer was compelled to re-dry the product every so often during storage. The cost of re-drying was charged separately and indirectly recovered from customers. Essentially, the retailer made a profit of at least UGX 280,000 per 100 kg-bag of freshly dried mukene. At some regional retail markets 100g of freshly sundried mukene was priced at US$ 5 - quite lucrative by any standards.

### Consumers

There are over 100 million potential mukene consumers within the East and Central African region. When purchasing sun-dried mukene from the local market, there are two principal factors of concern uppermost in the mind of a consumer. The cost per unit volume and the quality properties of the product: appearance, smell, levels of cleanliness, drying, sorting, fragmentation and lustre. The scrutiny of quality was comparatively more rigorous with mukene intended for human consumption than that for animal feed production. In most retail markets in Uganda, mukene for human consumption was sold in plastic cups whose contents weighed 100 g and cost UGX 700–1500, depending on distance from landing site. The same volume of mukene was sold for US$ 3 in the Democratic Republic of Congo (DRC) and US$ 5 in South Sudan.

### 5. CONCLUSION

The mukene value-chain in Uganda was multifaceted with different nationalities and gender participation in different chain segments. The gender disparity at the respective chain segment was influenced by socio-economic factors like labour intensity of task and capital investment. The processing segment was dominated by the womenfolk and male youth while the actual fishing and trade was the mainstay of adult males. However, each chain actor played an integral part in the mukene value-chain at the respective segment. Lack of policies, or inability to enforce them, has led to mismanagement within the chain. Consequently, formulation and subsequent implementation of harmonized policies for improved quality or standardized measurement tools, as strategy to curb malpractices is an uphill task. As expected some actors took advantage of the status quo to make huge profits and then move on to other enterprises. Understandably, there were some key-players who incurred more costs than others. For example, the boat-owner incurred the highest cost of inputs while the labourers offloading or loading mukene products on either boats or transport vehicles and some hired processors incurred the least input costs. The profit margins varied across the market outlets with the regional traders getting the lion's share while the fishers and processors earned least. However regardless of the profit margins, the various actors played irreplaceable roles and operated in tandem with market demands.

## 6. RECOMMENDATIONS

- Formulate and harmonize a mukene quality policy for the East and Central African (ECA) region to curb malpractices.
- Regional co-operation among respective inspection services and border post agencies should be forged to promote trade in improved mukene products.
- Conduct a detailed study within the ECA region for better understanding of the mukene value-chain dynamics.
- Create market platforms for chain actors to share challenges, risks and other relevant information about the mukene fishery.

## 7. REFERENCES

**Dekker, H.C.** 2003. Value-chain analysis in interfirm relationships. *Management Accounting Research.* 14. (1), pp.1–23. www.sciencedirect.com

**Kaplinsky, R.** 2000. Globalization and equalization: What can be learned from value-chain analysis? < *The Journal of Development Studies;* Dec 2000; 37, (2); ABI/INFORM Global p. 117. www.tandfoneline.com

**Masette, M.** 2008. The influence of Dagaa-based feed on quality of chicken egg production within L. Victoria Basin. Paper presented at *FAO Workshop on Fish Technology, Utilization and Quality Assurance in Africa* held at Agadir, Morocco 24-28 November 2008.

**Taylor, H.** 2005. Value chain analysis: an approach to supply chain improvement in agri-food chains. *International journal of physical distribution and logistics management.* 30 (10). www.ingentaconnect.com

# AMELIORATION DES REVENUS DES FEMMES DANS LA FILIERE PALOURDE: APPROCHE CHAINE DES VALEURS ET EXPLOITATION VENERICOLE DANS LE GOLFE DE GABES

## [IMPROVEMENT OF WOMEN'S EARNINGS IN THE CLAM FISHING SECTOR: VALUE CHAIN APPROACH AND CLAM FARMING IN THE GULF OF GABES]

by/par

Amine Ibn Chbili[1], Leila Hmida, Alessandro Lovatelli, Rakia Belkahia et Yvette Diei-Ouadi

### Résumé

Des données indispensables à l'amélioration de la situation de la population féminine, acteurs clés en amont de la filière palourde en Tunisie ont été collectées dans le cadre du projet TCP/TUN/3203. L'analyse de la chaine de valeur a montré que les femmes gagnaient 2 \$/kg, c'est-à-dire environ 5 pour cent de la valeur du produit final, alors qu'en général, 10 pour cent est la base minimale de part de la production primaire qui fonde des actions visant une juste redistribution de la rente en pêche. L'étude du contexte institutionnel a conclu à l'incapacité du groupement de développement et d'exploitation de la palourde (GDP), son illégalité et les mauvaises pratiques qui fragilisent la gestion durable de la ressource et maintiennent les femmes dans la précarité.

Se basant sur ces deux études, le projet a développé une stratégie incluant des options de schémas institutionnels. Il a aussi conduit le processus de refonte du GDP pour plus de représentativité des femmes dans le processus décisionnel lié à la gestion participative de la ressource, puis la création d'une association autonome de femmes collectrices pour renforcer leur pouvoir de négociation lors de la mise en marché de leurs produits.

Les tests pilotes de grossissement des palourdes de taille non marchande collectées involontairement ont montré un accroissement des revenus des femmes de 30 pour cent en suivant le protocole développé par le projet. Les informations relatives à l'exploitation du domaine maritime devraient guider tout processus de concession foncière aux femmes collectrices.

*Mots clés: Tunisie, Palourde, Femme pêcheur à pied, Chaine de valeur, Groupement, Transporteurs, Revenus, Grossissement, Taille commerciale*

### Abstract

Within the framework of the project TCP/TUN/3203 relevant data have been collected for the improvement of female population, key stakeholders in the upstream clam fishing sector in Tunisia. The value chain analysis showed that the women earned 2 \$/kg, representing about 5 percent of the end product's value while in general a minimal 10 percent share of the primary production is the reference justifying the implementation of actions of fair distribution of revenue in fisheries. The study of the institutional framework revealed a lack of capacity of the Groupement de developpement et d'exploitation de la palourde (GDP), its illegal status and malpractices, which weaken the sustainable management of the resource and keep the women in insecurity.

Based on these two studies the project developed a strategy which includes some institutional options. It has led to the process of reshuffling of the GDP to better represent women in the decision making process linked to the participatory management of resources, then to the creation of an autonomous association of women clam collectors, to strengthen their bargaining power during the sale of their products.

The pilot tests for ripening of undersized clam that have been collected unwillingly showed an increase in the women's income of 30 percent following the protocol developed by the project. The information related to the exploitation of the marine domain should guide any process of tenure concession to the women collectors.

*Key words: Tunisia, Clam, Woman fisher, Value chain, Group, Transporters, Revenues, Ripening, Commercial size*

---

[1] Agence de Formation et de Vulgarisation Agricole, CFPP Sfax, Tunisia. chbili_amine@yahoo.fr

# 1. INTRODUCTION

La collecte des palourdes en Tunisie est une activité en expansion depuis les années soixante à la suite d'une demande extérieure de plus en plus croissante des pays de la rive nord de la méditerranée en particulier l'Italie, l'Espagne et la France. La pêche à la palourde est principalement exercée par une population féminine de « pêcheurs à pied » assez jeune, marginalisée et dont le niveau d'éducation et de formation demeure assez faible. Par ailleurs, le rôle de ces femmes est malheureusement toujours peu connu et invisible en dehors du cercle communautaire local.

La revue de la littérature a montré qu'à défaut d'une activité supplémentaire génératrice de revenus en dehors de la campagne de pêche à la palourde, les collectrices sont exposées à de réelles difficultés et peinent à améliorer leur niveau de vie (Nouaili, 2006).

Il est aussi admis que la pêche à pied, de par son caractère artisanal et la précarité de la population qui s'adonne à ce genre d'activité voire les fermetures répétées des zones de collecte et l'absence de toute possibilité de diversification de leurs revenus, mérite davantage un appui conséquent et un soutient permanent. C'est dans ce contexte que s'inscrit le projet de coopération technique «Renforcement du rôle de la femme dans la filière pêche à pied de la palourde» financé par l'organisation des Nations Unies pour l'Alimentation et l'Agriculture, FAO et dont la tâche d'exécution revient au ministère de l'agriculture, des ressources hydrauliques et de la pêche.

L'objectif global de ce projet est d'appuyer les efforts déployés par le gouvernement tunisien en vue d'améliorer les conditions de vie et de travail des femmes pêcheurs à pied de palourdes, pour une production rationnelle et une utilisation responsable de la ressource « palourde ». Cet objectif sera atteint à travers :

- le renforcement du dispositif d'intervention en matière de vulgarisation et de formation basé sur une approche participative, ainsi que des échanges et de la collaboration entre les différents intervenants dans la filière ;
- la mise en valeur du savoir-faire acquis sur le tas par ces femmes en vue de les mener vers une organisation en groupements de productrices participant activement aux processus de prise de décision et de négociation des prix;
- l'amélioration des techniques de collecte et des conditions de débarquement visant une meilleure préservation de la ressource, ainsi que des aspects techniques de la vénériculture ou d'autres activités annexes (notamment à travers des microprojets productifs économiquement fiables).

Afin d'apporter des éléments d'appui et/ou d'ajustements aux actions et programmes stipulés dans le TCP/TUN/3203, deux volets ont été choisis, se fondant sur l'approche chaine de valeur pour générer des informations d'orientation et de développement d'un plan d'action d'amélioration du rôle de la femme dans la filière palourde en Tunisie. Ces deux volets portent sur :

- **Un volet socioéconomique** dont l'objectif est la valorisation de l'effort de collecte de la palourde, l'identification des interactions entre les collectrices et les différents acteurs de la filière dans la région de Zaboussa (Délégation de Graiba, Gouvernorat de Sfax) et l'initiation de mesures de renforcement du pouvoir des collectrices.
- **Un volet technique** vénéricole dont l'objectif est de suivre la croissance des palourdes *Ruditapes decussatus* indigènes du site, de faible taille commerciale collectées accidentellement, à travers des tests pilotes de grossissement dans la zone du projet afin d'introduire de nouvelles techniques aux femmes collectrices pour améliorer leurs revenus.

# 2. VOLET SOCIO-ECONOMIQUE

Ce volet a consisté en l'étude de la chaîne de valeurs, l'analyse du contexte institutionnel organisationnel et la mise en œuvre d'actions concrètes, dont notamment la refonte du GDP et la création d'une association de femmes collectrices.

*La chaîne des valeurs*

**Objectif et méthodologie**

L'intervention avait pour objectif l'analyse de la filière palourde dans la zone de Zaboussa à travers l' :

- Analyse de l'évolution de l'effort de pêche en fonction de l'état du stock et de l'exploitation de la ressource, le suivi et l'étude des méthodes de régulation et de contrôle.
- Analyse du rôle des intervenants et/ou des acteurs de la filière (fonctions, nombre, nature des relations entre les acteurs ainsi que le cadre juridique régissant la filière).
- Identification des utilisations du produit : description détaillée des circuits de commercialisation de la palourde.
- Analyse de la valeur ajoutée créée à chaque niveau de la chaîne ainsi que la répartition du revenu entre les groupes d'intervenants.

Pour accomplir la mission descriptive des spécificités de la filière en son amont « production » dans la zone d'étude et les interactions que les acteurs entretiennent entre eux, deux types d'enquêtes ont été réalisés :

- Des fiches d'enquêtes structurées auprès des pêcheurs à pied et des transporteurs : le présent travail se base sur la réalisation de séances de diagnostic participatif. Ces séances d'animation et/ou interviews ont été réalisées au niveau des sites de collecte.
- Des interviews semi structurées auprès du chef du groupement El IZDIHAR, les commerçants, les mareyeurs et le garde de pêche. Ce type d'interview se situe entre les causeries-débats ordinaires et les enquêtes classiques réalisées à l'aide d'une fiche d'enquête.

Pour mieux planifier le travail de terrain, des missions de supervision journalière de l'activité de collecte et de commercialisation de la ressource sur les deux sites de débarquement de Zaboussa et de Maouma ont été programmé durant les premiers dix jours de la campagne 2009 /2010.
Les constats faits durant ces missions ont aidé à relever les spécificités de la filière dans la zone et découvrir d'importants détails sur le quotidien des femmes, de leur arrivée sur le site de collecte jusqu'à la vente de leur produit.

Ce travail de pré-enquête a été mené aussi auprès de quelques pêcheurs à pied afin de déceler les éventuelles imprécisions au niveau des questions et pour éviter les questions « évidentes ».

Un échantillon raisonné de 130 collecteurs a fait l'objet d'enquête individuelle durant la période Novembre 2009 et Janvier 2010 (Tableau 1).

Pour les autres acteurs de la filière (les membres du GDP et les transporteurs) l'enquête a touchée tout l'effectif présent (six membre du GDP et 18 transporteurs).

**Tableau 1. Taux d'échantillonnage**

| Site de débarquement | Effectif total des pêcheurs à pied | Nombre d'enquêtes réalisé | Taux d'échantillonnage |
|---|---|---|---|
| **Zaboussa** | 196 | 65 | 33% |
| **Maouma** | 185 | 65 | 35% |
| **Total** | 381 | 130 | 34% |

Les questionnaires d'enquête finalisés ont été traduits en « formulaires » pour faciliter la saisie.

L'opération de saisie des données des questionnaires a été effectuée au fur et à mesure qu'un travail de terrain était effectué. Pour la réalisation de cette tâche, le logiciel EXCEL a été utilisé en spécifiant un codage pour les questions qualitatives. Le traitement des données a été effectué en deux étapes : la première étape consistait à épurer les données saisies et la seconde a été consacrée à l'exécution du programme de tabulation.

**Résultats**

Zone d'étude

L'étude des facteurs écologiques et physiques du littoral et la mise en place d'un réseau de surveillance sanitaire effectuée dans le cadre de la mise aux normes de la filière ont abouti à l'identification de 17 zones agréées de production de mollusques bivalves délimitées par des limites physiques et des coordonnées maritimes déterminées. Chacune de ces zones est affectée d'un numéro sanitaire.

La zone de collecte S5 (Figure 1), sujet de cette étude est située sur tout le littoral de la délégation de Graiba, Gouvernorat de Sfax, entre les îles de Kneiss (coté Nord) et le port de pêche de zaboussa (coté Sud) (MEDD, 2008).

Près de 350 femmes exploitent la zone S5 en tant que pêcheur à pied de la palourde. Ces collectrices sont originaires de quatre localités : Hchichina, Zaboussa, Khawala et Frichète, situéesà des distances de prés de 5 Km du rivage.

**Figure 1. Zone cible du projet (MEDD, 2009)**

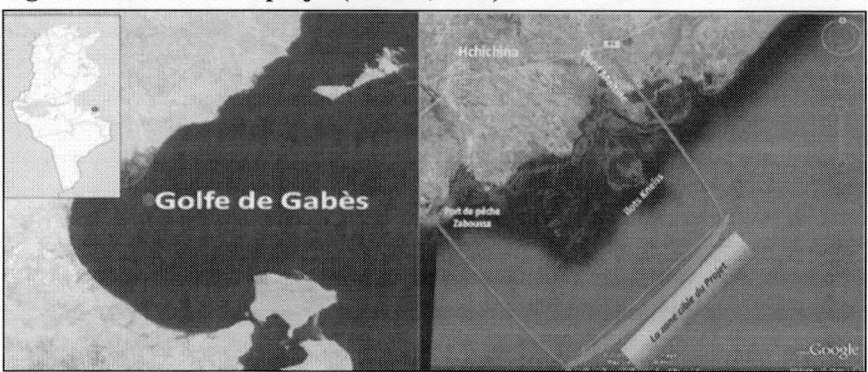

Les acteurs de production

- ***Les pêcheurs à pied***

Si l'on considère l'échantillon de collecteurs interrogés, 85 pour cent sont des femmes contre 15 pour cent d'hommes avec une moyenne d'âge de 30 ans.

L'enquête a montré que la majorité des collecteurs sont analphabètes ou ayant un niveau scolaire faible (85 pour cent des pêcheurs à pied sont soit « sans instruction », soit ayant une courte fréquentation de l'école coranique « Kôtteb »).

Quoiqu'aucune femme ne soit affiliée à la sécurité sociale, la collecte des palourdes constitue l'activité principale pour toutes les collectrices durant la campagne de pêche, contrairement aux hommes (parfois leurs époux) où l'activité de collecte constitue une activité supplémentaire à la pêche par embarcation.

En dehors des périodes autorisées à la pêche, seules 17 pour cent des collectrices interviewées affirment qu'elles réalisent un revenu familial supplémentaire en exécutant différents travaux agricoles entrant dans la production marchande en tant qu'ouvrières salariées. Le reste de la population (83 pour cent) exécute certaines activités agricoles aussi mais dans le cadre d'exploitation familiale servant à la production d'autoconsommation.
L'absence d'une activité supplémentaire génératrice de revenus des collectrices hors des périodes de pêche autorisées les rend plus vulnérables et réduit leurs moyens de subsistance.

- ***Les transporteurs***

Les collecteurs dans le site de collecte de Zaboussa et de Maouma parcourent une distance moyenne entre 4 et 7 Km pour se rendre aux lieux de collecte.

Selon la distance qui les sépare de la zone de collecte, la majorité des collectrices (92 pour cent) ont eu recours à un moyen logistique pour parvenir aux zones de collecte : Ce sont les transporteurs qui se chargent d'amener les femmes de leur lieu d'habitation jusqu'au rivage via des camionnettes et/ou des barques à voile par groupe de 15 à 20 femmes.

Les transporteurs à camionnette perçoivent 1 TND[2]/kg de palourdes collectées comme frais de transport pour les deux zones de collecte, alors que les transporteurs par barque font généralement alliance avec les transporteurs à camionnette qui leur assurent le remboursement de leur frais de transport de 1TND/Individu/jour.

Selon le témoignage des collectrices, ces transporteurs se sont intégrés dans le circuit de la filière et se sont rendus incontournables, en effet ils :

- ont des liens de parenté avec les groupes de femmes qu'ils transportent. Par conséquent, ils jouissent de la confiance des collectrices et surtout de leurs familles.
- veillent à les faire bénéficier de quelques prestations de services (prêts, courses…).
- détiennent les autorisations de pêche des collectrices.

Ainsi, l'on peut considérer que les transporteurs jouent le rôle d'intermédiaire entre les collectrices et le groupement et profitent de la vulnérabilité des femmes pour leur imposer une commission sur chaque kg de palourde collectée bien que le groupement ait été instauré afin d'enrayer ces derniers du circuit de la commercialisation primaire du produit.

- ### *Le Groupement GDP El Ezdihar*

Fondé le 20 Octobre 2004, le groupement EL EZDIHAR est dirigé par un comité composé de six personnes bénévoles : un président, un vice-président, un trésorier et trois membres.

Les membres du groupement sont tous des hommes qui connaissent de très près la filière et jouissent par conséquent du respect de toutes les collectrices et exercent des activités dans divers domaines tels que l'agriculture et le commerce.

Le groupement a mis en place deux abris en dalles de béton utilisés comme site de débarquement dans les deux sites de collecte, mais il souffre d'un grand manque de ressources humaines qualifiées et de dysfonctionnements financiers qui mettent en péril la planification de ces interventions en faveurs des adhérentes.

Malgré les importantes sources de financement du GDP qui perçoit des centres de purification et d'exportation de la palourde (CPE) 0,1 Dinars sur chaque kg de palourde vendu, l'environnement d'intervention du groupement se limite toujours à la commercialisation primaire du produit.

En effet le GDP veille à :

- Représenter les pêcheurs de palourdes de la zone dans les différentes manifestations qui les concernent.
- Informer les collectrices de l'ouverture et de la fermeture des sites de collecte.
- Aider à l'obtention des permis de pêche
- S'assurer de l'origine des produits.
- Etre présent durant l'opération de contrôle et de pesage de la palourde.
- Contacter les exportateurs.

Malgré que le GDP El Ezdihar est le premier responsable de la commercialisation primaire du produit, cette entité est affectée par un ensemble de facteurs qui compromettent sa viabilité, à savoir la faible compétence des responsables, l'absence des mesures de soutien ou d'encadrement, et surtout l'exclusion de la femme "pêcheur à pied" du processus de gestion de la ressource et du marché et par conséquent du circuit de commercialisation.

Circuit de commercialisation primaire du produit

La Figure 2 présente le circuit de commercialisation primaire de la palourde dans la zone du projet.

---

[2] 1 TND = 0.75 $.

**Figure 2. Circuit de commercialisation primaire du produit**

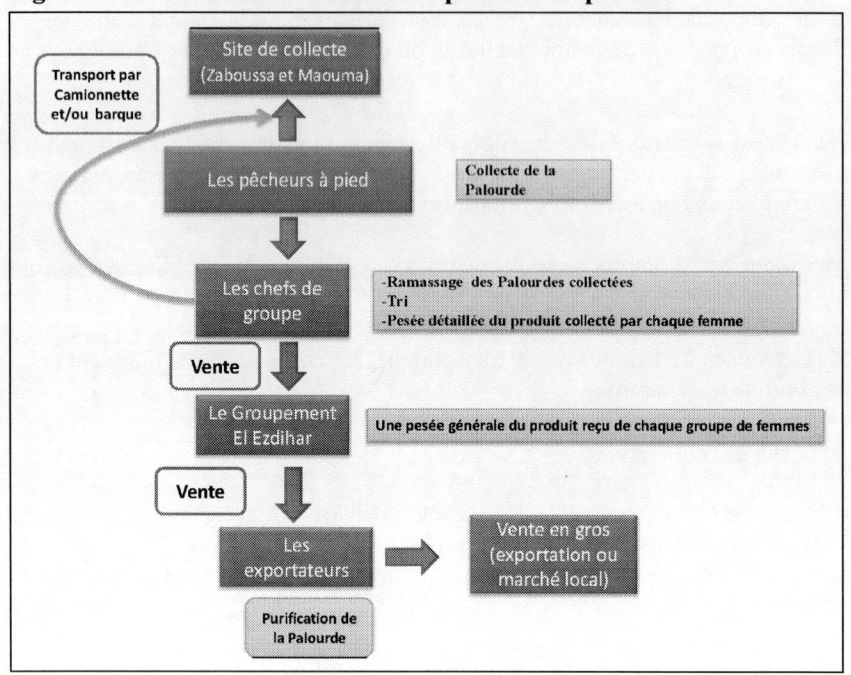

La valeur ajoutée des acteurs de la production

Les valeurs ajoutées des différents acteurs de la filière sont calculées selon les données fournies par les services responsables notamment la Direction Générale de la Pêche et de l'Aquaculture (DGPA) et sur la base des enquêtes semi directives avec les sujets cibles (collectrices, transporteurs, Groupement).

Pour le calcul du gain des collectrices et du GDP, il a été procédé au calcul de la moyenne de quelques indicateurs clés notamment le prix de la palourde ou encore le nombre de jour de travail autorisés sur la base des cinq dernières campagnes 2005–2010.

Toutefois, pour se rapprocher de la réalité et pallier aux insuffisances de la méthode basée sur les déclarations des transporteurs, qui ont tendance à gonfler volontairement les coûts et diminuer les revenus, l'on a essayé de calculer les revenus des transporteurs sur la base de certaines données telles que les distances parcourues /jour, l'amortissement (ans), la valeur actuelle du moyen de transport, le pourcentage d'utilisation du moyen de transport relativement aux autres activités, etc.

A travers le tableau (2) on peut constater que les activités des collectrices sont les activités les plus couteuses en termes de temps et d'énergie.

Les collectrices ont quand même l'avantage d'avoir une structure des coûts très faibles leur permettant d'enregistrer des gains positifs même si l'activité connait des problèmes. En plus, au cas où une femme désire abandonner l'activité de collecte, les charges de désinvestissement seront négligeables.

**Tableau 2. Caractéristiques de chaque acteur dans le maillon production**

| Acteur | Tâches | Degré de difficulté de l'activité [3] | RM (DT) |
|--------|--------|--------------------------------------|---------|
| Collectrice | Collecte de la palourde | Très difficile | 73,646 |
| Transporteur | Transport des collectrices Assiste au tri et à la pesée détaillée du produit | Difficile | 676,48 |
| GDP El Ezdihar | Assure la pesée générale du produit Vente du produit au CPE | Moyennement difficile | 1891 |

---

[3] Estimé en termes de temps et dépense d'énergie.

Variabilité des revenus mensuels (RM) des acteurs de production

L'étude de la variabilité saisonnière des RM, permet de confirmer la fluctuation significative du revenu des différents acteurs au cours du temps et prouve aussi l'existence d'une superposition des valeurs ajoutées mensuelles (VAM) entres les différents acteurs de la filière et ce en raison de la dépendance de ces derniers des quantités de palourde collectées mensuellement (Figure 3).

Le revenu du GDP dépasse celui d'une collectrice de 25,62 fois alors qu'un transporteur gagne 9 fois plus qu'une collectrice (Figure 3). Les transporteurs bénéficient de la non disponibilité/du faible nombre de matériel de transport et fixent par la suite des tarifs assez élevés ; alors que le GDP génère des revenus assez élevés grâce d'une part à la marginalité des charges supportées et d'autre part à sa position d'intermédiaire unique entre les femmes et le CPE.

**Figure 3. Variabilité saisonnière des RM pour la saison 2007/2008**

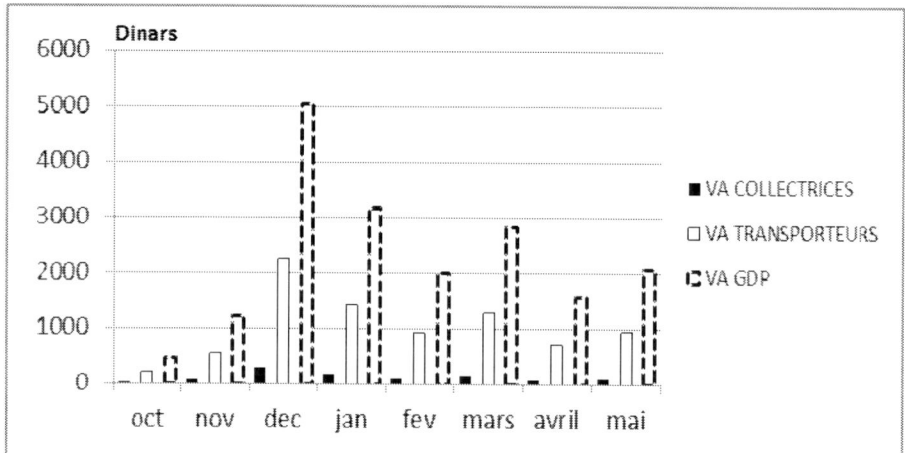

Comme le montre le graphique ci-dessous les courbes de variation de la valeur ajoutée pour les collectrices pendant les saisons 2006/2007 et 2007/2008 sont indépendantes ; c'est-à-dire qu'à partir du surplus réalisé pendant un mois, on ne peut pas prévoir le surplus réalisé au même mois la saison suivante. Cette constatation est illustrée par la divergence entre les deux courbes des deux saisons ainsi que par certaines observations.

Au cours de la saison 2006/2007, il n'y a pas eu d'activités de collecte pendant les mois d'Octobre et de Décembre alors que pour la saison suivante, la valeur ajoutée a atteint 27,109 DT en Octobre et 262,558 DT en Décembre.

**Figure 4. Variation mensuelle du revenu des collectrices**

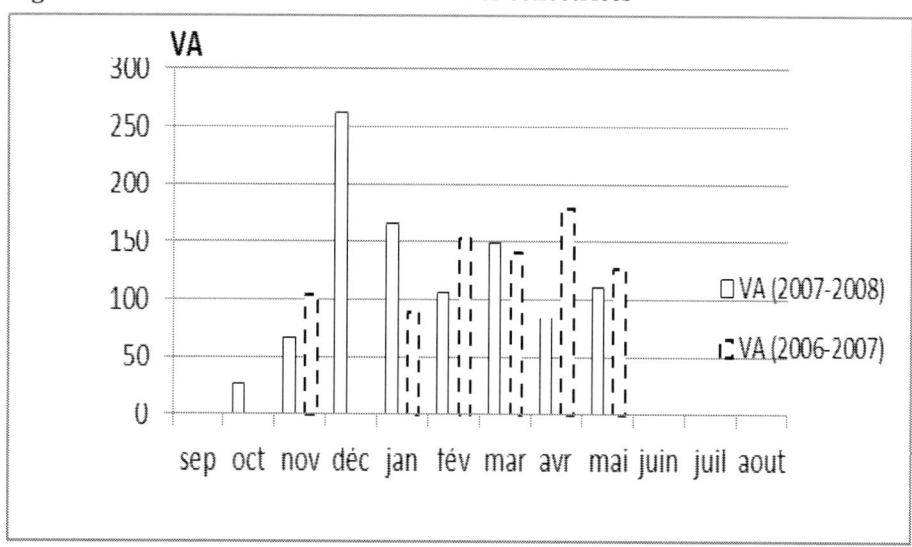

Les valeurs de la VAM admettent une moyenne de 99,348DT avec une variation de l'ordre de plus ou moins 67,337DT des deux côtés ; alors que pendant la saison suivante, la série admet un écart type de 72,071DT et une moyenne de 121,303DT. Par ailleurs, les VAM varient d'une saison à l'autre et d'un mois à l'autre de la même saison. Cette variabilité s'explique essentiellement par :

- Le nombre de jours autorisés à la collecte et qui dépendent de l'évaluation par les autorités de la situation sanitaire des sites de collecte.
- L'état du stock de palourde.
- Les variations saisonnières des prix à la production : Plus la demande des CPE augmente, plus élevés seront les prix de vente de la palourde et par suite, l'activité de collecte devient plus intense.

Les femmes sont exposées à un ensemble de contraintes qui accentuent leur vulnérabilité tels que l'analphabétisme, le non respect de la réglementation, l'arrêt de l'exercice de pêche pendant de longues durées, l'insuffisance et la fluctuation de leurs revenus ainsi que l'absence de toute forme d'encadrement, de formation ou d'autres sources de revenus complémentaires. En effet, un meilleur encadrement des femmes par le GDP ainsi que la diversification de leurs revenus est l'une des solutions judicieuses pour lutter contre la pêche illicite et la précarité de ces dernières.

### *Développement organisationnel et institutionnel des femmes*

L'un des principaux résultats du projet a été l'élaboration d'une stratégie nationale de renforcement du rôle de la femme dans la filière palourde et d'optimisation des revenus qui tracera le sillon des actions de développement y compris la dissémination aux autres régions concernées par l'activité de la pêche à pied de la palourde.

Le but de la stratégie est la promotion du développement organisationnel et institutionnel des femmes pêcheuses de palourde et la distribution équitable des revenus de la pêche.

Les travaux menés au sein du TCP/TUN/3203 ont montré qu'à divers degrés, chaque acteur de la filière est exposé à un ensemble de facteurs pouvant compromettre la durabilité de la ressource de même que ses activités.

Afin de remédier à ces défaillances qui mettent la durabilité de la filière en péril, l'équipe du projet a pensé à améliorer la situation actuelle à travers l'organisation d'une visite d'échange d'expériences et de connaissances au Maroc, la restructuration organisationnelle du GDP et la création d'une association féminine dans la zone cible du projet.

### La visite au Maroc

Pour inciter les collectrices à prendre conscience de leur situation et s'efforcer à défendre leurs propres intérêts, deux femmes leaders du site de Zaboussa ont bénéficié d'un voyage d'études et d'échanges de la part du projet. Elles ont ainsi pu être imprégnées de l'expérience de femmes d'une communauté de pêche au Maroc qui pratiquent la même activité qu'elles, mais sont mieux organisées, participent au processus décisionnel de tout ce qui touche leurs opérations, disposent de pouvoir de négociation et sont relativement épanouies.

Durant les sept jours de visite les deux participantes ont bénéficié d'une démonstration en technique d'élevage de palourde sous filet, d'une visite de la coopérative féminine de ramassage de moule à Douira et Tigri à Sidi Boulfdail et d'une démonstration sur le concept de fumage de poisson à l'Institut Supérieur des Pêches Maritimes.Ces deux femmes ont participé aussi à un atelier d'échange avec les membres du bureau de la coopérative de Sidi Moussa et une réunion avec les membres de l'unité genre et développement à l'Institut Supérieur des Pêches Maritimes.

Cette visite a été un élément déclic dans leur engagement ultérieur à mobiliser les autres pour se regrouper pour être plus fortes et mener des actions de cohésion face aux autres intervenants de la chaîne.

### La restructuration du GDP

Inspirée par les résultats des enquêtes, l'équipe du projet a jugé que le développement de la situation actuel passe par la restructuration du GDP à travers l'incitation d'une assemblée générale et l'intégration de trois femmes leaders dans son comité directeur pour une gestion participative du stock et pour améliorer leur représentativité dans le GDP et par conséquent renforcer leur pouvoir de négociation des prix.

Différentes actions ont eu lieu notamment:

- la tenue de séances de sensibilisation et d'animation auprès des collectrices de la zone pour les motiver à adhérer au GDP, à participer dans les élections et à déposer leur dossier de candidature;
- la tenue de séances de travail avec les autorités compétentes intervenant dans la filière afin de relayer les problèmes de terrain et convenir d'une date pour l'assemblée générale (AG) de bilan et restructuration du GDP;
- Appuyer le président du GDP à la préparation de son rapport financier;
- Information des autorités régionales de la situation financière et administrative du GDP (rapport financier à l'appui) et de la nécessité d'organiser une AG pour dépasser le blocage administratif lié à des irrégularités dans le bilan financier du GDP;
- Fixer une date pour l'assembler générale; et
- Lancement des procédures de l'AG conformément à la règlementation en vigueur à savoir la publication de la date de l'AG, la sensibilisation des dates du dépôt de la candidature au conseil d'administration, les pièces exigées pour participer à l'AG et la réception des dossiers de candidature.

Ainsi, avec certes quelques mois de retard sur le plan initial, le GDP a pu être restructuré pour être plus représentatif des utilisateurs de la ressource palourde.

**La création d'une association féminine**

En parallèle avec les actions de restructuration du comité directeur du GDP, l'équipe du projet a cherché à optimiser la participation des pêcheuses à pied dans l'amélioration de leur situation en encourageant la création d'une association féminine dans la zone du projet nommée l'Association des Femmes et de Développement à Graiba (AFDG).

Cette association aura comme rôle principal l'amélioration des conditions de vie et de travail des pêcheuses à pied, un rôle dont le GDP actuel est incapable jusqu'ici de jouer.

Formé par un comité de 7 jeunes pêcheuses à pied (âge moyenne de 25 ans) dont trois ayant un niveau universitaire supérieur, cette association sera le premier noyau du travail associatif des collectrices de palourde qui cherchera à travers ces programmes de terrain à contribuer à:

- Elever le niveau scolaire, professionnel et culturel des femmes;
- Promouvoir la santé maternelle des femmes;
- Améliorer la qualité de vie de la population;
- Valoriser les produits agricoles locaux;
- Promouvoir l'autonomisation économique des femmes;
- Représenter les femmes auprès des structures nationales et des partenaires au développement;
- Protéger les intérêts des femmes travaillant dans le domaine de la pêche maritime;
- Conserver les ressources naturelles et rationnaliser leur utilisation; et
- Généralement faire tout ce qui pourrait contribuer à servir les intérêts de l'association.

## 3. LE VOLET TECHNIQUE VENERICOLE D'AMELIORATION DES REVENUS : TEST DE GROSSISSEMENT DE LA PALOURDE

### *Objectif et methodologie*

Afin d'améliorer l'environnement du secteur vénéricole et la rentabilité des revenus des femmes collectrices, des mini projets de grossissement de palourdes ont été réalisés dans le site pilote de Zaboussa. En effet, 10 pour cent des palourdes collectées involontairement de taille inférieure à la taille commerciale autorisée (35 mm) sont rejetées en mer d'une façon aléatoire ; ceci constitue une perte dans l'effort de pêche et par conséquent dans le revenu. Le projet a jugé judicieux d'ensemencer ces dernières dans le cadre du travail dans des parcelles de grossissement afin de préserver le stock naturel, limiter la surexploitation et améliorer la rentabilité des collectrices. Les tests pilotes de grossissement visaient à collecter les données authentiques et informations fiables à mettre à la disposition des femmes collectrices pour mieux les orienter dans la valorisation des palourdes préalablement non commercialisables.

Les tests d'élevage de la palourde dans la zone de Zaboussa ont été répartis sur deux sessions :

- ***1ère série*** entre mars-aout 2010 : qui consiste à installer 7 parcelles de grossissement afin de suivre la croissance des palourdes ensemencées dans deux zones différentes (zone herbier non exploitée par les pêcheurs à pied et zone claire exploitée souvent par les pêcheurs à pied) avec deux densités différentes (150 pièces/m2 et 300 pièce/m2).
- ***2ème série*** entre septembre-décembre 2010 : Cette deuxième phase du projet a consisté à des essais de grossissement en fonction de la taille d'ensemencement des palourdes dont le but est d'approfondir les résultats obtenus lors du 1er essai (période d'étude mars- aout 2010) et améliorer le protocole de grossissement pour les femmes pêcheurs à pied.

Les palourdes ensemencées lors de cet essai (2ème phase) proviennent toutes des parcelles de la première partie du projet c'est-à-dire des zones herbiers et des zones claires confondues. Ce choix est respecté car une telle opération nous permettra d'une part d'apprécier la croissance effective de toutes les palourdes ensemencées après 7 mois de grossissement, de pouvoir les partager en classe de taille et de préserver ainsi le stock naturel. Les individus faisant l'objet de cette 2ème partie d'étude sont donc des palourdes sauvages prélevées du même site.

Après tamisage des palourdes sur tamis 25 mm et 15 mm pour séparer les différentes classes de taille 2 classes ont été identifiées :

- 1ère classe de taille > à 35 mm, taille commerciale ;
- 2ème classe de taille comprise entre 30 et 35 mm.

Afin d'identifier la croissance des palourdes dans les différents essais, il a été on a procédé à des mesures biométriques moyennant le pied à coulisse et la détermination de la croissance physiologique des palourdes en estimant le gain en chair, sur un échantillon de palourdes (n = 100) collecté bimensuellement.

***Résultats***

Les résultats de la première série d'étude (mars-aout 2010) ont permis de faire deux constats :

- Le premier constat concernant les palourdes issues de zones herbiers qui présentent une croissance hautement significative (Figure 5) par rapport à celles issues des zones claires (Pvalue=0,026, Comparaison de k échantillons indépendants, test Kruskal-Wallis, logiciel XLSTAT Pro). Cette différence est principalement due à la présence des herbiers de posidonie qui offraient aux espèces qui cohabitent, un habitat, une nurserie et une zone de reproduction.

Le taux de croissance en longueur enregistré dans cette zone est de l'ordre de 1,21 mm/ mois.

- Le deuxième constat est que le taux de croissance des palourdes ensemencées dans les parcelles à 150 ind./m$^2$ est significativement supérieur à celui observées dans les parcelles à 300 ind/m$^2$ (Tableau 3).

**Tableau 3. Taux de croissance des palourdes en fonction de la densité d'ensemencement**

| Parcelle Zone herbier | Densité (ind./m$^2$) | Taux de croissance(%) |
|:---:|:---:|:---:|
| P4 | 150 | 54 |
| P7 | 300 | 35 |

Tenant compte des résultats obtenus lors de la première phase de ce projet la décision a été prise de travailler lors du 2ème essai de grossissement (septembre-décembre 2010), seulement dans les zones à herbier de posidonie avec des densités de 150 ind/m$^2$.

Après avoir ensemencé les palourdes dans les nouvelles parcelles un suivi de leur croissance a été alors effectué.

**Figure 5. Suivi de la croissance en longueur et en poids**

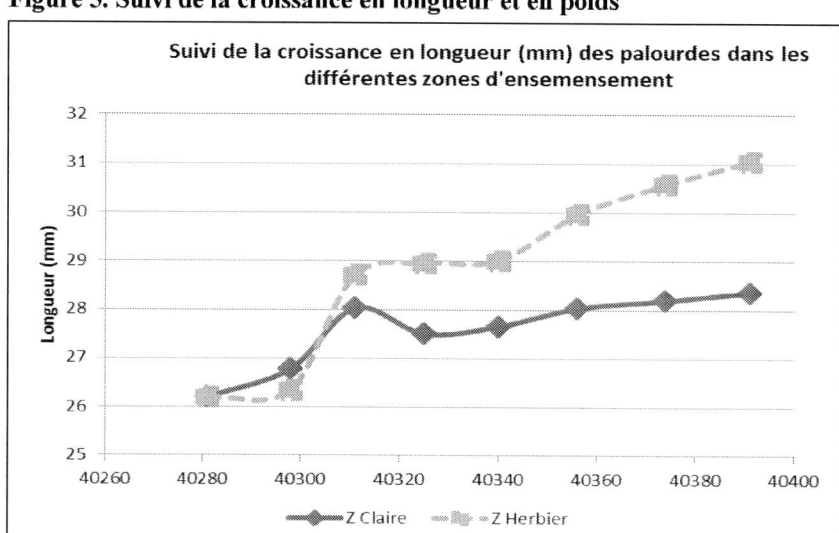

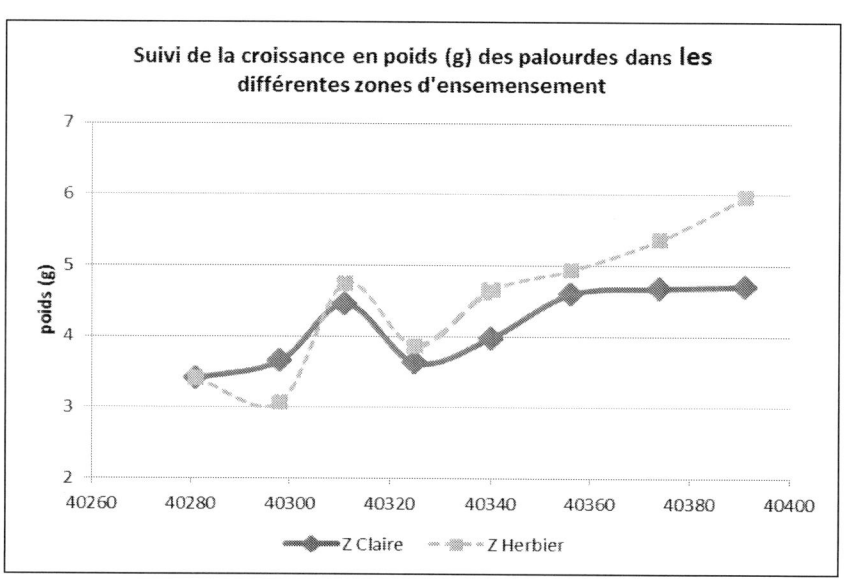

Comme le montre la Figure 6, quelque soit la taille d'ensemencement (parcelle A et B), les palourdes enregistrent une croissance significative aussi bien en taille (longueur) qu'en poids, un gain de 0,43mm/mois (toutes parcelles confondues) est alors enregistré. Suivant cette démarche, il faudrait au minimum 7 mois de grossissement pour que les palourdes ensemencées atteignent la taille commerciale. Le taux de croissance enregistré entre avril et octobre 2010 (lors de la première phase de cette étude) est nettement supérieur (1,21 mm/mois) à celui enregistré pendant cette période (0,43mm/mois). Cette différence est probablement due aux efflorescences planctoniques pendant la saison printanière (avril) et estivale (août) qui enrichissent l'eau en planctons principales proies pour les bivalves filtreurs tels que les palourdes.

**Figure 6. Suivie de la croissance et le poids moyen des palourdes pour les différentes tailles d'ensemencement**

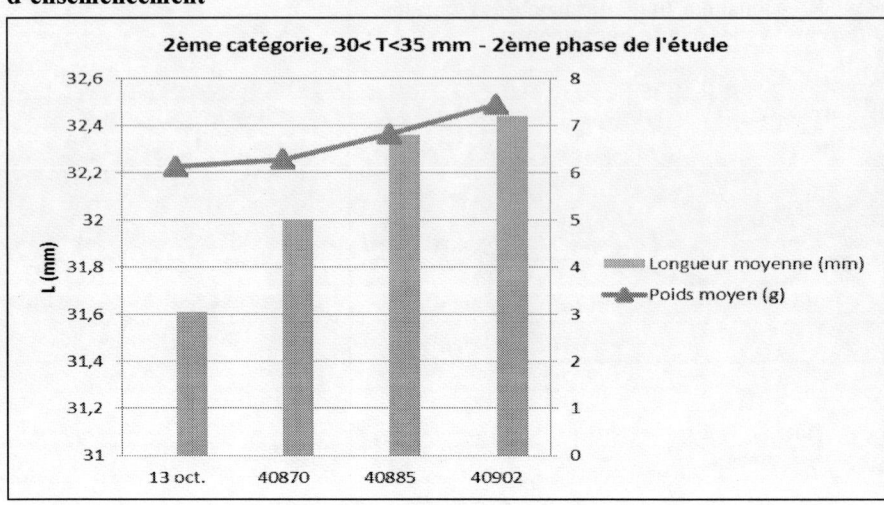

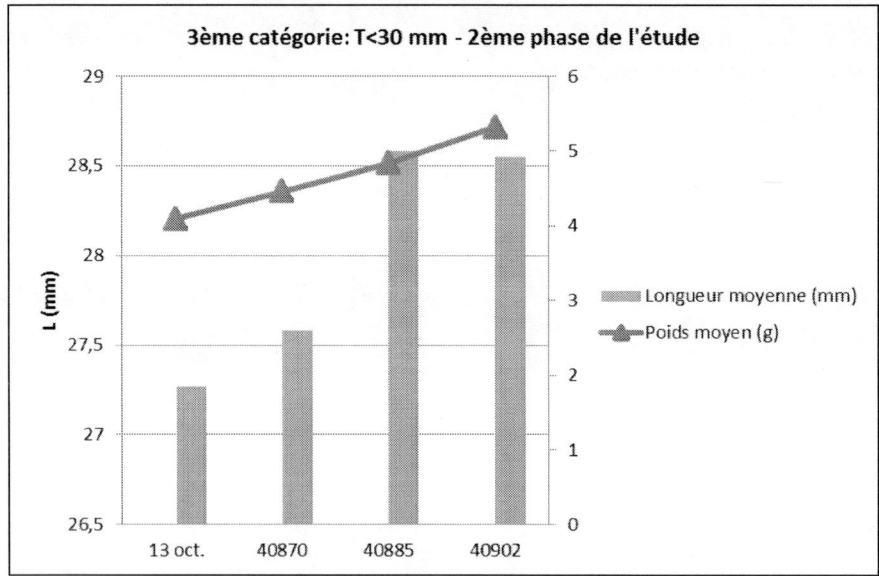

Les résultats des tests ont permis de calculer le bénéfice qui peut être engrangé par les collectrices (Tableau 4).

**Tableau 4. Estimation du revenu annuel supplémentaire gagné par femme par campagne**

| | |
|---|---|
| Nb kg collectés/ femme/campagne | 212,45 kg |
| % palourde collectée (T<35 mm) /femme /campagne | 10% |
| Nb kg ensemencé/ femme/campagne | 21,24 kg |
| Nb kg retenu en fin exercice | 42,48 kg |
| Prix moyen/kg | 3,5 Dinars |
| Revenu annuel supplémentaire/femme | 150 dinars |
| Augmentation du revenu annuel | 30% |

Sachant qu'une femme pêcheuse à pied gagne en moyenne 520 dinars par campagne, une telle activité augmente son revenu annuelle de près de 30 pour cent. Ce qui lui permet d'améliorer sa situation financière d'une part et de couvrir aisément les frais de sa sécurité sociale durant toute l'année d'autre part.

En effet, un tel scénario est très accepté par la législation tunisienne surtout que les parcelles seront sous la gouvernance du GDP. L'analyse de la spécificité de l'institution, du domaine public maritime (DPM), de la réglementation de pêche ont démontré que la meilleure forme juridique d'exploitation du DPM par le GDP consiste à déposer une demande d'autorisation d'exploitation d'une pêcherie fixe au sens de la législation relative à l'exercice de la pêche et ses textes d'applications.

Cette autorisation d'exploitation d'une pêcherie fixe est subordonnée à la satisfaction de certaines conditions prévues dans la législation en vigueur.

### Conclusion

Suite aux deux essais de grossissement de palourde un manuel de procédures a été développé dans lequel ont été soulignés des aspects devant permettre aux femmes de mieux conduire le grossissement afin de renforcer leurs rôles dans la filière et améliorer ainsi leurs revenus. Ils portent notamment sur :

- La période propice pour les essais de grossissement de palourde.
- La zone de grossissement adéquate.
- La charge adéquate d'ensemencement en termes de nombre d'individus par m2 et de kg par $m^2$ selon la taille des naissains
- Le mode opératoire d'ensemencement.
- Le nombre de femmes intervenant par parcelle : Suivant le nb de kg pêchés par femmes.

Afin de diversifier le revenu des pêcheuses à pied via la vénériculture et de pérenniser leur autonomie, un scénario a été monté par l'équipe du projet pour l'acquisition des parcelles de grossissement de palourde en se basant sur les normes techniques d'élevage acquises lors des essais pilotes et sur les contraintes socioéconomiques de la zone que l'étude diagnostique a révélées.

Le défi de cette approche est de joindre les résultats de toutes les études et les travaux réalisés dans le cadre du projet pour avoir un scénario réalisable qui répond aux critères socioéconomiques de la zone et biologiques de l'espèce sans être loin du cadre juridique Tunisien.

Le scénario proposé consiste à l'installation de deux grandes parcelles (une à Maouma et l'autre à Zaboussa) où la gestion technique et administrative serait sous forme de coopérative gérée par le nouveau GDP. Cette proposition est fondée sur les acquis, résultant du contexte actuel de la filière dans la zone du projet, qui sont essentiellement la présence préalable :

- D'un GDP restructuré
- Des groupes de femmes : chaque groupe est transporté par un seul transporteur, les membres (collectrices) pratiquent l'activité de pêche ensemble et ont des liens de parenté très proche.
- Une volonté de la part des collectrices de diversifier leurs revenus d'une part et améliorer leur pouvoir de négociation via le stockage de leur produit non vendu d'autre part.

L'objectif de ce scénario est d'avoir 18 parcelles (une parcelle pour chaque groupe : 5 à Zaboussa et 13 à Maouma) dont chacune aura une superficie selon la formule suivante :

Superficie de la parcelle = nombre de femmes/groupe x 42 $m^2$

*La durée effective du grossissement pour atteindre la taille commerciale est en fonction de la taille d'ensemencement :*

- Taille comprise entre 25 et 30 mm : maximum de 8 mois.
- Taille comprise entre 30 et 35 mm : maximum 4 mois.

Et tenant compte du fait que la collecte n'est autorisée que pour 7 mois, il est impératif de diviser les parcelles de grossissement en 14 unités selon le schéma suivant :

**Figure 7. Schéma d'une parcelle de grossissement de palourde**

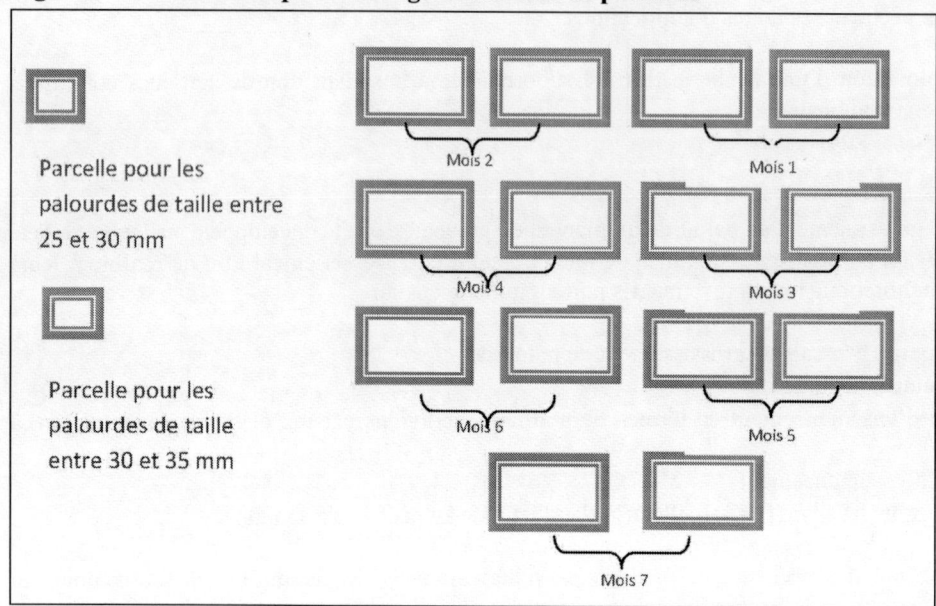

## 4. CONCLUSION ET RECOMMANDATIONS

A travers les études réalisées et les actions menées sur le terrain le projet a pu formuler une série de recommandations aux autorités, agences de développement et tout acteur de la filière palourde, illustrant les conclusions retenues et les potentialités de développement réalisable selon le contexte socio-économique de la filière et des acteurs de production.

Le contexte socioculturel avec les pesanteurs y afférents ont ralenti le rythme d'exécution des activités. De plus le vent de la révolution qui a soufflé en Tunisie à la fin 2010 a contraint à reporter certaines interventions. Malgré cela des acquis conséquents ont été réalisés qui méritent, de même que le processus de mise en œuvre du projet, d'être partagés avec d'autres pays où des communautés de pêche (notamment des femmes) sont confrontées aux mêmes problématiques.

Il est plus que jamais nécessaire de sécuriser ces acquis. Dans l'immédiat, il faut s'atteler au développement de la capacité du nouveau GDP pour qu'il exerce avec efficacité professionnelle son rôle. De même, une assistance directe à l'association des femmes créée s'avère impérieuse. Il convient également de souligner que la durabilité de la ressource est un impératif qui va de pair avec l'épanouissement de ceux qui l'utilisent. Par conséquent, les actions de renforcement du pouvoir de négociation, la redistribution juste et équitable de la rente tout au long de la chaine des valeurs, la promotion du grossissement de palourdes ainsi que l'introduction d'espèces non valorisées (ex : le couteau ; *Solen marginatus)*, et à terme, la mise en place d'un petit centre de purification devraient être considérées avec la plus grande attention.

## 5. REFERENCES BIBLIOGRAPHIQUES

**Belkahia, R.** 1997. «Appui à la filière pêche à pied de la palourde », Direction Générale de la Pêche et de l'Aquaculture.

**DGPA.** 2009. Rapport des statistiques de la pêche en Tunisie de 2005, 2006, 2007, 2008, 2009

**MEDD.** 2008. Ministère de l'environnement et du développement durable, Direction générale de l'environnement et de la qualité de la vie. Projet de l'étude de faisabilité environnementale de la zone de la Skhira : Rapport de la phase III.

**MEDD.** 2009. Ministère de l'environnement et du développement durable, Agence de Protection et d'Aménagement du littoral : Projet de la préparation d'un plan de gestion des îles Kneïss et préparation de sa mise en œuvre : Rapport définitif du bilan socio-économique et environnemental.

**Nouaili, R.** 2006. Contribution à l'étude de la filière et de la pratique de pêche de la palourde en Tunisie. Master de recherche. Institut National Agronomique de Tunisie. P 15-19, P40.

**Siala, W.** 2003. Essai d'analyse de la filière palourde tunisienne et ses perspectives de développement. Projet de fin d'étude. Institut National Agronomique de Tunisie, 118 p.

# A HOLISTIC APPROACH TO RESOURCE SUSTAINABILITY: THE INTERVENTIONS IN LAKE SARDINE FISHERIES IN THE UGANDA PART OF LAKE VICTORIA

## *[UNE APPROCHE HOLISTIQUE DE LA DURABILITÉ DES RESSOURCES: LES INTERVENTIONS DANS LES PECHERIES DE LA SARDINE DU LAC DANS LA PARTIE OUGANDAISE DU LAC VICTORIA*

by/par

Yvette Diei-Ouadi[1], Margaret Masette and Edward Rukuunya

**Abstract**

Interventions based on systematic post-harvest loss assessment have been conducted in the 3 East African riparian countries. In Uganda, upstream and downstream interventions were aimed at fostering the role of Lake Sardine/*Rastrineobola argentea* (called Mukene in Uganda) in contributing to animal protein supply, as a trade commodity and as a key species in the food web for a balanced aquatic ecosystem. Though a clear-cut conclusion is premature at this time, it would seem that the impact of these interventions in Uganda will be positively appreciated along the Lake and have a wider scope, provided some sound follow up measures are taken and due consideration is given to factors such as, (i) the political will for enforcement of standards and regulations; (ii) promotion of an acceptable credit scheme; (iii) a coherent regional approach to effectively tackle the " plague" of the quality-blind cross-border markets and (iv) sharing the documented information on technology viability with the other 2 countries.

*Key words: Lake Victoria, Sardine, Post-harvest, Quality*

**Résumé**

Les interventions basées sur l'évaluation systématique des pertes post-capture ont été effectuées dans les 3 pays riverains de l'Afrique de l'est. En Ouganda, des interventions en amont et en aval visaient à promouvoir le rôle de la Sardine du Lac/*Rastrineobola argentea* (appelé Mukene en Ouganda) dans la contribution à l'approvisionnement en protéines animales, en tant que denrée commerciale et espèce clé dans le réseau trophique pour un écosystème aquatique équilibré. Bien qu'une conclusion sans équivoque soit pour le moment prématurée, il semblerait qu'un impact de ces interventions en Ouganda sera positivement appréciée le long du Lac et sur une portée plus large, sous réserve que soient prises de bonnes mesures de suivi et que soient prises en considération des facteurs tels que: (i) la volonté politique pour l'application des normes et règlements (ii) la promotion d'un schéma d'octroi de crédit acceptable (iii) une approche régionale cohérente pour lutter efficacement contre la "peste" des marchés transfrontaliers insensibles à la qualité et (iv) le partage de l'information documentée sur la viabilité de la technologie avec les 2 autres pays.

*Mots clés: Lac Victoria, Sardine, Post-capture, Qualité*

## 1. INTRODUCTION

*Rastrineobola argentea* (local names Omena/Mukene/Dagaa) is one the 3 main fish species in Lake Victoria, a water body shared by Kenya, Uganda and Tanzania. The importance of fisheries to the economy of these countries cannot be overstated. The Lake fisheries resources contribute about 7 percent of the GDP in each of the riparian states. As an export commodity, the high commercial value Nile perch has been the focus of significant private investments and development progammes in the past 2 decades in the three East African nations. However, recent trends in production show that a decline of this species, has resulted in a shift to exploitation of small fish species namely *Haplochromis* ("Nkejje"), and *Rastrineobola argentea* on Lake Victoria and Lake Kyoga and other small fish species like *Neobola bredoi* ("Muziri") and *Brycinus nurse* ("Ragoogi/Ndolo") from Lake Albert and the Albert Nile. The potential of these small fish is very high in terms of commercial catches in Uganda, contributing 80 percent by weight on Lake Albert and 40 percent on Lake Victoria. At the same time, small fish are the main prey (food) targeted by the Nile perch and other predator species. The small fish are harvested for direct human consumption (40 percent) and form a key crude protein

---

[1] FAO, Products Trade and Marketing Service (FIPM), Viale delle Terme di Caracalla, 00153 Rome, Italy. yvette.dieiouadi@fao.org

ingredient for industrial as well as cottage processed human and animal feeds. This shift however, has not translated into tangible socio-economic development as observed during the boom era of Nile Perch in terms of infrastructure development, private investment and technical know-how, yet the two species play a vital role in domestic and regional trade, rural employment, poverty reduction and sustaining food security of the population of the riparian countries. The small fishery industry is dominated, or employs, many rural women among the fishing community mainly in artisanal processing and trade.

While this contribution cannot be overemphasized, there have been ever growing concerns in the way these two Lake Victoria species have been managed. The poor fishing practices, mishandling and poor processing of catches and market malpractices threaten their sustainability. The case of *Rastrineobola argentea* is of particular concern because of its importance as a food source for the Nile Perch (hence a high risk for the depletion or collapse of this commercial species if exploitation of *Rastrineobola argentea* is not properly managed).

A few years ago FAO in collaboration with respective Fisheries Departments within the region, coordinated a study under the post-harvest loss assessment programme (PHLA). The results of the study indicated that the sardine-like species (*Rastrineobola argentea*) incurred the highest post-harvest losses (Akande and Diéi, 2010) in two of the three riparian countries. Kenya was more interested in tilapia fisheries than omena losses. Figures of 35 percent in Uganda and 20–40 percent in Tanzania were reported, which translated into significant economic loss and some lost opportunities for food availability.

It was therefore fundamental to integrate these sources of information in the plan for interventions along the whole supply chain, including feed processing and marketing enterprises. The findings of the PHLA programme have been strengthened by another study in Uganda by the FBRC (Masette, 2008) in collaboration with the poultry farming sector, which happens to use substantial quantities of the dried mukene in the manufacture of animal feed. This particular study revealed that using mukene-based feed that had been adulterated with foreign matter contributed to a significant loss to poultry farmers as it negatively impacted on the performance of the chicken layers and on the quality of eggs produced. It further concluded that whatever the end-use, mukene must be better handled.

Quite often, addressing post-harvest-related issues has been a standalone and piece-meal initiative with little acknowledgement of its role as a resource management tool, and frequently prescriptive solutions are considered, notwithstanding a stock of key information that often leads to sustainable interventions. As a result, they may be technically effective but not necessarily cost effective and socially acceptable, or might have missed the root causes of post-harvest losses, hence a risk of reoccurrence. The current paper presents the key initiatives in each of the Lake Victoria riparian countries, especially in the Uganda part of the Lake. In considering these aspects a stepwise approach was used in Uganda to address deficiencies for sustainability of mukene as well as other fish resources, like Nile perch, that depend on it for their existence. In addition, small operators depending on mukene for their livelihoods; also contribute to national food and nutrition security (Kikafunda *et al.*, 1998) and regional trade (Odongokara, *et al.*, 2005). It mostly describes the process within the technical cooperation assistance from FAO to the Government of Uganda.

## 2. THE METHODOLOGICAL APPROACH ADOPTED BY EACH OF THE THREE EAST AFRICAN COUNTRIES

This approach stems from the findings of the systematic loss assessments and further initiatives which probed into the multifaceted context of the way Lake sardine is utilized. The lessons learned from these led to the following important points on which decisions in each country were based:

- The most common coping strategy on losses incurred by the fishers was to recoup on subsequent catches/supplies. For example when fishermen incurred a loss on a consignment, the first coping strategy they had in mind was to try and get more catch, thereby increasing the fishing effort;
- While the mukene post-harvest operators incurred most losses owing to the magnitude of mismanagement of the catches, it was also pinpointed that upstream factors, especially in connection with the weak law enforcement measures, contributed to a certain extent to the high losses observed in the sub-sector. The inability to enforce laws perpetuated the use of obnoxious fishing methods and harvesting of immature and juvenile fish species, including some *Haplochromines*.
- Losses were observed at all points of the chain although the magnitude varied in each segment. However, losses during drying were more common, while some additional processes were pinpointed in some other countries. For instance the storage of dried product in Tanzania and packaging in Uganda were also of great concern with widespread adulteration in the latter.

- was a direct link between production and magnitude of losses for mukene processed and packaged for animal feed. In this country the proportion of mukene utilized for fish feed production was abnormally high, which makes it an absolute priority, to shift the trend towards direct human consumption if the objective of significantly curbing post-harvest losses is to be attained. Data from the Fisheries Department (DFR, 2005) indicated that 80 percent of the landed sundried mukene was meant for animal feed (poultry, aquaculture and piggery). The widespread mishandling and adulteration practices have triggered stigmatization among consumers and impacted on the demand for human consumption of this relatively cheap fish, despite the increasing price of meat and other animal products. This leads to a cycle as indicated below (Figure 1) that can only be broken if adequate measures are taken for better handling of Lake sardine:

**Figure 1. Schematized cycle of mukene use in Uganda**

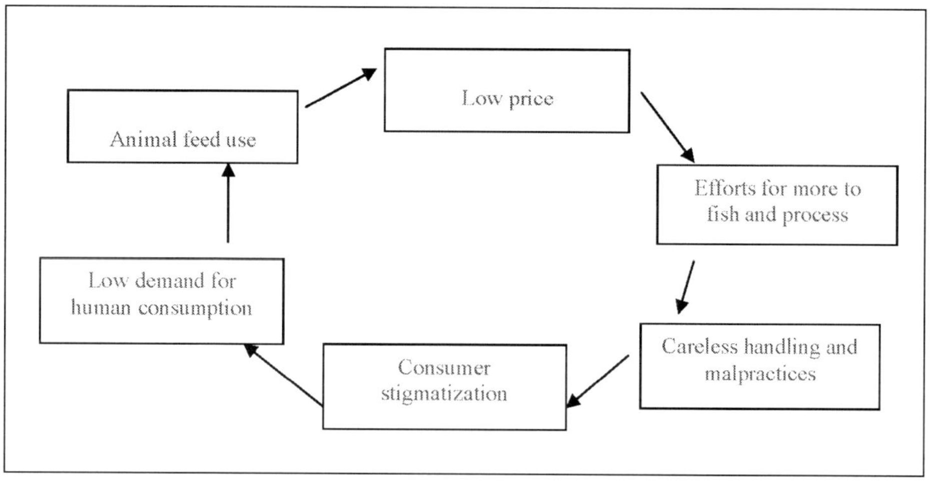

The above information paved the way for several measures – interim or longer term - in the riparian countries:

- Given that Kenya was more concerned with fresh tilapia losses and losses in some specific fishing gears in the Indian Ocean, and taking into account the fact that utilization of omena for human consumption is more advanced than in the other countries, there has been no specific follow up action on this species. Instead, the subsequent in-depth analysis and actions were on tilapia fisheries. This approach is in line with the rationale of the fish loss assessment framework, suggesting that prioritization of loss reduction interventions within limited/scarce development resources be based on concrete data.
- In Tanzania the loss assessment report has been instrumental in triggering the assistance from a bilateral cooperation agency to provide value-adding machinery. Storage tests to determine the best shelf life for less rancidity and without significant change in colour of dagaa was proposed to be carried out.
- Given the overwhelming concern in the Ugandan part of the Lake, as presented during the Jinja workshop jointly organized with Lake Victoria Fisheries Organization (LVFO) and which marked the end of the regional PHLA programme, the Government of Uganda requested technical assistance from FAO to address the issues facing the mukene fishery. The project "Increase the supply of *Rastrineobola argentea* for human consumption" was then approved and started in April 2010.

## 3. THE TECHNICAL COOPERATION PROGRAMME TO ASSIST UGANDA IN TACKLING SUSTAINABILITY OF MUKENE RESOURCES

The main purpose of the project "Increase supply of mukene (*Rastrineobola argentea*) for human consumption" was to assist the government in preparing the policy, strategy and management plans for mukene fisheries in order to comply with LVFO directives and to obtain maximum economic benefits within a sustainable fisheries framework. While dealing with post-harvest issues affecting this fishery, it also addresses upstream (fishing) as well downstream (market, feed manufactures and use) segments.

The participatory process during the inception phase then visits to key stakeholders generated further information based on the following:

1. feed manufacturers were increasingly concerned about the poor quality of mukene offered on the market, which nonetheless, destroyed their feed processing equipment. This has led to some poultry farmers to

switch to substitution of mukene products by soya pellets; mukene use in fish feed is, however, still a critical requirement;

2. there is a good demand, especially in regional untapped markets for "clean" mukene, but in most instances the small scale fishers were incapable of meeting it. This was not only linked to their weak technical capacity, but also the limited resource to invest in raised racks for improved drying and the fact that they are caught in a poverty trap with advance payment from financiers which prevented them from selling at the best price. A case with a medium scale processor and exporter of dried "clean" mukene products to the regional markets is self explanatory of the latter illustration. This operator who cannot meet the high regional demand engaged unsuccessfully into partnership negotiations to buy from the fishermen the fresh mukene at the market price of dried product. This was simply because these fishers had already committed themselves through loans from financiers, to repay with products based on the price imposed by those financiers. Interestingly it was planned within the project that the Government will mobilize the savings and Credit Cooperative Organizations (SACCO) so that the fisherfolk and processors can access credits.

The main upstream measure has been the introduction of catamaran technology to transfer some fishing effort from the shallow areas, where there is a considerable unsustainable by-catch of juvenile fish (immature mukene, Nile perch, tilapia and Haplochromines) known to be around 75 percent (Legos and Masette, 2010), to offshore. In addition, generation of relevant scientific information including promulgation of mesh size regulations and involvement of fishers through BMUs in the enforcement process were also introduced. Pilot demonstrations with proven catamaran technology were conducted with local boat builders trained in the construction of these vessels according to the best practices. The economic viability of this technology was assessed. While effort is being put into the design of canoes and more selective fishing gears, it is expected that Government also acts firmly for law enforcement in order to discourage the use of chemicals and harmful methods of fishing.

**Picture 1 Testing of a catamaran boat constructed**

With regard to utilization of mukene, interventions focused on improving existing processing and developing new products/value-added products, including assessing economic and operational feasibility, conducting market testing of economically viable products and their promotion in an effort to establish commercial partnerships.

Several products were developed from mukene as a strategy to increase the supply for human consumption. The new products included **powders** flavoured with local ingredients like ginger, lemon and onion, **chips** also locally flavoured, IQF and bulk frozen, deep-fried, sesame coated , sweetened and deep-fried. In addition, trials to cater for the feed industry were also conducted which included production of forms of silage using acid or enzymatic processes.

**Pictures 2, 3 and 4. Examples of new mukene-based products developed (powder, deep fried and chips)**

Using standard methods of determining shelf-life of fishery products, including chemical, microbiological and sensory testing, the keeping quality of economically viable and socially accepted products was established. The deep-fried whole mukene lasted for 3 months while the powdered version lasted for only two months at room temperature (28 °C).

One other important area where the project planned interventions is the packaging materials, currently being addressed and standardization of the traditional/local measurements of dried mukene.

The market test was evaluated on various mukene-based products which included crisps, powder and pancakes with salt only and similar products sweetened with sugar. Apart from general acceptability of the products, other sensory variables like appearance, sweetness, saltiness, bitterness, flavour, taste, size, packaging and price were tested. The products were market tested at the two project landing sites of Kiyindi in Buikwe District and Kasekulo in Kalangala district. In addition, six peri-urban centres including some primary schools around Kampala and Makerere University were included in the study. Results indicated that product acceptability varied with type of product, although salted products were more acceptable to the majority of consumers than similar products that had been sweetened with sugar. However, primary school children with ages varying from 5-12 years liked almost all the products regardless of colour, texture and taste. Products with potential commercial viability included most salted versions and powders. Other products, like crisps, required further improvement to enhance acceptability among consumers.

Draft standards and regulations were set in preparation for formulation of policy guidelines. Some of these efforts were in support of the feed manufactures to ensure sustainable supply of high quality product, but again the success will definitely depend on the actual implementation of these guidelines and sanitary regulations regarding the products standards.

Inclusion of extraneous matter in raw materials, either intentionally or accidentally, has negative consequences on feed formulation, which in turn affects the nutritional health of the animal and the quality of products e.g. egg shell. In the food retailing sector of Europe and America, adherence to set sanitary and regulatory protocols ensures uninterrupted supply of high quality fishery products from third countries. So far, it has worked well for Nile perch and Nile tilapia products from Uganda. Emulation of similar guidelines for the mukene fishery, intended for either human consumption or animal feed production, would undoubtedly improve the way mukene is handled and processed. It is envisaged that with improved handling and processing protocols, the quality of the corresponding mukene-based final products will improve, as well as the income of the fish operators. The current sanitary status of the mukene distribution chain from fishing ground to market outlets is abysmal. The proposed guidelines are designed to involve all stakeholders along the value chain.

## 4. CONCLUSION AND RECOMMENDATIONS

The process that led to the interventions in the Lake sardine fishery can be food for thought in addressing problems facing other fisheries with similar interconnected social, economic and environmental interests at stake. Not only has it helped better understanding but most importantly has been instrumental in shaping the actions in each of the 3 East African countries.

The approach in Uganda deserves attention as it is resource management-centered. Even if it is too early at this stage to sense the impact of the interventions within the Uganda TCP, it would seem that wider benefit can be derived from it, given its integration of upstream and downstream measures. The choice available to fishers in terms of technologies and markets fostered by the economic viability should strengthen the intent of making mukene an important trade commodity, as it remains a key species for the food web and stability of the aquatic ecosystem. Uganda's multispecies fishery dominated by many small fish species has a lot to benefit from this intervention and many lessons have been learnt. Therefore the stakeholders within and beyond the Lake sardine fishery will perceive the interests at stake. Improved long term returns would depend on a set of conditions:

- The political will for enforcement of standards and regulations is set up
- The promotion of acceptable credit schemes by SACCO for adoption by fishermen and processors of the catamaran technology and the products developed and also to free them from the tight arrangements with financiers that prevent fair pricing of their products;
- A regional approach to effectively tackle the "plague" of the quality-blind cross-border markets. There is need for a concerted process among the food/fish inspection and trade promotion services of buyer and seller countries to develop a sound roadmap to put an end to any practice which undermines quality promotion efforts.
- The use of catamarans in the EAC region: the viability data need to be shared with Kenya and Tanzania so as this technology can be a reference in Lake Victoria if sustainability of fishing operations is to be met.

# 5. REFERENCES

**Akande, G.R. & Diei-Ouadi, Y.** 2010. Post-harvest losses in small-scale fisheries, Case studies in five sub-Saharan African countries. *FAO Fisheries and Aquaculture Technical Paper* 550.

**DFR.** 2005. Annual report and background to the budget.

**Kikafunda J.K., Walker A.F., Collet, D. & Tumwine, J.K.** 1998. Risk factors for early childhood malnutrition in Uganda. *Journal of American Academy of Pediatrics.* 102    (4) p. 45

**Legos, D. & Masette, M.** 2010. IND017UGA - Testing of different processing methods for Mukene for human consumption and fish meal in Uganda.
http://sfp.acp.int/en/allprojects?title=mukene&term_node_tid_depth=All&term_node_tid_depth_1=78

**Masette, M.** 2008. The influence of Dagaa-based feed on quality of chicken egg production within L. Victoria Basin. Paper presented at *FAO Workshop on Fish Technology, Utilization and Quality Assurance in Africa* held at Agadir, Morocco 24–28 November 2008.

**Odongkara K., Kyangwa, M., Akumu. J., Wegoye, J. & Kyangwa, I.** 2005. Survey of the regional fish trade. *LVEMP Socio-economic Research Report* 7. NARO-FIRRI, Jinja.

# POTENTIAL SOCIO-ECONOMIC BENEFITS OF POST-HARVEST FISH TECHNOLOGY PLATFORMS (PHFTP) APPROACH IN SMALL-SCALE FISHERIES DEVELOPMENT

## *[POTENTIELS AVANTAGES SOCIO-ECONOMIQUES DES PLATEFORMES DE TECHNOLOGIE POST CAPTURE DANS LE DEVELOPPEMENT DES PETITES]*

by/par

Yahya I. Mgawe[1], Isaac Flowers and Eric Kekula

## Abstract

Small-scale fisheries are a major source of income, employment and food fish supply in many countries in sub-Saharan Africa, and have great potential to contribute more towards the achievement of Millennium Development Goals (MDGs). However, almost fifty years have passed since most of the countries in the region gained their independence. Since that time different stakeholders have stressed the importance of improving this sub-sector towards a competitive "industry", but the situation has not improved much.

Different approaches and development initiatives that have been tried to address key issues in small-scale fishery have rendered mixed results. In most cases, the efforts have been piece-meal interventions focusing on isolated segments of inter-woven challenges. This has resulted in producing poor dividends in the value chain. Based on experience, a key challenge in small-scale fisheries development is organizational, a factor that hampers any effective technical assistance.

In view of this situation, there is now a good possibility of promoting a Post-Harvest Fish Technology Platform (PHFTP) approach. This is an integrated approach drawing on the Community Fishery Centers (CFC) concept. The approach focuses on creating an early action or entry strategy to mitigate the inter-woven challenges in the industry. The presupposed benefits include provision of an effective organizational platform for dealing with issues such as: fusion of improved technologies, infrastructure, sources of finance and market dynamics. It is also a platform for building the institutional capacity of fishers' organizations to meet their development challenges on a sustainable basis.

The underlying philosophy is that effective early action would enhance the interest of fishers to cooperate in dealing with their common challenges and thus trigger broad small-scale fisheries development. The experience with regard to employing the PHFTP approach in Liberia has provided basic frames of reference for its substantial appraisal, and thus has been used in this paper to articulate the functioning and potential social-economic benefits of the approach.

***Key words: Post-harvest Fish Technology Platform (PHFTP), Fishery Based Organization (FBO), Micro-finance, Strategic marketing, Physical evidence, Servicescape, Competitive advantage***

## Résumé

La pêche artisanale est une importante source de revenu, d'emploi et d'approvisionnement en nourriture de poissons dans de nombreux pays de l'Afrique Sub-saharienne, et a un grand potentiel pour contribuer davantage à la réalisation des Objectifs du Millénaire pour le Développement (OMD). Cependant, presque cinquante ans se sont écoulés depuis que la plupart des pays de la région ont acquis leur indépendance. Depuis ce temps, différentes parties prenantes ont souligné l'importance de l'amélioration de ce sous-secteur vers une "industrie" compétitive, seulement, la situation ne s'est pas beaucoup améliorée.

Différentes approches et initiatives de développement qui ont été tentées pour aborder des questions majeures de la pêche artisanale ont donné des résultats mitigés. Dans la plupart des cas, les efforts déployés ont été fragmentaires, avec des interventions axées sur des segments isolés de défis entremêlés.. Cela a conduit à de maigres dividendes dans la chaîne de valeur. Par expérience, un défi majeur au développement des petites pêches est organisationnel, un facteur qui entrave toute assistance technique efficace.

---

[1] Ministry of Livestock & Fisheries Development, (Fisheries Education & Training Agency), PO Box 83, Bagamoyo, United Republic of Tanzania. ymgawe@yahoo.com

Compte-tenu de cette situation, il y a maintenant une bonne possibilité de promouvoir une approche Plateforme Technologique post capture(PTPC). Il s'agit d'une approche intégrée s'appuyant sur le concept des Centres Communautaires de Pêche (CCP). L'approche met l'accent sur la création d'une action précoce où d'une stratégie d'entrée afin d'atténuer les défis entremêlés dans l'industrie. Les avantages présupposés comprennent la mise à disposition d'une plateforme organisationnelle efficace pour traiter les questions telles que: la fusion des technologies améliorées, l'infrastructure, des sources de financement et de dynamiques de marché. C'est aussi une plateforme de renforcement des capacités institutionnelles des organisations de pêcheurs pour répondre à leurs problèmes de développement sur une base durable.

La philosophie sous-jacente est qu'une action précoce efficace permettrait de renforcer l'intérêt des pêcheurs de coopérer pour faire face à leurs problèmes communs et donc de déclencher ainsi un grand développement des entreprises de pêche artisanale. L'expérience en ce qui concerne l'utilisation de l'approche de la PTPC au Libéria a fourni des références de base pour son évaluation importante, et donc a été utilisée dans ce document pour expliquer le fonctionnement et les potentiels avantages socio-économiques de l'approche.

*Mots clés: Plateforme Technologique post capture (PTPC), Organisation paysanne de base (OPB), Microfinance, Commercialisation stratégique, Évidence physique, Facilités de Services, avantage compétitif*

## 1. INTRODUCTION

Fisheries are important to the economy of Liberia contributing about 3.2 percent to the GDP and they provide a means of employment and livelihood for over 33,000 people who are engaged on a full-time basis and hundreds of thousands more on allied industries. Also, fish is a cheap source of animal protein, income and a potential source of foreign exchange.

Over 60 percent of the total domestic fish catch is landed by the small-scale fishery, with landing sites distributed throughout the country, especially along the coastline. The small-scale or artisanal fleet comprises about 3500 canoes of different sizes but dominated by small dugout canoes commonly known as *Kru* canoe which can take 1–3 fishers onboard, and other relatively big planked canoes that can take up to 15 crew known as *Fanti* canoes. The former is propelled by oars and or sail using mostly hand-lines whereas the later is powered by 15–45 hp outboard engines using gillnets, ring and purse nets.

As in the case of many other developing countries, collection of fish landing data from small-scale fisheries in Liberia is difficult, however, it has been estimated that landings from the sub-sector range between 15,000–20,000 tonnes/year.

Most of the fish landed by small-scale fishers is processed in local communities. Processing is mostly limited to fish smoking by using simple oil drum and cylindrical mud ovens. The practice causes massive fuelwood utilization and is thus becoming an environmental threat in terms of mangrove depletion and coastal forest clearing. Also, the oven causes increased operational costs to small-scale fishers as the wood is purchased from local dealers.

Failure to use ice in fish handling, both at sea and ashore, contributes to high quality loss in the Liberian small-scale fishery. Although fishers try hard to salt-dry most of their spoiled fish, which could be over 30 percent of total fish landing, the price for such product, mainly used as condiment, is very low. Generally, existing bottlenecks in the supply chain cause high post-harvest fish losses which threaten food security and hinder fish operators from securing greater post-harvest benefits from their rich fishery.

Again, just like in many other countries in sub-Sahara Africa, the small-scale fishery sector in Liberia is faced with other challenges including; low levels of technology, inadequate capital associated with lack of credit facilities, poor extension services and marketing bottle-necks. These kinds of challenges can hardly be addressed by using a single solution approach, rather the situation calls for integrated or holistic solutions.

In view of the prevailing situation, the Liberian government decided to implement a comprehensive project aimed at addressing most of the challenges in the sub-sector under the project entitled Food Security through Commercialization of Agriculture (FSCA). The project is mainly being funded by the Italian Government with technical support from FAO. The goal of the FSCA project is to develop the agricultural sector (which includes fisheries) into a modern and commercially vibrant sector.

The specific objective of the project is; *"Agricultural productivity, marketed output and incomes of project beneficiary Farmer Based Organization (FBOs) increased on a sustainable basis, resulting in improved livelihoods and food security of FBO members".*

Indeed, the FSCA project has implemented a number of activities to address the situation. These have included promotion of the post-harvest fisheries sub-sector by using the integrated PHFTP approach, which draws from CFCs' guidelines (Ben-Yami and Anderson, 1985) and experience obtained from similar interventions in Chad (Ndiaye and Diei-Ouadi, 2008).

The initiative centered on enhancing fishers' participation to address pertinent issues in Liberian small-scale fisheries including; organizational development, dissemination of improved technology, infrastructure, source of finance and capacity building for sustainable livelihood.

## 2. OBJECTIVES

The main objective of the study was to appraise the PHFT approach, as a continuation of the quest for searching effective strategies in addressing multiple challenges in small-scale fisheries. Specifically, the study focused on the following aspects:

- To identify major challenges faced by small-scale fishers in Liberia, especially those engaged in post-harvest sub-sector.
- To introduce and appraise potential benefits of the PHFT approach.
- To outline requirements for successful application of the PHFTP approach under similar situations to that of Liberia.

## 3. METHODS

This study was based on qualitative research methods, which included review of available literature, semi-structured interviews (check-list on Appendix 1) and direct observation through participation during implementation of project's activities in communities by using the PHFTP approach.

*Study site*

The study took place at Banjor Beach in Montserrado County and in Grandcess City in Grand Kru County (Figure 1). Montserrado, which includes the capital City Monrovia, is the most populous county in the country providing the largest domestic fish market whereas Grand Kru located in the Southern part of the country enjoys a relatively rich fishery but with poor infrastructure, constraining fish marketing.

**Figure 1. Map of Liberia showing study sites**

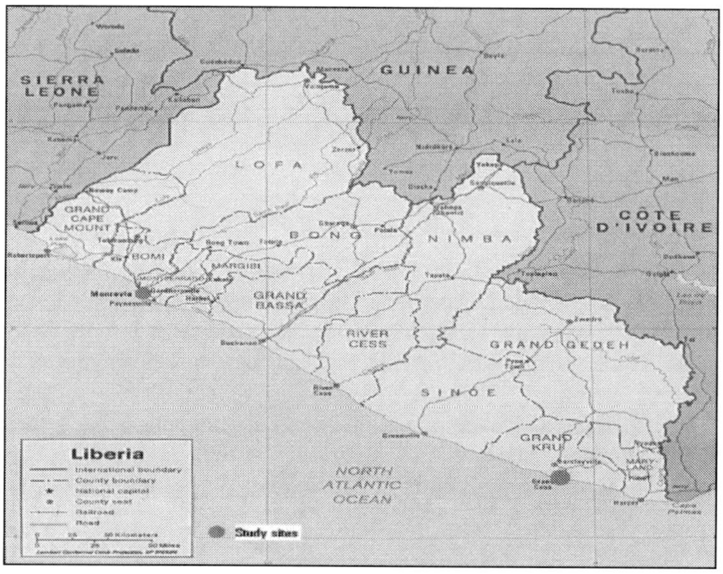

*Number of respondents*

The study involved 168 fishers representing about 70 percent of FBO members (Table 1).

**Table 1. Names of FBOs and total number of their members**

| No. | Fishery Based Organization (FBO) | Location | Members |
|-----|-----------------------------------|----------|---------|
| 1. | Good Father Fishery Based Organization (GFFBO) | Grand Kru | 38 |
| 2. | Ma-nya-nubo FBO | Grand Kru | 35 |
| 3. | God Will Provide Fishery Based Organization | Grand Kru | 40 |
| 4. | Katarward Fishery Based Organization | Grand Kru | 42 |
| 5. | Nimean Fishery Based Organization | Grand Kru | 24 |
| 6. | Banjor Beach Fishery Based Organization | Montserrado | 38 |
| 7 | King Gray Fishery Based Organization | Montserrado | 24 |
| | **TOTAL** | | **241** |

*Source:* FSCA project documents.

## 3. RESULTS

The results have been presented in three broad categories; challenges faced by small-scale fishers, potential socio-economic benefits of PHFTP and requirements for successful introduction of the platform.

*Challenges faced by small-scale fishers*

Challenges in small-scale fisheries in Liberia are mostly the function of the prevailing business environment. As such this could be either general business environment or the operational environment of the business.

**General business environment**

The general business environment for industry includes trends in resource availability, policies and politics, economics, social, technological, legal aspects and other such factors that could have a strong influence on the performance of fishers.

With regard to the general business environment in the case of SSF in Liberia; the situation could be said to have improved over the past six years following the end of the civil war and successful installation of a democratic government. As a matter of fact, the Liberian policies and economic reforms are very supportive when it comes to SSF development. This is happening at the time demand for fish in the domestic, regional and global market is increasing exponentially as a function of increasing population and increased awareness of nutritional superiority of fish over food from other types of animals. However, the main challenge in this context is how to develop the industry especially in terms of accessing rewarding markets.

**Operational environment**

The challenges associated with the operational environment seem to be more pronounced since these are of immediate concern compared to the ones associated with general business environment. The traditional challenges under this category include lack of capital associated with inadequate credit facilities, low level of technology, lack of entrepreneurial skills, poor and high cost of transport, lack of appropriate infrastructure to support the sector and marketing dynamics.

Indeed, several countries in the region have attempted to address these challenges through adoption of various development strategies, including the cooperative approach; however, the result is yet to be laudable. The major problem has been the fact that in most cases, the efforts have been piece-meal interventions focusing on isolated segments of interwoven challenges in small-scale fisheries. In view of this situation, there is now a good prospect of promotion of the PHFTP approach, already experimented in Africa (Ndiaye and Diei-Ouadi, 2008).

The PHFTP is an integrated approach for small-scale fisheries development that aims at resolving many of the challenges in general business and operational environments. The approach is considered to be more effective in mitigating the interwoven nature of the challenges because it helps to build organizational and technical capacity of fishers in communities.

*Potential socio-economic benefits of the PHFTP approach*

The PHFTP approach is an integrated platform for building institutional capacity of fishers' organizations to meet their development challenges on a sustainable basis. It is based on the premise that many of the challenges faced by small-scale fishers, especially from fish production to marketing, can be successfully confronted if individual fishers have an adequate organizational support base and are backed by entrepreneurial skills and technical support.

The PHFTP approach stems from the Community Fishery Centre (CFC) concept developed by FAO way back in 1970s. Indeed, various models of CFC have been employed to promote the commercialization of fisheries in rural areas in many countries in sub-Saharan Africa and beyond. Examples may include; The Gambia (Tankur and Kemoto artisanal cooperatives), Ghana (Fishermen service centers) and fishers cooperative societies in many other countries. In most cases, these CFC centers focus(ed) on providing services such as ice making, mechanical repair of fishing boats and engines, a collection points for fish transport to markets, and a base for fisheries extension activities. Likewise, they had secondary objectives of improving cash incomes, slowing rural urban drift, and enhancing supply of food fish.

In many countries, however, the traditional CFC models, centered on providing various physical facilities and services, have not delivered the expected returns. At most, they have proved to be overcapitalised, under-utilised, economically non-viable, providing only minimal benefits to village communities and incurring excessively high production and marketing costs.

Learning from the dismal performance of many CFCs across the region, there was an obvious need to refine the strategy, hence development of the PHFTP approach. This is a rather flexible intervention focusing on building participatory organizational capacity to deal with issues of common interest to fishers. In the case of Liberia, the organizational development initiative is focused on entrepreneurial skills, fusion of improved technologies, infrastructure, and source of finance.

The thrust is in building the capacity of individual fishers within fisher-based organizations. While it could be premature to articulate concrete the commercial viability of the intervention noticeable achievements have been made in many socio-economic facets as follows:

## Organizational development

Various training programmes were developed and implemented that aimed at building the institutional capacity of FBOs. Currently, the FBOs have been able to organize themselves and exhibit basic managerial skills, based on their plans and by-laws. Meetings are convened and important decisions made and implemented. In some communities, the FBOs have gone to the extent of establishing networks or clusters where two or more FBOs work together to mitigate their common challenges. Certainly, this development has improved social cohesiveness in communities, which is a very important livelihood asset in rural communities.

## Entrepreneurial skills

Building of entrepreneurial skill was another area that was facilitated through the intervention. The main thrust was placed on developing business plans as the most important tool or road map which ensures the success and sustainability of any kind of enterprise. An exemplary excerpt from Grandcess's GFFBO management plan is as follows:

**Table 2. Excerpt from Grandcess's Good Father Fishery Based Organization**

| No | Components of a business plan | Excerpts from GFFBO business plan | |
|---|---|---|---|
| 1 | Executive summary | A short stand alone section which summarize each section of their business plan | |
| 2 | Vision and mission | *Vision* | To become lead supplier of fish in Liberia |
| | | *Mission* | To provide excellent fish handling, processing, storage and marketing services to fishers in communities |
| 3 | Business description | Their business is at an early stage of development but the FBO banks on their combined practical knowledge and long history of experience in fishing industry. Also, being located near Guinea and Cote d' Ivoire they believe they can easily capture the regional fish market through cross border trade. | |
| 4 | business environment analysis | Political stability, economic development and increased demand for fish in domestic, regional and international market coupled with improved infrastructure and communication systems provide conducive conditions for promoting fish business in Grand Kru county. | |
| 5 | Business goal and objectives | *Goal* | Secure improved sustainable livelihood in community |
| | | *Objective* | Increase income of FBO members |
| | | *Output* | Increased quantity of good quality fish produce at minimum cost |
| | | *Activities* | Construct oven base and trays of standard Chorkor ovens in community |
| | | *Inputs* | Clay mud, water, wood, chicken wire, plywood, carpenter's tools |
| 6 | Product and services | The FBO plan to use the PHFTP in providing ice and fish storage space at a fee. Also, to provide marketing service by collecting/purchasing fish from members and sell it collectively before sharing the revenue proportionally. | |
| 7 | Market strategy/plan | They have segmented the market and targeted a few locations, including in Monrovia and other major towns where populations are growing rapidly. Also, the regional market through cross-border trade. Later on they will procure a vacuum packing machine targeting the increasing number of super markets. The market mix elements including Product, Pricing, Placement and Promotion have been analyzed in the plan. They have noted that developing services without any clear cut means of selling those goods and services would not help. Hence, they have decided to appoint sales agents in major towns. | |
| 8 | Technology | The FBO is banking on PHFTP as a infrastructural advantage providing physical evidence and service scope. The facility includes improved smoking ovens, ice production and insulated containers for fresh fish handling and storage. | |
| 9 | Competition and competitive advantage | Stiff competition is not eminent given the rich Grand Kru fishery and the willingness of surrounding communities to also use the PHFTP at Grandcess. At most the FBO has positioned themselves to collect increased quantities of fish from its neighbours to potentially increase their market share and benefit. | |
| 10 | Operation plan | The men do the fishing whereas women members of the group handle, process and market the product. They also purchase fish from other fishers for value-addition before selling out to target markets. The management team is responsible for providing and supervising services being provided by the PHFTP. | |
| 11 | Financial, projection and use of proceed | Sources of revenue include; fish sale, sea transport service, selling of ice, interest from extended credit facilities, shares, storage and marketing services. | |
| | | On the other hand the costs include; fuel for generator and boat engine, engine oil, maintenance, purchase of fish, contribution to community development, labour and depreciation | |
| 12 | Management plan | Organization chart, elaborative by-laws and decision-making procedures are articulated in the plan. | |

*Source:* GFFBO business plan.

In the example of a business plan presented above, it could be noted that fishers consider the establishment of their FBO as a conduit for service provision. Profits from services being provided through their joint activities are apportioned within the community, effectively providing a form of social welfare for members.

## Fusion of technologies

With regard to fusion of technologies the PHFTP approach has proved to be an effective education and training instrument where fishers can access continued learning. In the case of Liberia, a number of training workshops have been organized cutting across multidisciplinary fields including technical, administrative and business areas. The effort has helped to build FBOs' institutional capacity to meet their occupational challenges in the process of promoting community development.

A vivid example in this context is the dissemination of improved technologies such as *Chorkor ovens* and *insulated containers* in Liberia. These technologies were innovated over twenty five years back by the Food Research Institutes in Ghana and Senegal respectively, but could not effectively reach fishers in Liberia until the PHFTP approach was employed.

## Infrastructure

As part of the intervention, four fish handling and processing buildings have been constructed in four different communities for joint use in meeting the changing market demands (Figure 2). The facilities include; water supply (boreholes), fish handling, processing and storage and display rooms as well as small (0.5 tonnes/24 hours) ice producing machines. Also, they possess washrooms, offices, and meeting rooms. The plan is to fence the compound at a later stage in order to protect the facility from animals, pests and unauthorized persons.

**Figure 2. Generalized front view of PHFTP**

Through provision of the facilities communities have got access to important services such as reliable water supply for their hygienic fish handling and domestic use generally improving standards of living.

Specifically, the facilities were meant to be an entry point for building up organizational cohesiveness among fishers in communities. Members in communities get access to services being provided by the facility, at a fee. The money accrued is ploughed back to sustain and expand the facilities.

Likewise, given the physical surroundings of the facility, where the chain approach is used, with fish being handled, processed and stored creates an appealing physical environment for improved delivery of quality fishery products and other services. Hence, fishers could use the physical evidence of the platform as a brand strategy to create competitive advantages when it comes to targeting rewarding markets.

## Sources of finance

Another important aspect in terms of socio-economic benefits of the PHFTP approach is in mitigating the problem of inadequate capital and lack of credit facilities in fishing communities. In Liberia, the PHFTP has also demonstrated the potential of the approach as a unifying driver, when it comes to sourcing finances and cultivating a saving culture.

For example, after undergoing training on basic accounts and basic business management skills the FBO at Grandcess, the GFFBO, managed to improve their operations. They managed to procure a canoe, 25HP outboard engine and assortment of fishing gear in addition to meeting other variable costs in one operational year. The variable or operational cost for GFFBO includes; fuel, food, depreciation for canoe, engine and fishing gears. Also, it includes payment for fishers on duty. All these happened after their original canoe capsized and was lost in the Atlantic Ocean.

Currently, this FBO has embarked on a project to construct decent houses for members in line with their expressed objective of attaining improved sustainable livelihoods. Indeed, their financial performance leaves no doubt with regard to their ability in realizing their objective (Table 3).

**Table 3. Gross profit of GFFBO**

| Month | Gross profit |
|---|---|
| July 2010 | 17 050 |
| August 2010 | 50 195 |
| September 2010 | 22 400 |
| October 2010 | 73 400 |
| November 2010 | 2 000 |
| December 2010 | 90 000 |
| January 2011 | 32 500 |
| February 2011 | 100 100 |
| March 2011 | Capsizing of canoe |
| April 2011 | 30 400 |
| May 2011 | 29 050 |
| June 2011 | Off-season for their type of gear |
| **TOTAL** | **LD 447 095** |
| **TOTAL** | **US$ 6 387 071** |

*Source:* GFFBO Board report, July 2010 to June 2011.

Similarly, women empowerment is another potential benefit of introducing PHFTP in communities. Women are often hardest hit as they struggle with the burden of trying to cope financially at the household level but with the PHFTP in place it has been proved that they can come together and use the common service to become socio-economically empowered.

In most of the FBOs in Liberia women form over 40 percent of the members and their active participation can easily be visualized. For example, the FBO at Banjor Beach is being led by a woman whereas in Grandcess women FBO members take charge of all fish being landed by their male colleagues on credit for processing before they can submit the revenue at an agreed price.

With this kind of mutual support women have been able to support their children's education and provide food and income to their families. In the process they are empowered to reduce vulnerabilities such as extreme poverty in communities, child mortality as well as combating HIV/AIDS and other diseases in line with MDGs.

When it comes to market dynamics, the PHFTP approach is very helpful in synchronizing the traditional marketing mix (4Ps) for goods (Product, Price, Placement and Promotion) as well as the 3Ps for services (People, Physical evidence and Processes), creating a competitive advantage.

Collectively, they would be able to enhance their bargaining power realizing high prices for their fish. In terms of promotion, the physical evidence coupled with word of mouth can form a basis of promoting their product, even before engaging in conventional means.

On placement (distribution), having a PHFTP could lead to reversible selling, whereby increasing number of customers would come to the community instead of the other way round, reducing hurdles associated with selling to distant customers.

Similarly the platform caters for the "People" aspect of the service marketing mix. In most cases, the group would appoint those of with basic entrepreneurial skills to handle the marketing function as service spanners (front liners) and thus buffer the problem of inadequate skill among other group members. On the other hand, documentation of operational processes, which is one of the important characteristics of PHFTP, would help to build FBO's image for delivery of good service to customers.

In conclusion, the PHFTP approach provides an improved working environment and more important, is an easy way of bringing fishers to work together in overcoming their common challenges. It helps them communicate with customers, control production of good quality fish products and empower fishers to continually seek improvement in their work to serve changing customer demands. All these would enable fishers to target rewarding markets and capture a large market share to secure greater post-harvest benefits for improved sustainable livelihoods.

### *Requirements for successful introduction of PHFTP*

Based on lessons learnt from Liberia, it is rational to expect that the same benefits could be experienced in other developing countries especially in sub-Saharan Africa, where the quest to boost economic prosperity cannot be overemphasized. But an obvious question arises, what are the requirements for successful application of the PHFTP approach?

Quite often small-scale enterprises, fishers included, face distinctive problems based on the type of trade they are involved in. Hence, existing problems in respective communities have to be identified before further steps are taken with interventions.

In the same frame of mind, before the introduction of PHFTP in a community it is recommended that a participatory situation analysis be conducted to identify problems and potential solution. The analysis should also aim at identifying potential stakeholders who are ready to use the PHFTP productively and on sustainable basis.

The situation analysis has to be followed by the preparation of a strategic plan pointing out the vision, mission, goal and objectives of stakeholders. It should contain a critical analysis of the business environment relative to their strengths, weaknesses, opportunities and threats (challenges). Similarly, the plan has to point out explicit strategies to be employed together with an outline of activities to be performed.

Once the introduction of the PHFTP in a certain community has been analyzed and its feasibility established, then the process can continue taking note of the following critical points.

### Group formation and dynamics

Group formation and dynamics is a key to ensuring the success and sustainability of an FBO and their PHFTP. A group usually starts when people face a common need and decide voluntarily that the best, or the only, way to satisfy that need is for them to join together and pool their own resources, and thus own, control and patronize their own enterprise. The central concept underlying the formation of the group is that of self help. Group formation, given time, eventually develops into a cooperative society.

Probably the most important question is: who comes together to form a group? People having the same interest, problem and need come together to solve their problems. FBO implies: "together we stand divided we fall" or unity is strength.

Based on the Liberian experience, it is better to start a group with a few reliable people than to start with a larger number containing many doubtfuls. Again, those forming a group should avoid making extravagant promises to prospective members about what the government or NGOs are going to do for them. They should not make it seem as if group formation is going to work magic in their lives without their active participation.

With regard to group dynamics, this requires setting a clear goal and ensures group maintenance. The goal of the group should be set by members rather than being imposed. This is the only way to ensure that activities are directed towards achieving the objectives. For example, the goal of the Good Father Fishery Organization (GFFBO) in Grand Kru County is to improve the sustainable livelihoods of members. In line with this goal, one of their immediate objectives is construction of "zinc" roofed houses for each member in six years time. In the meanwhile they are busy rallying their efforts to get there.

Likewise, the group should be maintained through effective administration in such a way that members would feel comfortable belonging to the group. The group maintenance should aim at creating an atmosphere of mutual trust, respect, free opinion and democratic space. Indeed, meetings have proved to be a very important aspect in the organization and running of FBOs. It is through meetings of members that groups have been controlled.

It must be emphasized that financial reporting is one of important items during meetings. Although most FBO could face difficulties in coping with the relatively complex double entry system of accounting and reporting, even a simple income-expenditure form of reporting tends to help. Through simple system of bookkeeping and reporting the FBOs in Liberia have managed to calculate gross profit (Table 3) and monitor their cash flows. Regular financial reporting builds trust and helps in upholding group harmony.

In organizing meetings, members should be involved by encouraging many of them to speak their opinion. That is the best way to have democratic participation in decision making and to get the benefit of everyone's views on important matters. In the case of Liberia, women have been encouraged to take key positions in FBOs in order to ensure their effective participation and enhancing gender balance. However, it ought to be noted that meetings should be arranged in such a way that members' precious time is not wasted.

Rules and regulations is another critical element in fostering group dynamics, all FBOs in Liberia have set their own rules and regulations to guide the running of their organizations. These rules though not uniform across the FBOs generally follow basic guidelines stipulated by the government through the department of Cooperatives.

**Capacity building**

In order to be successful there is a need for strengthening FBO's capacity towards production, value-addition, marketing and good governance practices. This requires proper planning and implementation of a series of training workshops relevant to the various subjects. The training should aim at empowering FBO members to take responsibility for their own production and marketing and the sustainability of their organization.

**Extension service**

The philosophy of an extension service revolves around the need to promote development of rural people. The work itself involves carrying out activities that could help uplift the socio-economic condition of the people. In the case of Liberia, extension workers have acted as advisors and technicians helping FBO members to improve technical and social aspects of their business. They have communicated new ideas and improved technology in order to improve production and standard of living in communities.

Extension workers have acted as a link between FBOs and government institutions as well as Non Governmental Organizations. Similarly, they are assisting, as advisors; in the administration of the FBOs through consultations.

The main lesson learnt from Liberia is that extension workers need to possess certain qualities before they can be effective in helping FBOs. For example, three extension workers attended a train of trainers' training organized by the project in Monrovia before one could be dispatched to Grand Kru and two retained to work with Banjor Beach FBO within the city. The one who went to Grand Kru did commendable work in disseminating improved a fish smoking oven in communities whereas those retained in the city were transferred to different places and replaced by new personnel who could not deliver.

Accolades given by FBO members in Grand Kru to their extension officer suggest that it is very important for extension workers to learn the culture and social values of the people in communities. Equally, they should be knowledgeable, approachable, honest, sympathetic and emotionally stable. And when it comes to training or introduction of new technology, they should be at the forefront in demonstrating how to go about doing it rather than making speeches to fishers.

**5. CONCLUSION AND RECOMENDATIONS**

The post-harvest fisheries sub-sector includes all activities at all stages from capture to consumption. In real life it involves a number of groups playing different roles in handling fish on board, unloading, processing, storing and in distribution. Again, the sub-sector is inter-linked with other socio-economic services in fishing communities, including educational, health and other undertakings. Hence, the sector provides diversified employment opportunities and is a major source of sustainable livelihoods in fishing communities.

Challenges of small-scale fisheries development, compounded by rapid technological changes and market dynamics, exceed the capability of fishers to effectively perform when working on an individual basis. Hence, there is an urgent need to assist them in forming a solid bond among themselves. The PHFTP approach is indeed an effective way that can hold fishers together under their service-oriented organizations.

Certainly, the PHFTP approach is not the only solution to small-scale fishery development but provides an integrated vision on how to address the inter-woven challenges faced by fishers in communities. It has a great promise in stimulating small-scale fisheries development by acting as a platform for organizational development, production of good quality fish and linkage with rewarding markets. The experience from the Liberian initiative has set another example that is likely to influence adoption of similar strategies in future therefore it is important to internalize requirements for its successful introduction.

## 6. REFERENCES

**BBPFBO.** 2010. Banjor Beach Progressive Fishery Organization model business plan. Monrovia: BBPFBO.

**Ben-Yami, M. & Anderson, A.M.** 1985. Community fishery centres: Guidelines for establishment and operation. *FAO Fish. Tech. Pap.*, 264:94 p.

**GFFBO.** 2011. Good Father Fishery Based Organization's business plan. Grandcess: GFFBO

**Ndiaye, O. & Diei-Ouadi, Y.** 2008. Approche plateforme technologique post-capture du poisson: point d'entrée pour la résolution des questions technologiques et socioculturelles en pêche artisanale. In report and papers presented at the second workshop on fish technology, utilization and quality assurance in Africa. Agadir, Morocco, 24–28 November 2008. *FAO Fisheries and Aquaculture Report* 904. ISSN 2070-6987.

**CHECK LIST FOR SSI**

1. What is the name of your organization?
2. When was the FBO formed?
3. How many members are in your FBO?
4. What is the composition of your organization (men and women)?
5. What kind of resources do you have (human/physical/financial) as an organization?
6. What reasons prompted you to join the FBO?
7. What benefits have you accrued from the organization?
8. What are potential benefits that you are expecting to gain from the FBO?
9. What are the major strengths of your organization?
10. What makes your organization strong?
11. What are the major weaknesses of your organization?
12. What makes your organization weak?
13. What are the opportunities in having an FBO?
14. What are the major challenges faced by those fishing at sea?
15. What are the major challenges faced by those handling, processing and marketing fish?
16. What are the major threats to the FBO?
17. What kind of documents have you prepared (example; strategic plan/business plan/ by-laws/ cash book)?
18. What are your short-term plans?
19. What are your long-term plans?
20. Which form of support do you get from the government?
21. How are you going to sustain the FBO?

# ALTERNATIVE COMMUNICATION TECHNIQUES FOR TRAINING IN FISHING COMMUNITIES

## [TECHNIQUES DE COMMUNICATION ALTERNATIVE POUR LES FORMATIONS DANS LES COMMUNAUTES DE PECHES

by/par

Luisa Arthur[1]

**Abstract**

This article, rather than the result of research, presents a practical experience in the training of fishing communities, where the participants are often illiterate, but do have practical knowledge of their productive activity, such as fishermen, processors, sellers and workers in production units. It gives examples of how sketches, drawings and cartoon strips can be used in written materials and of how the use of "corporal expression" and simple theatre plays can help to convey complex ideas.

*Key words: Fishing communities, Training materials, Corporal expression*

**Résumé**

Cet article, plutôt que le résultat de recherche, présente une expérience pratique dans la formation des communautés de pêches, où les participants sont souvent analphabètes, mais ont une connaissance pratique de leur activité, tels que les pêcheurs, les transformateurs, les vendeurs et les travailleurs dans les unités de production. Cela donne des exemples de comment des croquis, dessins et bandes dessinées peuvent être utilisés dans les documents écrits et comment l'utilisation de « l'expression corporelle » et de pièces de théâtre simple peuvent aider à exprimer des idées complexes.

*Mots clés: Communautés de pêche, Matériel de formation, Expression corporelle*

## 1. INTRODUCTION

Governments and institutions like FAO, WHO and others, are focussing on objectives like poverty reduction of the most disadvantaged populations, e.g. through improved handling and processing techniques to avoid post-harvest losses or to add value to processed fishery products or even in quality assurance or traceability systems; all of these, in general, involve a training component.

We could say that "training is not an event, but a process that makes you think" (Figure 1) and that can involve all the participants in a continuous process of working together to improve the safety and quality of the products in the country and so reduce post-harvest losses (Figure 2).

**Figure 1. The chain**

Our Teachers          Our lessons          Dissemination

---

[1] Ministry of Fisheries, Maputo, Mozambique. luisa.arthur@gmail.com

**Figure 2. Proposed system of training to improve sanitary quality (GHP/ GMP) and to reduce post-harvest losses (INFOSA, 2009)**

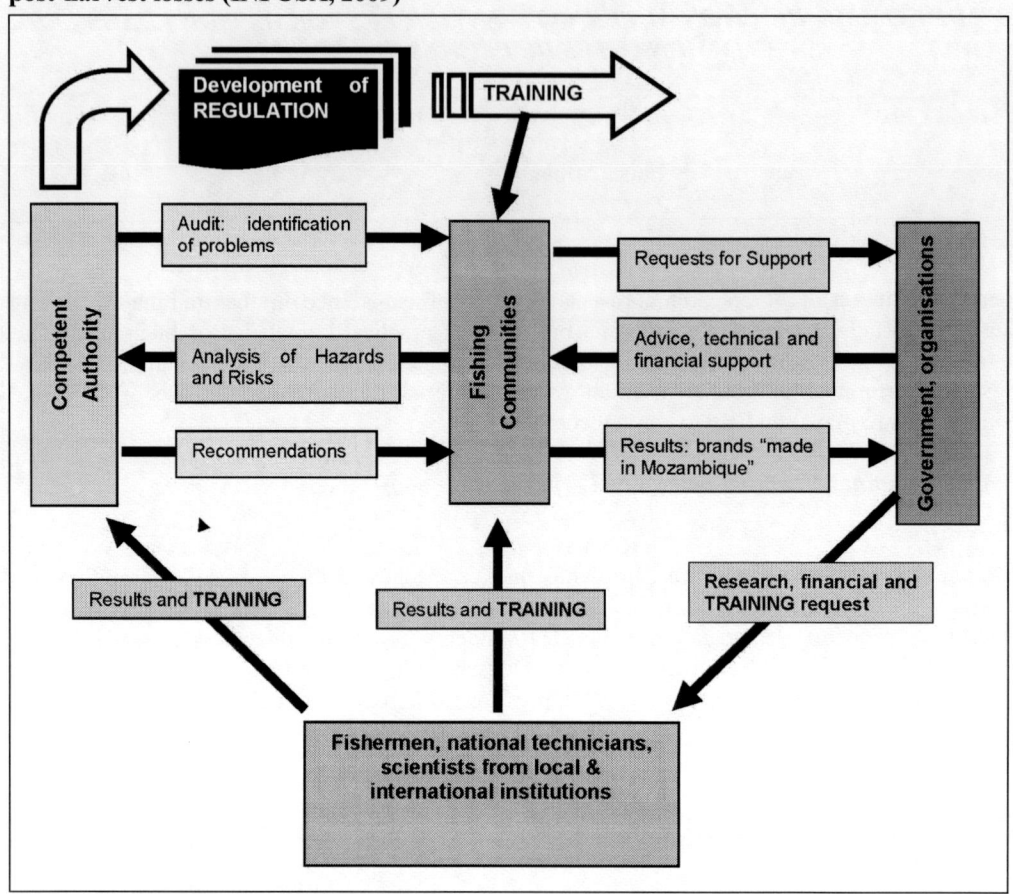

Training can be done at different levels of education, such as:

- at university level, training professionals, who not only carry out their direct activities, e.g. in the production of fish products with quality and the highest possible profitability, or even in research or quality control laboratories, where they test more effective methods, but who also train others;
- at high school level, training technicians with a more practical knowledge, who can have a direct effect on the production lines of the food chain, from harvesting to marketing, and also on training activities and in the dissemination of technical information;
- at basic level, where professionals, such as fishermen, factory workers and sellers have a direct effect on the production and trading chain, e.g. when handling fish, and their actions are e.g. the basis for quality assurance of fish products.

The diversity of methods, manuals, means and training skills and of the techniques to study the different topics (subjects) is immense and is being adapted to the various target groups of the training activity and also to the circumstances and the goals one wants to achieve and to the existing level of technology; thus, different training methods and assessment criteria are being developed, adapted to the cultural context and the traditions of the society where the training is taking place. But there is a factor that is always present, even though in different forms, which are the communication skills of the facilitator. In countries of Southern Africa, oral transmission of knowledge has been very much used up to the present day. In a society in which the communication and conveyance of information is mainly done orally, "corporal expression" - also known as "mime" - takes on an important role and this can sometimes lead to different forms of theatre.

## 2. OBJECTIVES

This article, rather than the result of research, intends to present a practical experience in the training of fishing communities, where the participants often are illiterate, but do have practical knowledge of their productive activity, such as fishermen, processors, sellers and workers in production units.

## 3. METHODOLOGY

There are many techniques for the transmission of information, but this article focuses on the method of transmission of information in training processes by means of corporal expression or theatre.

A "facial expression" can transmit better than a thousand words a feeling or an opinion about e.g. dislike caused by the smell of a fish in a state of deterioration or by the taste of the food that we are being served (Figure 3 and 4).

**Figure 3. The smell, the taste and the texture evaluation**

If during e.g. a training activity, while transmitting a fact, different tonalities or nuances of voice and even facial expressions are added to the talk, most of the time the effect is to attract more attention from the participants, because:

- The topic becomes clearer,
- The learning process becomes "funny" and makes the topic more attractive and maybe easier to understand.
- The topic becomes more intriguing, captivating the attention of the participants, resulting in a more effective learning process, etc.

**Figure 4. I do not like the tomato soup**

If those training activities are completed with gestures, using the hands or even the whole body, the explanation of a technical topic can become much easier. In that way the training process becomes more efficient.
However, if these variations of the voice, facial expressions, gestures, etc. are exaggerated, they may also have the opposite effect on the group of participants

The training courses for facilitators in the projects were organised around the presentation of theoretical topics, but especially around practical activities, many of which were developed in working groups that presented the results to their colleagues.

Another type of activity consisted in using the posters to exercise their role as facilitators. The main goal was to make the facilitators understand those topics, so we could have some assurance that they would be able to communicate them correctly to the group that they were going to train next.

## 4. RESULTS

Within the context of a project with the *"Escola de Pesca"* (Fisheries School) from Mozambique[2], and later a joint FAO and INFOSA project in Angola and Mozambique[3], training guides were developed for workers and fishermen and sellers of fish products in small fishing communities. The need was then felt to also develop material to train the trainers (to be called "facilitators" in this article), who would be transmitting the technical information to people who are often illiterate.

Many of these facilitators themselves had had a more practical education (at high school level) and to make it easier for them to understand the subjects, it was decided to develop Training Guides in which the topics were presented graphically or by means of a drawing, e.g. to explain the chemical processes of deterioration, bacterial activity, or the role in preservation played by salt in the salting of fish or by ice in refrigeration.

We also developed posters, to be used by the facilitators during the training activities that they would have to carry out. Those posters were simplified copies of slides that had been used during the training of the facilitators and in which graphical drawings and cartoons were used to make the explanation of e.g. chemical processes easier.

Those posters, which became known as "African PowerPoints", were merely slides, printed in size A2 on laminated material that can be hung on trees or used on board of boats and they make it possible to organise courses in small fishing communities and to convey information by means of those posters, as if the facilitator had sets of slides for each of the various topics.

We also developed leaflets with cartoon strips to be handed over to the participants in the training activities in the fishing communities. The main subjects covered were: Hygiene, HACCP, Traceability, Icing, Salting and Drying, Smoking, Safety in Aquaculture, Safety in the Market, Shrimp Processing and Sanitary Plan of Molluscs.

Figures 5 and 6 show some examples of drawings and cartoon strips.

**Figure 5. Fish dries more quickly in the wind**

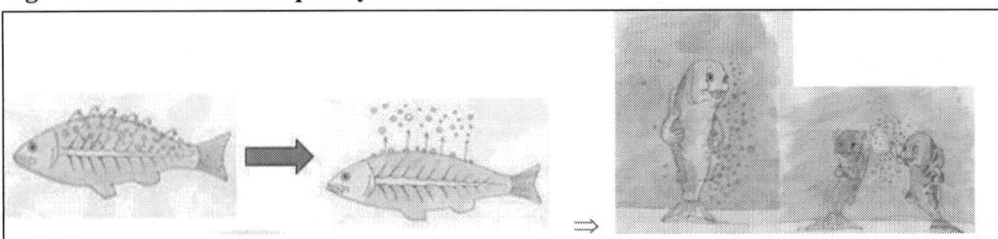

**Figure 6. Your hands remain dirty; no matter if you use a modern toilet or a simple latrine, you always have to wash your hands**

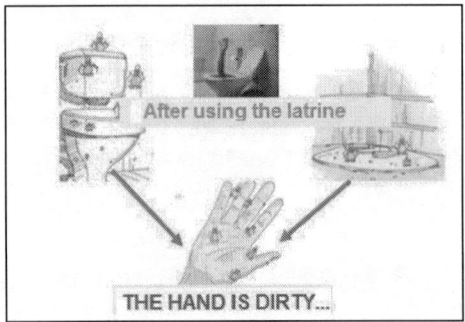

---

[2] Financed by ICEIDA.

[3] Financed by CFC (Commune Funds for the Communities).

Many of the facilitators, even though they showed a satisfactory acquisition of technical knowledge during the training courses (that, apart from theory, included demonstrations and practical activities), while exercising their activities as facilitators in the community, transmitted the technical information in a theoretical way and were often unable to answer concrete questions by the participants. They showed a lack of confidence in the knowledge they had acquired and it looked as if they had only learned the topics by heart and had a limited command of the underlying principles or foundations of the different methods of salting and smoking, of the deterioration processes, of organoleptic evaluation, etc.; apart from that, they only seemed to have a vague notion about how to teach those topics to other people.

In an overhaul of the organisation of the training, the time dedicated to it, the practical activities carried out, etc., the activity of training the facilitators how to train was introduced - and to train under the circumstances prevailing in the communities, which often means:

- Having no electricity and therefore no audio-visual means such as PowerPoint, video etc.;
- Being in an open space (e.g. under a tree), where you have to raise your voice to be heard in the middle of all the normal sounds of the surrounding community;
- Working with big groups that move around, e.g. every time a ship arrives, the women go to buy fish and then come back to listen to the facilitator;
- Working with mothers carrying their babies and often the facilitator has to help the mother and carry her baby for her to be able to participate in certain training activities, etc.

During the training courses for the facilitators the need was felt to explain certain topics, such as chemical phenomena, in a more practical way, because they often have no more than high school. And as we mentioned about the development of the Guides, the same sketches and cartoon strips that were used to explain complex phenomena that occur e.g. in the salting of fish, were dramatized in the form of "corporal expressions" and simple theatre plays (drama) were organised, using the participants themselves as actors.

That is how the use of "corporal expression" and of "simple theatre" arose as a means of training the facilitators and for them to use the same techniques later in their own training activities in the fishing communities.

In a project financed by FAO/IFAD, in São Tomé and Príncipe[4] a guide was prepared for the facilitators in which the PowerPoint presentations contain certain slides that can be explained with the use of corporal expressions or simple theatre plays, which were later used to also train the trainers in the communities themselves, whose role it would be to convey the knowledge that they had acquired to other people in their own community (Figures 7 and 8 show examples of slides and of the corporal expressions or theatre plays that can be used to convey the technical information).

---

[4] FAO/IFAD Regional Capacity–building and Knowledge Management for Gender Equality: IFAD/ FAO Grant Programme (Gender and Youth Expert). EOD: 01/06/2010 – NTE: 01/03/2011. FAO/IFAD Grant (GCP/GLO/233/IFA).

**Figure 7. Example of a slide and theatre about "Hygiene"**

| HYGIENE | | |
|---|---|---|
| **SUBJECT** | **SLIDE** | **CORPORAL EXPRESSION OR THEATRE** |
| **How do detergents clean the equipment?** | The soap CATCHES the dirt and with the help of water drags it away 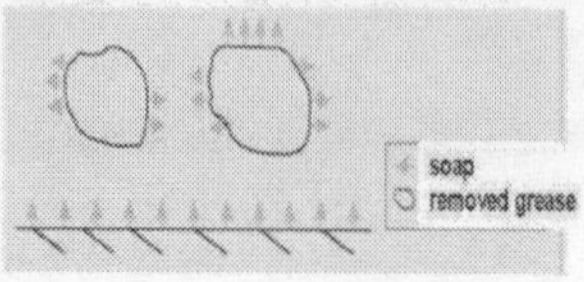 | → 2 people (A e B) spread over the room (which represents the surface of a food product) are the dirt<br>→ 2 other people (C and D), also spread over the room, are the bacteria<br>→ 1 person (E) is the water and runs across the room, with arms spread wide and pushes A, mas does not manage to push (or drag) B, C, D<br>→ 1 person (F) is the soap (detergent), gliding and dragging his feet across the room and catching B e D by the hand, but does not succeed in catching C<br>→ Person E (the water), comes again with arms spread and drags F, B and D<br>→ Only C remains, whom the water and even the soap do not succeed in removing from the surface of the food product<br><br>**NOTE: the soap (detergent) catches <u>some</u> of the dirt and the water drags it away, including the soap itself. The water also makes it easier for the soap to spread.** |

**Figure 8: Example of a slide and corporal expression about "Salting"**

| SALTING | | |
|---|---|---|
| **SUBJECT** | **SLIDE** | **CORPORAL EXPRESSION OR THEATRE** |
| **How does the salt preserve the fish?** | Effects of salt–1  <br>Os iões de sal entram no músculo do peixe provocando a saída da água constituinte.<br><br>When the salt enters the meat of the fish, the water leaves | → 1 person (A) sitting in a chair represents the water of the body of the fish,<br>→ 1 person (B) represents the salt, that wants to sit down in the same chair as A, and therefore B pushes A out of the chair. |

The printed materials already benefitted approximately 500 people, in the training courses for facilitators and in the short courses in the fishing communities themselves, and many more copies must have been printed though the electronic versions that were distributed. The materials and the technique of corporal expression have already been used in six African countries - Angola, Namibia, Malawi, Mozambique, Sierra Leone, São Tomé and Príncipe.

This method led to development of a Guide for facilitators and of a set of posters ("African PowerPoints") that are being used by "teachers" in the communities themselves.

In those communities school-going children became involved in the training processes because, being more open to change, the theatre plays attracted them and when the course was over, they started conveying information e.g. about Best Practices in Hygiene to younger children.

To complete these training materials, in the context of the above mentioned project in São Tomé and Príncipe, we also developed:

- Several videos showing the theatre plays that were presented.
- A radio course, in which the listeners have a manual to be able to follow the spoken text. The material of this course could eventually be adapted to television and thus facilitate the conveyance of information.

## 5. CONCLUSION AND RECOMENDATIONS

The use of "corporal expression" is very important, especially in training activities for large groups of people with a lot of practical experience but with a low level of schooling, and in open spaces.

This article presents examples of some of the topics dealt with in PowerPoint slides and a proposal of "corporal expressions" or theatre plays that can be used by facilitators to make the explanation of the topic of the slide clearer and easier.

## 6. REFERENCES

**CFC, FAO, INFOSA, IPA, IDPPE.** 2008–2009. Training material from the Project: "Aumento da eficiência nos mercados do sector de pequena escala" em Angola e Moçambique.

**INFOSA**. 2009. Technical reports for the fish industry.

**Arthur, L.** Final report of the Regional Capacity-building and Knowledge Management for Gender Equality: *IFAD/ FAO GRANT PROGRAMME.* (GENDER AND YOUTH EXPERT). EOD: 01/06/2010 - NTE: 01/03/2011. FAO/IFAD Grant (GCP/GLO/233/IFA). *Sao Tomé and Principe: Thematic "hands-on" training of fisher women.*

# A SEAFOOD TRADE CORRIDOR APPROACH TO DRIVING ECONOMIC PERFORMANCE[1]

## *[UNE APPROCHE DU CORRIDOR COMMERCIAL DE FRUITS DE MER COMME MOTEUR DE PERFORMANCE ECONOMIQUE]*

by/par

Mike Dillon[2], John Heap, Rory Dillon and William Davies

**Abstract**

A country's wealth depends on the price that exported goods can command in open markets, the ease with which routes to market can be opened up and the efficiency with which goods can be produced.

Seafood Trade is critical to many developing countries economies and this paper looks at how added value export can be driven using buying clusters connected to developing and developed countries who wish to increase their seafood trade.

Within this paper the strategy for productivity (including infrastructure) development is to underpin innovative approaches to trade development using a model known as a 'trade corridor'. Case studies are looked at including a developing country (Indonesia) and a developed country (Canada). This is then extended to look at how all trade corridors of commodities are transported into a market cluster.

The Corridor added value vision can include new technology, branding using sustainability and improved logistics.

*Key words: Trade Corridor, Seafood Supply Chain, Industrial Clusters*

**Résumé**

La richesse d'un pays dépend du prix des marchandises exportées dans les marchés libres, la facilité avec laquelle les voies d'accès au marché peuvent être libres et l'efficacité avec laquelle les marchandises peuvent être produites.

Le commerce de fruits de mer est important pour de nombreux pays en développement et ce document présente comment la valeur ajoutée peut être pilotée à l'aide de groupes d'achat connectés aux pays en développement et aux pays développés qui souhaite augmenter leur commerce de fruits de mer.

Dans ce document la stratégie du développement de la productivité (y compris l'infrastructure) est de soutenir les approches novatrices au développement du commerce en utilisant un model connu comme un « corridor commercial ». Des études de cas sont envisagées incluant un pays en développement (l'Indonésie) et un pays développé (le Canada). Ceci est alors étendu à l'examen du comment tous les corridors commerciaux de marchandises sont transportés dans un groupe de marché.

La vision de la valeur ajoutée du Corridor peut inclure une nouvelle technologie, l'image de marque grâce à la durabilité et, l'amélioration de la logistique. .

*Mots clés: Corridor commercial, Chaîne d'approvisionnement de fruits de mer, Groupes industriels*

## 1. INTRODUCTION

The Trade Corridor vision that is explored in this paper considers the linkage of supplier and buyer. This can often be spread across the developed and developing worlds. In the strategy to increase productivity there is an important use of trade to facilitate wealth creation.

---

[1] This paper is based on a report produced for UNIDO. The authors thank them for their contribution.
[2] IAFI, 10 Scartho Road, Grimsby DN33 2AD, United Kingdom. mikedillon2010@hotmail.co.uk

At the national level, the value created arises from both the total volumes of trade and the value created by each unit. For some products or services, increasing volume may not be possible (because the market is restricted in some way) and it becomes even more important to concentrate on adding value to each unit of production.

Improving productivity allows a country to support higher wages; a strong currency and attractive returns to capital and with them support a higher standard of living and a reduced level of poverty.

The Trade Corridor approach has been designed in part to address some of the issues faced by food processors operating within global supply chains - those of food safety, quality and social compliance throughout the supply chain.

The current systems of inspection and approaches of aid donors to increase developing country compliance focus too heavily on achieving international standards and accreditation and too little on increasing trade. Technical assistance is provided in an ad hoc way which cannot easily be handed over to developing country governments.

The trade corridor approach aims to make gains from trade development real by bringing in buyers from export markets. Co-operation through a trade corridor structure can lead to trade partnerships which increase exposure for firms in emerging markets and lead to incremental added value. Trade corridor technical assistance such as economic impact assessment and sector benchmarking allow governments to increase trade capacity.

Within this paper the use of case studies are looked at including a developing country (Indonesia) and a developed country (Canada). This is then extended to look at how all trade corridors of commodities are transported into a market cluster.

## 2. DEVELOPING COUNTRY SUPPLIERS

Within a global economy, manufacturers and retailers must deal with suppliers and traders who are based within different time zones, continents and ensconced in different cultures. Those operating in developed country markets will now have to deal with the challenges of working with suppliers in developing countries. For this reason, the problem of development economics and problem of supply chain management begin to overlap: the problem of underdevelopment in large parts of the world is now inexorably linked to the problem of insuring suppliers consistently offer high quality products at low cost. The capacity of supplier countries is directly relevant to the experience of consumers and manufacturers in developed countries: not only because of social issues such as the treatment of workers early in the supply chain will affect, but also the quality of the product and the flexibility of supply chains. These factors will be linked to factors such as the level of skilled labour and access to capital within that country. Likewise, the success of new industrialised countries, such as the Asian tigers, in raising average incomes has been in overcoming barriers to trade and developing the skills to manage and become competitive within complex supply chains. Projects which can serve both of these needs will become more and more important for both the development community and private businesses in the future.

In the UK market, almost 70 percent of the total seafood supply is imported (Marine and Fisheries Agency, 2007) and the top ten import markets include China, Thailand and Mauritius (Seafish, 2008). Developing countries sometimes have competitive advantages in these markets due to access to particular species of fish. Developing countries also often have lower labour costs that lead to lower prices in the market place. In moving from national and regional supply chains to international supply chains, premium seafood manufacturers have been able to lower production costs at the cost of increasing transaction costs (Matthews, 1986, pp 906-907) i.e. seafood processors have had to come up with strategies for mitigating the costs of forming new trade relationships. Retailers have to protect themselves from reputation damage, potentially caused by the unethical business practices of new suppliers. If fish is discovered to be caught unsustainably, by workers who have been mistreated or using practices which compromise food safety, then present and future sales could be significantly diminished. Other considerations include the timeliness, quality and price of goods exchanged. There are several strategies employed in mitigating these risks, most important of which are: private auditing systems, national and supra-national regulation on food safety as well. Private, governmental and aid led strategies are also in place to provide supply instead of demand led approaches to supply chain management.

This paper contends that existing approaches are sub-optimal and significant gains can be made by exploring innovative forms of supply chain management. Demand led systems based around auditing and private standards, such as ETI, GlobalGAP and ISO enable businesses to monitor and police supplier behaviour. However, a standards-based approach has been criticised in terms of both effectiveness and its unintended consequences for development. Standards often act as barriers to trade for developing countries as they impose large upfront costs on suppliers (Ellis and Keane, 2008). Legalistic approaches to ensuring food safety and sustainability sometimes result in a lack of engagement this can cause suppliers to work outside rather than within rules structure to achieve the desired results: Freiberg (2008) offers a case study of Ghanaian Baby-Veg production, which though launched in partnership with UK retailers with bespoke monitoring systems went bankrupt when managers were revealed to be fictionalising accounts.

Technical assistance programmes are designed to support businesses within economies that are excluded from international markets due to technical barriers to trade. Within the food sector this often involves training workers and managers to meet international standards required for export. In this way the aid spending complements private spending by investing in regions which are excluded from the free market.

However trade related technical assistance has been seen as an imperfect solution to the problems raised above (this is discussed in detail in Section 3).

The trade corridor approach seeks to carve out an approach which is mutually beneficial to producers across the supply chain by addressing limitations in inspection based systems and the traditional technical assistance designed to allow developing country producers to comply with international standards.

## 3. DEFICIENCIES IN CURRENT APPROACHES

The accepted approach toward trade development and trade-related technical assistance is flawed. Aid agencies have systematically mischaracterised the proper role of market forces and export demand in development. These bodies have focused on a trade liberalisation approach which has identified a lack of openness to trade as the limiting factor to growth. Current programmes looking to reduce tariffs for Less Developed Countries (LDCs) such as Aid for Trade and EBA (the "Everything But Arms" initiative) pursue openness directly. However, successful trade development needs to focus directly on demand: developing relationships with key stakeholders in target markets.

UNIDO has stated that trade liberalisation "has not resulted in a form of integration that will support sustained and socially inclusive development" (2007, §56). However, by focusing on a supply led strategy to trade capacity building, the trade development community is tacitly endorsing this trade liberalisation approach. Projects which focus only on improving infrastructure - such as inspection laboratories - rely on export markets to react quickly to signals of supply safety and quality. However, it is not clear that this is sufficient to deliver increased trade. UNIDO's mandate states it seeks to help developing countries "compete, comply and connect": and our new approach is suggesting that connecting markets should be a real focus for trade development.

Reviews of current trade-related technical assistance state that it is 'poorly coordinated and prioritised' (World Bank, 2006, p60). As indicated above some of these problems are to do with a wrong headed approach; however there is also a lack of thoroughgoing project management. For projects to be successful they must be have specific, measurable and time bound goals. Project teams must understand the current capacity of the plant, supply chain or economy they are analysing and the gap between the target and actual state of affairs. A root cause analysis must be conducted to identify constraints to closing this gap. Projects must then be selected on merit: with those which have most impact prioritised. We identify four specific deficiencies that are often found in TCB initiatives, below:

First, too often goals for trade development projects are vague and nebulous. Alternatively, they concentrate on medium term outputs instead of true strategic goals such as increasing trade or employment. For example, project success is measured by the accreditation of new inspection laboratories rather than evidence of increased trade or employment, which such compliance infrastructure is intended to deliver.

Second, data on capacity or performance is not well integrated into project design. The gap between the current state of affairs and the desired state of affairs is not calculated in a way which can be measured, controlled and improved. There is often a need to generate project specific information on say, logistics costs or energy costs per tonne exported.

Third, project selection can also be problematic with decisions made on the whims of international consultants or national governments. Often national governments are in the position to fund only a limited number of infrastructure improvement projects within a sector. Without tools to consistently measure the impact of proposed projects governments must make planning decisions in the dark.

Finally, most development projects develop some means of setting goals, selecting projects and measuring impacts. However, when this is done in an ad hoc way, which is heavily reliant on outside experience or expertise, it cannot be replicated. For this reason technical assistance fails to perform a sustainable handover to those working in country. Continuous improvement cannot be established.

## 4. A NEW APPROACH

The value chain is central to our approach to trade development.

The 'value chain' describes all the stages of getting a product from its source to the customer: these include transport, transformation and trade.

The value stream denotes a section of the chain managed by one company.

The value chain concept is useful in emphasising the interconnectedness of organisations and activities. These factors are particularly important for products that target high quality markets.

Mapping the value chain allows stakeholders to think about where value is added, where inefficiencies exist and which strategies can provide benefits for buyers and suppliers. This creates the potential for increased trade: individual organisations within the value chain must then be facilitated to exploit this potential by addressing and improving their own value streams.

The trade corridor then provides a forum for these sector wide strategies to be discussed and implemented.

The trade corridor approach seeks to use top down and bottom up approaches to achieve greater value. Business to business solutions will be implemented to increase market access, develop desirable new products and adopt more effective business practices. Top down planning of new infrastructure will utilise best estimates of impact and good data on industry capacity to allocate resources optimally.

There is a need to build and support effective trade partnerships from the bottom up. This requires a mechanism by which trust can be established. This is particularly important in overcoming the risk-averse tendencies of many buyers who, contrary to what might be expected in a 'perfect market' often prefer to remain with current networks of less competitive but established suppliers. The risk (and perhaps the changeover cost) of moving to a new supplier is too high: we aim to lower it through industrial co-operation.

In many markets there are individuals and firms who influence or control key elements of the market. It is then wise to adopt a strategy of 'key account management' in which relationships are formed and maintained with individuals/ firms in target markets, to allow specific marketing of products informed by their specific needs and preferences. Because the various stakeholders are brought together, there is opportunity to identify and address specific barriers to trade.

However, building trade capacity also involves top down planning from government: for example, to change transport or testing infrastructure. The trade corridor approach seeks to make technical assistance about helping decision markers rather than making decisions. Technical assistance should provide tools to increase the effectiveness of governmental planning: economic impact assessment tools which estimate how projects will affect export revenue and employment, establishing productivity centres that can generate reliable data on sectoral performance and creating trade corridors to act as an industry think tank.

## 5. CASE STUDIES

### CANADIAN

The seafood clusters of the Humber and Nova Scotia were linked up through a programme of partnership along a trade corridor between both countries. This Trade Corridor was specifically designed using the model agreed with UNIDO. Government to Government links were established, embryonic steering and working groups have been put in place. An initial visit by the Trade Corridor CEO and logistics chair enabled key issues to be included in the preliminary analysis, ensuring proper briefing for trade opportunities was undertaken. The enterprise agency in the key area, wishing to drive economic development, facilitated the industrial programme with clear performance indictors based on existing trade and views of options for Canada. The cluster model enabled different sizes of firms, levels of requirement and competition for products to be built in.

The cluster then supported a group of local industry on a Trade Mission to Cape Breton in June 2010. During the visit the team from Grimsby met a range of organisations from Government, Academia and Industry as this was the best approach to embed trade. The results of various discussions during the visit have improved links between the UK and Cape Breton. A number of projects are currently in discussions following the event. Trading relations have already been implemented between the two geographic areas and the Trade Corridor would act to continue and strengthen these relationships. Cape Breton and the Humber are both strong seafood areas with the need to expand international trading activity. The successes for the industry from the recent trade visit in terms of successful orders in both seafood and packaging along with other potential leads, highlight the future trade prospects that would benefit from the stimulus of a trade partnership group (The Trade Corridor).

In Canada the link to supplying seafood to Humber is either via Reykjavik and scheduled sailings or into Liverpool and across a land bridge of good road and rail links. The supply of seafood into the Humber following the event has been increased from all the companies involved in the event and the next steps in the corridor is to capitalise the inertia from these results into longer term trade and economic growth. Trade Statistics showed a 19 percent increased up to August 2010 against the same period in 2009 for seafood entering the port of Immingham in the Humber. The trade of Canadian seafood over the previous 12 months was 5.8 Million kilograms. The increased volume through Immingham port from the Trade Corridor programme is calculated to be an extra 20 containers with a proposed value of just over £1.5 million in 3 months.

The set up of any Trade Corridor needs clear Terms of Reference and discussion of its membership in the relevant industry clusters. Upon its establishment a meeting programme and pattern were included to suit the needs of membership and trading. The cluster are currently tracking the impact of setting up this trade corridor and wish to replicate this with other partners they wish to increase trade with including Norway, Iceland, Faroe Isles, China, Indonesia and India. UNIDO are also driving this programme within Pakistan and a delegation involving CEO`s and academics and the NPO are at the World Conference to pilot this approach.

**Figure 1. Cumulative total 2009 vs 2010**

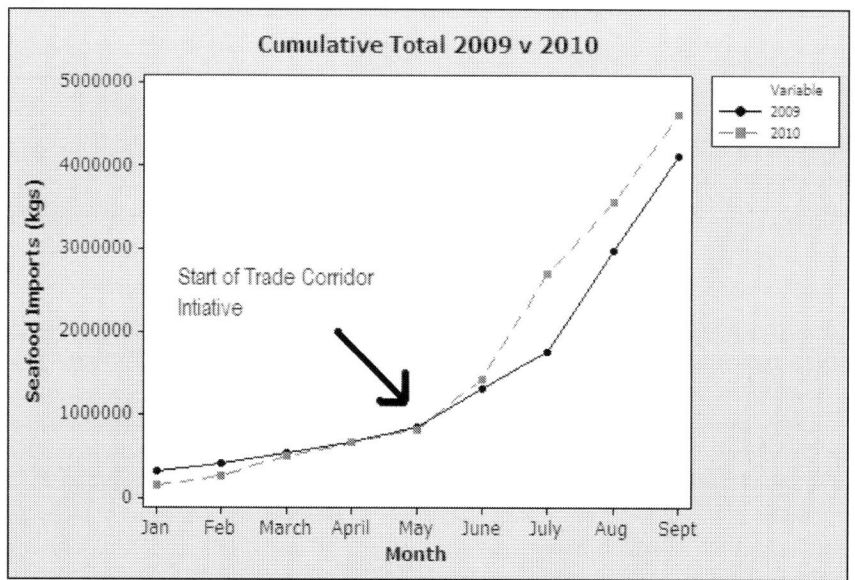

## INDONESIA

The Indonesian case study within this section demonstrates the impact of technical assistance implemented under a trade corridor approach. Below is a value chain diagram for the import to the UK of Indonesian yellowfin tuna. By examining how the value of the product changes as it moves through the chain it is possible to identify strategic changes which could increase value added for consumers and revenues for businesses. Four key lessons from the value chain are presented below. Section 6 indicates how interventions under a trade corridor vision can be implemented to address these problems.

**Figure 2. Value Chain: Indonesian-UK Chilled Yellowfin Tuna, 2007**

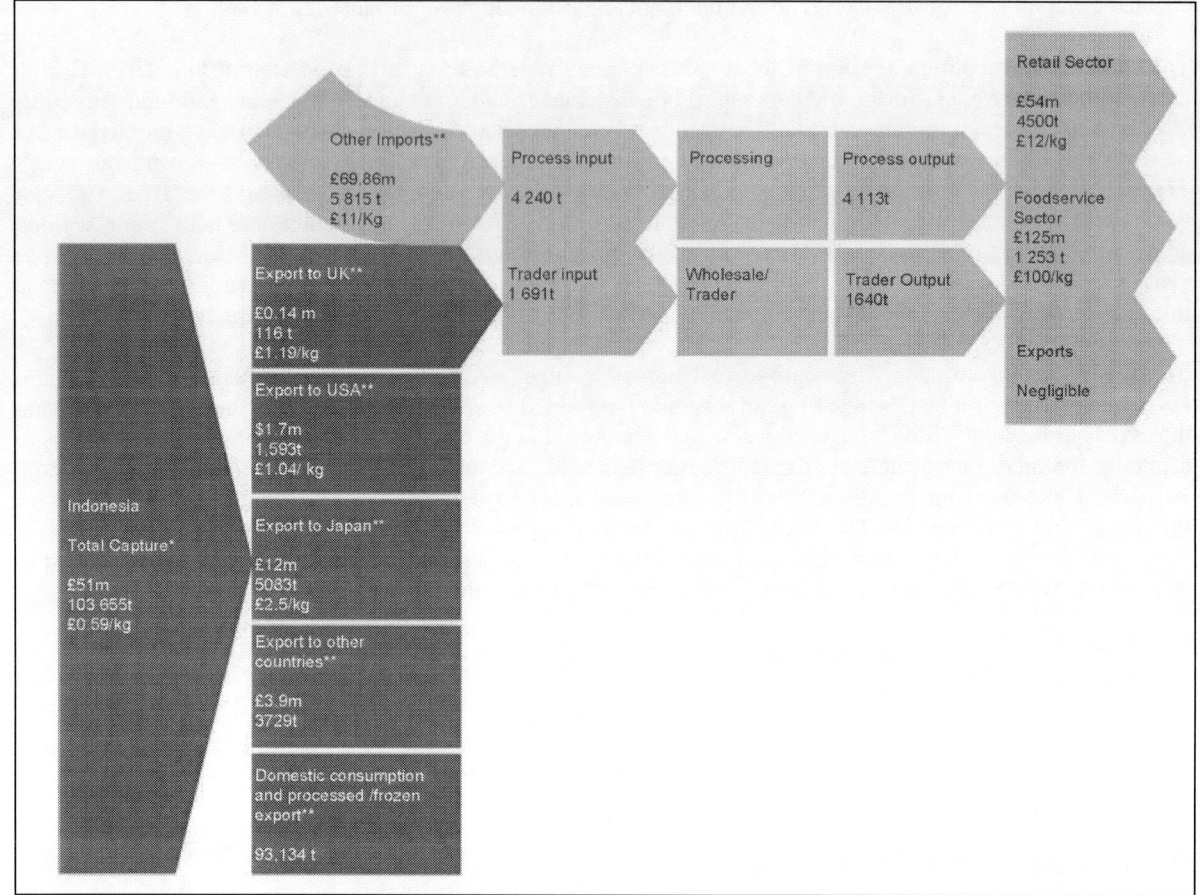

\* Ministry of Marine Affairs and Fisheries (2007).
\*\* United Nations Statistics Division (2008–2).
*Sources:* Garrett & Brown (2009).

This diagram indicates several factors within the supply chain which offer scope for increased productivity and revenues for stakeholders in Indonesia and within export markets.

First, Indonesian tuna is low value at the point of capture - significant value is lost at the first stage of the supply chain.

Processors and traders within Indonesia complain that product quality limits their competitiveness. Hygiene and handling failings on boats and within aquaculture units mean non-compliance with international standards. This lowers prices and prevents access to highly regulated foreign markets such as the EU.

Fishermen complain that they face the highest costs of maintaining standards but fear that these costs will not be reflected and recovered in the market place. They also indicate that factors such as high fuel costs and high feed costs mean that those involved in capture cannot afford investment, such as vessel upgrades, and must engage in destructive practices such as long sorties which cause product to perish.

**Figure 3. Cost breakdowns for key elements of the value chain**

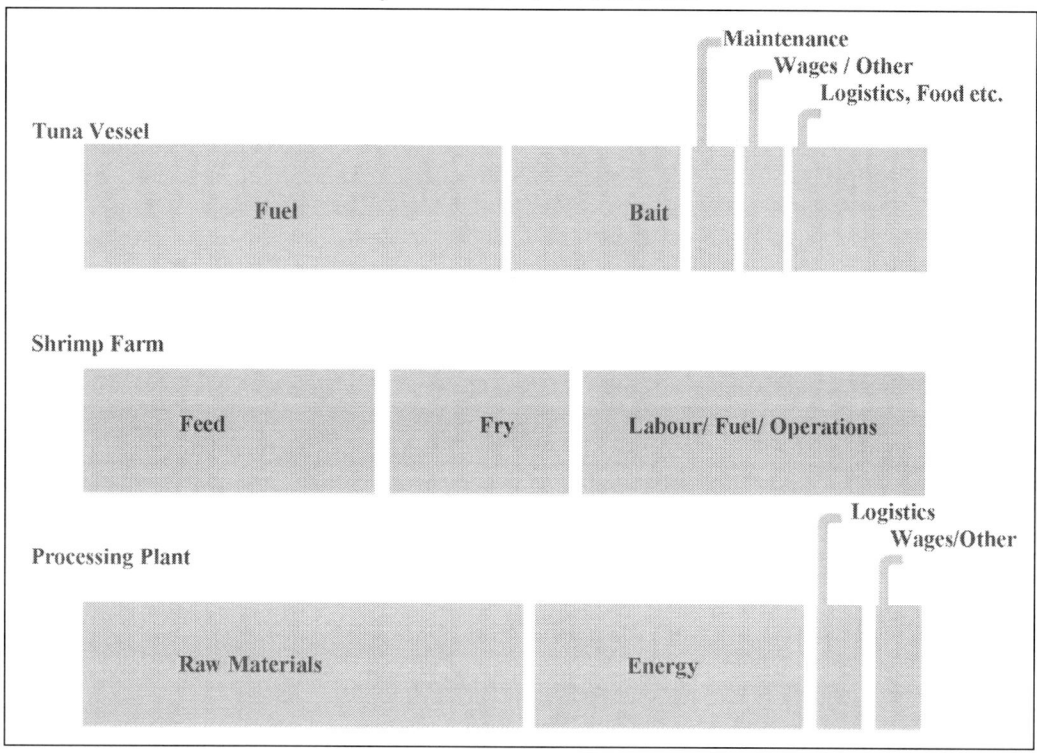

The chart above presents data collected within a UNIDO mission to Indonesia (Dillon et al, 2008). Data was collected in focus groups from workshop meetings with tuna, shrimp and processing associations in Jakarta, Surabaya, Bali and Bungus.

Second, only a small proportion of tuna captured enters the higher value export market, only 10 percent of yellowfin caught is exported (chilled, whole). This is indicative of a misallocation of goods and markets with high value products such as tuna and shrimp not funnelled towards highest value international demand.

A significant problem is the handling issues which mean product does not meet an export grade. However, the Indonesian chain also faces several structural constraints to functional market led behaviour. As discussed above, a lack of compliance to standard 'good practice' at capture stages and throughout the chain acts as a barrier to trade. Within the EU, compliance infractions have resulted in market bans for Indonesian product as well as increased costs from tariff charges and mandated testing on arrival.

These compliance issues are of paramount concern to processing and trading firms within the UK as transaction costs due to disrupted supply chains could impact on profitability and the reputation of the business with its customers.

Indonesian firms also report that a poor logistics infrastructure increases costs and lowers chain productivity. Fishermen and processors are keen to absorb skills and technologies which can reduce their use of scarce resources such as fuel. They also feel they could benefit from better packaging and new product developments more closely linked to buyer demands. In this respect, it is illuminating that UK processors report lack of awareness as a factor that impedes uptake of Indonesian goods.

The trade corridor approach seeks to address these weaknesses in the supply chain by directly linking suppliers and buyers with a view to both increasing trade in the short term and incentivising investment in structural constraints over the long term.

Indonesian imports achieve a low price point. This can be seen clearly in the value chain diagram above.

Some of the constraints described above reduce the market demand for Indonesian goods: productivity and compliance issues for Indonesian processors as well as perceived risk or lack of awareness for importers.

However, Indonesian seafood is also marketed largely as a wholesale commodity product. It is minimally packaged and minimally processed for secondary processing in export markets. In the yellowfin tuna value chain illustrated above, tuna will be exported as gutted whole rounds (chilled or frozen); it will be further processed in the UK where the majority enters the retail market packaged as ready meals.

The higher price points in the chart above often relate to positioning of products within the value chain. This is achieved by adding features which are desirable to stakeholders downstream. These could be features demanded by processors and traders such as further processing (i.e. fillets and loins) packaging or improved systems such as traceability and sustainability labelling. It could also be features demanded by retailers - development of ready meals with appropriate packaging.

## 6. WHAT IS A TRADE CORRIDOR?

The trade corridor is an organisation which seeks to strategically link trade between key clusters. This is particularly important for strengthening trade between countries at different stages of development as business often identifies emerging markets with increased risk. The trade corridor system aims to lower these risks by increasing communication between potential partners, by conducting research into potential problems and by mobilising funding to address long term supply chain obstacles.

The Grimsby Institute team have worked with the United Nations Industrial Development Organisation on added value export programmes and industrial development for the past 15 years. Field observations by the team had revealed that existing aid for trade strategies did not result in significant export. The team had therefore begun investigating an alternative strategy in the UN programme in Pakistan between 2005 and 2008 in collaboration with the Ministry of Trade, Industry and Fisheries. This resulted in an alternative approach to trade related programmes encompassing national project planning, national productivity centre and economic impact measurement linked to EC systems. This emerging approach was subsequently validated by the findings of the 2007 Aid for Trade ministerial conference (UNIDO, 2007). The Pakistan programme was agreed in 2008 and acted as the basis for the pilot programme on a Trade Corridor subsequently developed with Indonesia.

The UK-Indonesia Trade Corridor Strategy emerged organically from the work of the international team at the Grimsby Institute. It was suggested by the then Indonesian Ambassador to the UK that ongoing work between GIFHE and the Indonesian government be continued under a banner of 'the trade corridor' to drive existing assistance strategically to increase trade. The approach aims to create a 'strategic link' between partner countries to increase profits for all stakeholders.

In September 2007, the interested parties signed a letter of agreement inaugurating the co-operative relationship. The links between NEL Council, GIFHE and the Indonesian government were cemented with the signing of a Memorandum of Understanding on 30/05/08. During 2006-2007, the trade corridor project established working and steering groups composed of industrial and government stakeholders from both ends of the corridor. The UK working group has met 8 times in 3 years. There have been 6 trade delegations from and to Indonesia, including two ministerial visits to the UK. The programme has received coverage in the print and television media in Indonesia, as well as the Grimsby Institute presenting to over a hundred delegates at the Indonesian Cold Chain Conference. Research activities have commenced with mapping the movement of Indonesian imports in the UK and modelling the cost/benefit trade off of accessing UK markets through the Humber ports.

## 7. TRADE CORRIDOR: ADDRESSING VALUE CHAIN WEAKNESS

The trade corridor acts to tackle the supply chain problems listed above. It does this in two distinct modes. The first is as an incubator of trade partnerships and the second is as a think tank driving government planning. This allows Indonesian stakeholders to tackle both the upstream and downstream constraints to increased added value: working with importers to increase trade and with government to improve infrastructure.

An essential part of the trade corridor process is to find an appropriate partner cluster that can provide the right kind of supply. The selection of the North East Lincolnshire (NEL) seafood cluster as the partner for Indonesia meant that export expansion could be directed at a premium market, which was of a large enough capacity to absorb increased volumes. North East Lincolnshire is the centre of seafood trade for the UK with over 70 percent of seafood consumed in the UK being processed there (Competitiveness, 2008). The Humber has the highest regional density of processing units in the UK, with 19 percent of all sites and 27 percent of processing employment (Brown, 2008, p66). With a location quotient of 42 percent it is one of the most concentrated clusters in the UK and perhaps the biggest seafood hub in Europe (Griffiths, 2009).

Increasing linkages between secondary processors in the UK and primary processors in Indonesia will increase value added simply by capitalising on the best practice already in the country. Those firms producing at an international standard can overcome end of chain blockages, such as lack of awareness amongst importers, to achieve premium market prices. The trade corridor has been proactive in fostering business partnerships. It has conducted a supplier-to-buyer event to allow face to face networking. It has also contributed to research: through a survey, UK buyers indicated which factors stopped them buying Indonesian seafood and what added value features they would be interested in. This acts to increase the constraints to increased exports outlined in Section 3.

Trade partners can work with suppliers to improve the amount of value added activity which is carried out within Indonesia. The Trade Corridor (TC) group has explored strategies which would be beneficial both to UK buyers and Indonesian suppliers. Higher price points for products can be achieved by adding features that are desirable to stakeholders downstream. These could be features demanded by processors and traders such as further processing - fillets, loins etc, packaging or improved systems such as traceability and sustainability labelling. It could also be features demanded by retailers - development of ready meals, appropriate packaging. These activities move Indonesian fish incrementally from low value commodity to added value products.

The second role of the trade corridor is to exploit knowledge within the supply chain to drive government and supra government spending on infrastructure. Sometimes the trade corridor can drive business solutions to infrastructure problems: for example, UK logistics firms improved linkages with Indonesia simply by reviewing and replacing existing partners in the region. However, long term problems such as poor compliance and insufficient port and road infrastructure demand government level planning.

The Grimsby Institute has sought to establish a Trade Corridor vision that delivers not just a greater focus on trade partnership but also embeds continuous improvement within the supply chain. The UNIDO programme conducted within Indonesia has recommended greater utilisation of sector benchmarking and economic impact assessment.

Too often infrastructure development is hampered by poor central planning. Project goals are not clear, there is a lack of good data on market demand and sector capacity and there is poor understanding of the likely impact of different projects.

The economic impact model developed for Indonesia has been created to provide a planning tool for the Indonesian government. By projecting the impact of different project plans on employment and export revenue they can make informed decisions on which programmes to prioritise.

Good planning demands a detailed insight into current productive capacity and efficiency. Setting up a productivity centre is therefore key in driving sector benchmarking programmes and in building the right infrastructure for potential productivity gains.

This new approach puts the Indonesian government in a position to increase sector competitiveness in the long term. The trade corridor will continue to highlight new opportunities to add value. The National Productivity Centre (NPC) will indicate where efficiency savings can be made. And, the economic impact assessment tool will utilise this data to estimate macroeconomic effects of different decisions. Interventions will take place within this framework. Long term problems such as logistics, cold chain and compliance would be reviewed in this context. Innovative solutions such as the education of logistics professionals would be evaluated before implementation.

## 8. CONCLUSION

This report has outlined a new approach to trade development which focuses on the value chain and fosters partnerships through trade corridors. Traditional approaches have focused on meeting international standards rather than ensuring trade volumes or value increased as a result. Technical assistance is also constrained by a lack of strategic planning which stops developing countries from being able to continually improve their value chain interventions. The trade corridor approach aims to meet these challenges by developing trade partnerships with buyers in an associated export market. Increased communication between buyers and suppliers can lead to win-win strategies to increase trade volumes and add value. Better communication between industry and government can improve infrastructure spending, especially when reinforced with technical assistance, which provides tools for economic impact assessment and sector benchmarking. The case study in Indonesia demonstrated key problems in the value chain: suppressed value due to poor handling and poor compliance, inefficient production and logistics increasing costs and lack of awareness and perceived risk by importers. It then showed how the trade corridor could mitigate these problems: increasing trade by communicating with key stakeholders in export markets, working with buyers to add value and conducting research to quantify costs of inefficient port and cold chain arrangements. This new approach can help developing countries by letting the private sector drive how and where technical assistance is provided. The immediate impact of increased trade was seen in both Canada and Indonesia. The reason for sustained seafood trade of approx 20 percent to the UK port should be evaluated.

## 9. RECOMMENDATIONS

- East African Nations (Kenya, Tanzania and Uganda) should consider forming a trade corridor piloting approach including the export association, artisanal suppliers and the government. These counties have specific concentrations of fish supply which will benefit via a corridor. The existence of the strong UFPEA and Regional Export association in alliance with inspectors and Government policy makers will make the trade corridor approach useful.
- North Africa-Morocco- we discussed the approach in the World Congress in 2009 - further in Cochin at the Aquaculture congress with FAO in 2011- it was believed to be of interest for larger suppliers of pelagic species.
- Artisanal level trade can piggy back on the larger corridor programme to access new markets.
- Project funding should be sought to adapt the approach for the African network and a pilot sought to drive the programme.
- Net increase of Trade in 3 months was seen to be at 17 percent for range of products, including unusual species (yellowtail flounder).
- By-products should be investigated as part of the added value chain.

## 10. REFERENCES

**Brown, A.** 2008. *"2008 Survey of the UK Seafood Processing Industry"* for Seafish: [Online] available at, http://www.seafish.org/resources/publications.asp?c=Economics%20and%20Business, accessed 15/10/09.

**Competitiveness.** Jan, 2008. The Humber seafood clustering efforts in Yorkshire, UK (Case Study for the Commission of the European Communities Enterprise and Industry Directorate-General).

**Dillon, M., Esser, J., Heap, J. & Dillon, R**. 2008. Increasing Trade Capacity of the Fisheries Sectors in Indonesia UNIDO Project Ref: TE/GLO/08/035.

**Ellis, K. & Keane, J.** 2008. for the Overseas Development Institute "A review of ethical standards and labels: Is there a gap in the market for a new 'Good for Development' label?" Working Paper 297. Available online at, http://www.roundtablecocoa.org/documents/ODI%202008_297–good–for–development–label–ethical–standards–trade.pdf, accessed 11/11/09.

**Freidberg, S.** 2008. *"Postcolonial Paradoxes: The Cultural Economy of African Export Horticulture"* in Nutzenadel, Alexander & Trentmann, Frank (Eds) *Food and Globalization: Consumption, Markets and Politics in the Modern World* Oxford: Berg.

**Garrett, A. & Brown, A.** 2009. Yellowfin tuna: A Global and UK Supply *Chain Analysis* (March) [Online], available at http://www.seafish.org/resources/publications.asp, accessed on the 31/03/2009.

**Griffiths, W.** 2009. *"Introduction to Food Sector"* at the European Productivity Conference, Grimsby, UK.

**IMF & the World Bank.** 2006. Approved by Mark Allen and Danny Leipziger. *Global Monitoring Report 2006: Strengthening Mutual Accountability – Aid, Trade, and Governance*, Washington DC: World Bank.

**Indonesian Statistics Agency.** 2007. Statistics of Marine and Fishery Jakarta: Ministry of Marine Affairs and Fisheries, available at http://statistik.dkp.go.id/download/StatistikKP_2007/index.htm, accessed 09/09/2009.

**Marine and Fisheries Agency** 2007. *"UK Sea Fisheries Statistics"* [Online]. Eds. Barratt, Craig & Irwin , Craig. available at, http://www.mfa.gov.uk/statistics/documents/UKSeaFishStats_2007.pdf, accessed 10/01/2010.

**Matthews, R. C. O**. 1986. "The Economics of Institutions and the Sources of Growth" *The Economic Journal* 96. (384), pp 903–918.

**Seafish** 2007. *"The Economic Impacts of the UK Sea Fishing and Fish Processing Sectors: an Input-Output Analysis"* [Online] http://www.seafish.org/upload/file/economics/FINAL–%20Input%20output%20report%20%20,full%20report.pdf [accessed 04/04/2009].

**Seafish** 2008. *"2008 Trade Summary"* available at, http://www.seafish.org/upload/file/market_insight/2008%20Trade%20summary.pdf accessed 10/01/2010.

**United Nations Industrial Development Organization (UNIDO)** 2007. Background paper: How can Aid for Trade transform LDCs? Least Developed Countries (LDC) Ministerial Conference, Vienna. 29–30 November 2007.

**United Nations Statistics Division.** 2008–2. *"UN Comtrade Database"* available at http://www.bps.go.id/sector/employ/index.html, accessed 05/04/2008.

# ENHANCING FISH MARKETING THROUGH ICT:
# EXPERIENCES OF EFMIS PROJECT IN LAKE VICTORIA

## *[AMELIORATION DE LA COMMERCIALISATION DU POISSON A TRAVERS LES EXPERIENCES TIC DU PROJET EFMIS DANS LE LAC VICTORIA]*

by/par

Richard O. Abila[1], William Ojwang, Andrew Othina, Carolyne Lwenya, Robert Oketch and Raphael Okeyo

## Abstract

Small-scale fisheries in developing countries often fail to perform optimally due to lack of vital market information, leading to inefficiencies in market operations, inequity in sharing benefits and substantial post-harvest losses. Increased access to information is therefore the key to improving performance of the sector towards poverty reduction. The mobile phone, which is the fastest growing communication media in Africa, is most suited for addressing market information gaps.

Although the Lake Victoria region is well covered by mobile phone networks, there had not been a systematic means of collecting, packaging and disseminating market information. *Enhanced Fish Market Information Service* (EFMIS), an ICT project based on mobile phones, was developed and piloted in Kenya to address this problem. The objective of the project was to enhance fish trade and incomes for the fisher community easily, cheaply and faster through improved access to market information. Through the system, data from about 165 fish landing sites and inland urban markets was continuously relayed to a central database where it was appropriately packaged into a format that users could access in real time by sending a query through mobile phone SMS. The system operated for 24 hours every day and gave an automated response usually within 10 seconds. EFMIS also disseminated information through electronic bulletins.

This paper discusses the EFMIS project, focusing on the design of the market information system, achievements, impacts and implementation challenges. The EFMIS model can be appropriately adapted for application in other small-scale fisheries, paying attention to the lessons learnt from this pilot activity.

*Key words: Commerce, Lac Victoria, Information, Poisson, Projet EFMIS*

## Résumé

La performance optimale des petites pêches dans les pays en développement fait défaut du fait du manque d'une vitale information du marché. Ceci mène à des déficiences des opérations commerciales, une inégalité dans le partage des bénéfices et des pertes post capture significatives. Un accès accru à l'information est par conséquent clé pour améliorer la performance du secteur vers la réduction de la pauvreté. Le téléphone mobile qui est le plus rapide moyen de communication en Afrique, est plus adéquat pour aborder les déficits en information commerciale.

Bien que le Lac Victoria soit bien couvert par les réseaux de téléphonie mobile, il n'y a pas eu de moyens systématique de collecter, organiser et disséminer l'information commerciale. Le projet TIC basé sur les téléphones mobiles « Amélioration du service de l'information commerciale du poisson (EFMIS) » a été développé et piloté au Kenya pour résoudre ce problème. L'objectif du projet est d'améliorer le commerce du poisson et les revenus par un accès facile, peu onéreux et rapide des communautés de pêche à l'information commerciale. A travers ce système des données d'environ 165 sites de débarquement de poisson et marchés urbains riverains étaient continuellement relayées a une base de données centrale où elles sont adéquatement organisées dans un format accessible en temps réel aux utilisateurs, en envoyant une demande par un texto téléphonique. Le système a fonctionné pendant 24 heures chaque jour et a donné une réponse automatique habituellement en l'espace de 10 secondes. EFMIS diffuse aussi l'information à travers des bulletins électroniques.

---

[1] Kenya Marine and Fisheries Research Institute, PO Box 1881, Nkrumah Road, Kisumu 40100, Kenya. abilarichard@yahoo.com

Ce document discute du projet EFMIS, se concentrant sur la conception du système d'information commerciale, les réalisations, impacts et défis de mise en œuvre. Le modèle EFMIS peut être adapté de manière appropriée pour application dans toute petite pêche, en accordant une attention aux leçons apprises de cette activité pilote.

**Mots clés: Trade, Lake Victoria, Information, Fish, EFMIS project**

## 1. INTRODUCTION

### Theoretical context

The perfectly competitive market model is traditionally used in economics as a standard for attaining efficiency and equity (Mann, 1966; Roberts, 1987). It presupposes and entails an economically efficient allocation of resources. Since each trader maximizes profits by equating the given price to its marginal cost, competitive prices correctly reflect both consumer demand and the cost of resources. The competitive market model is characterized by large numbers of buyers and sellers, low barriers to entry, product homogeneity and complete knowledge of alternative choices on the part of producers and consumers (McNulty, 1967; Kotler, 1989). An efficient market will establish prices that relate transport, processing, and storage costs respectively to the provision of services in space, form and time dimensions. Competition should thus ensure that prices and marketing margins fully reflect the costs of resources used.

The complete knowledge of alternative conditions in the market is one of the cornerstones of the perfect market system. Market information is the main factor influencing sellers and buyers' decisions and choice in the market, such as; what, how much, what price and when to sell or buy (Wikipedia, 2011). However, such information is not readily available to most small scale fisheries in developing countries, a situation which gives undue advantage to those with access to information. Players operating at the upper end of the value chain (e.g. middlemen) tend to have greater access to market information and take advantage of fishers at the lower end of the value chain (Kambewa *et al*, 2007).

Information deficiencies and resulting inequities in the fish value chain are important issues that need to be addressed for sustainable development of small-scale fisheries. Effective market information systems can make a significant difference to the fishing community in terms of incomes, costs and post-harvest losses. Various approaches have been applied to improve access to market information in different regions; however, in some cases the cost of providing information is prohibitive. A common approach in addressing information gaps is by physical congregation of all players in a centralized auction market system, thereby enhancing information exchange and competition (Kambewa *et al*, 2007). However, the costs of setting up and maintaining a central auction can be astronomical in rural situations, considering the diverse locations and limited infrastructure of fishing areas and landing sites.

Other means of providing market information include; radio broadcasts, newspapers and bulletins, however, these often cannot keep pace with fast market transactions. While there is a lot of future potential in the use of the internet, most parts of Africa are still not online. The mobile phone is one media attracting increasing interest as a tool for quick dissemination of information. Africa has the fastest-growing mobile phone market worldwide, which is already being applied in many ways for profitable and non-profitable ventures. The penetration of the mobile phone is far greater than that of the internet in Africa, especially in rural areas, making it one of the most accessible communication tools. In East Africa, mobile phones are increasingly being used to access agricultural information including for weather, advice on inputs, reporting crop diseases and agricultural commodity prices, among other applications. The use of mobile phones has much potential for fish marketing as it can deliver information in real time to enable fast decision making.

### Information constraints in Lake Victoria

Lake Victoria is Kenya's most important fishery resource, producing about 120,000 MT of fish valued at about US$ 70 million and earning US$ 50 million from fish exports annually, which constitutes about 90 percent of the countries fishery output (Abila, 2008). The lake has a multispecies fishery of which three are commercially important, namely; Nile perch (*Lates nilotics*), Tilapia (mainly *Oreochromis niloticus*) and *Dagaa* (*Rastrineobola argentea*). The lake supports over 50,000 fishers, 300,000 local fish processors and traders, 7 fish processor and exporter firms and many fisheries organizations, including; 30 small fisher cooperative societies, 300 beach management units and over 350 women fish trader associations (Abila *et al*, 2008; Odongakara *et al*, 2009; Lwenya *et al*, 2008).

Although much of the lake region has been well covered by mobile phone networks for the past decade, fishers had not adequately taken advantage of this technology to market fish (LVBC, 2007). There was no systematic means of collecting, synthesizing and disseminating information on fish prices, demand and supply at various levels of the market chain in a useful way. This resulted in a situation where there were significant fish price disparities between markets and middlemen took advantage of this to buy fish from fishers at less competitive prices. It also caused considerable inefficiencies in market operations, and when fish landings were high combined with rains, led to substantial losses in fish volumes and value. Fish market information problems have been identified and cited in several previous studies. The current management plan for Lake Victoria fisheries developed by the Lake Victoria Fisheries Organization has identified increased access to fish market information as one of the best ways of improving performance of the sector and reducing poverty (LVFO, 2008; LVBC, 2007).

## 2. ENHANCED FISH MARKET INFORMATION SERVICE (EFMIS)

To address information deficiencies, an innovative IT project, based on mobile phones, was developed and piloted for Lake Victoria fishery in Kenya from June 2009 to June 2011. The project, named *Enhanced Fish Market Information Service* (EFMIS), aimed to empower the fishing community with useful fish market information to improve their bargaining position and increase incomes from fish trade. Through this service it was expected that pricing would be more transparent, fish prices improve, marketing costs reduced and that post-harvest fish losses would decline. The successful implementation and positive results of the pilot attracted a lot of interest in the region, consequently a proposal was developed to up-scale it to the national level. The EU accepted to support the up-scaled electronic fish market information system as from March 2011, which will serve the entire Kenya's fisheries sector, including the marine fisheries, all other fresh water lakes (Turkana, Baringo and Naivasha) and aquaculture, as well as fish markets across the country. The design and software systems of the national system would draw on the experience of the pilot project.

### *Objectives of EFMIS*

The objective of EFMIS was to enhance fish trade and incomes of the fisher community through improved access to market information easily, cheaply and faster. The project aimed to establish an electronic system of information collection and dissemination that would enable fishers, fish traders and consumers to access real time fish market information using their mobile phones.

### *Implementation framework*

EFMIS was conceptualized, developed and coordinated by a small team at the Kenya Marine and Fisheries Research Institute (KMFRI). The project was implemented in a coordinated framework bringing together a number of institutions from government, NGOs, community-based fisheries organizations and the private sector. The key collaborators were; the Department of Fisheries, the Beach Management Units, Cooperatives Societies, Women Fish Traders Associations and the Association of Fish Processors and Exporters of Kenya (AFIPEK). From the private sector, the leading mobile phone networks in Kenya and one SMS/ internet firm were involved as service providers (Figure 1).

The project was supported by the International Labour Organization through its Cooperative Facility for Africa (Coop[Africa]) Challenge Fund program, a regional technical cooperation program mainly funded by the Department for International Development of the United Kingdom (DFID). ILO provided US$ 83,950 through a competitive grant facility for a one-year implementation phase from June 2009 to May 2010, which was eventually extended to June 2011. KMFRI and other collaborators provided counterpart contribution in terms of facilities, project staff time and services valued at US$ 43,773.

**Figure 1. Partnership arrangement of EFMIS**

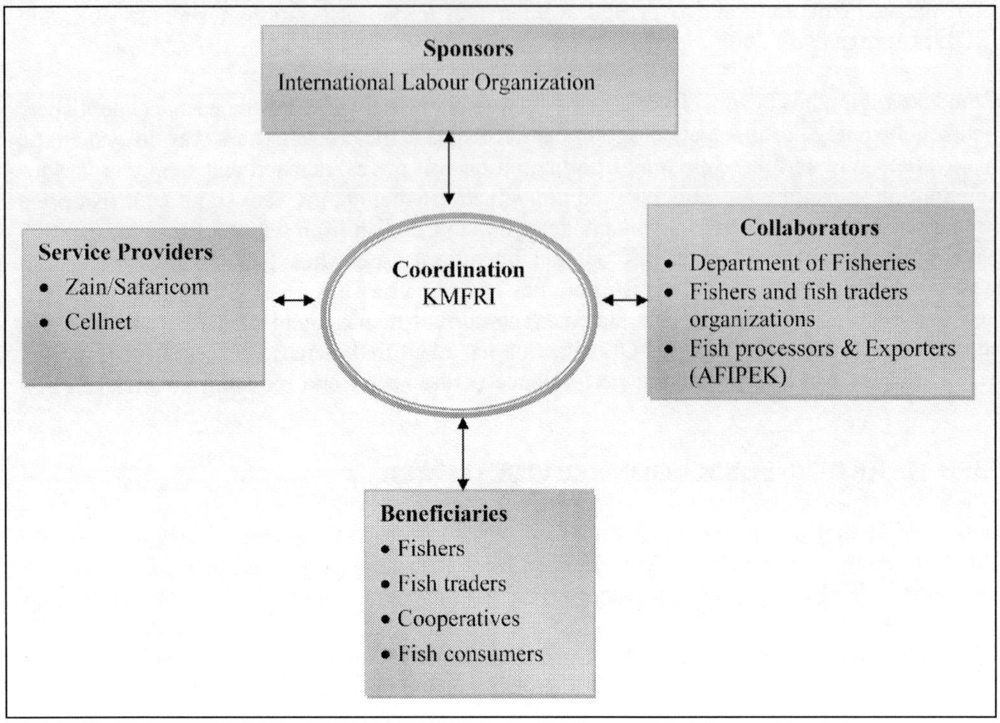

For the new up-scaled national project the EU has provided a grant of Euro 120,000 awarded through competitive bidding under the Assistance to Micro and Small Enterprises Program (ASMEP) in support of the Private Sector Development Strategy. The funds are channeled through the Micro-enterprise Support Program Trust (MESPT) established jointly by the Government of Kenya and the EU.

*Effecting the partnership*

EFMIS owes its success to the active participation of several institutions either as implementers, service providers and users of the system. To get all the participants on board, stakeholder workshops and meetings were organized that were attended by fishers, fish traders, fish exporters, cooperatives, government representatives, NGOs, and the media, among others. The apex of this was a highly publicized ceremony to launch the project which drew the attention of potential users of the service.

The project procured specialized services from the private sector competitively by tender in order to recruit a competent firm offering the most cost-effective services. As a result two highly reputed communication firms in Kenya were contracted to support communication services for the project, which included the supply of low-cost mobile phones for data recorders, provision of talk time and maintenance of the 24-hour automated SMS and internet data response system.

In addition, resources were expended on publicity materials and events so as to increase public awareness about project. The publicity materials included; two versions of wall posters, brochures, flyers and T-shirts, which were distributed to the fisher communities and other targeted groups. In addition the project organized exhibitions during agri-business shows and other forums held across the country, including twice each at the Kisumu Agribusiness Show and the Nairobi International Trade Fare. During such events EFMIS conducted "live" shows where people were able to observe data coming in, packaged and entered in the system and response to incoming queries. The EFMIS data centre itself attracted a lot of interest as demonstrated by high profile visits.

## 3. EFMIS SYSTEM DESIGN

*Overview of the system*

EFMIS was implemented through a stepwise process of continual learning and improvement, at each stage building on the previous achievements and correcting errors that occur. The project established a system for generating, packaging and disseminating key market information from about 165 fish markets, including 150 at fish landing sites and 15 inland markets in major urban areas across the country. The system had capability to handle data on four key variables that influence fish market decisions, namely; (i) Fish prices at landing sites and inland markets (ii) quantities of fish at landing sites and inland markets (iii) number of fish trucks at landing sites (iv) basic weather information (e.g. whether wet or dry).

The EFMIS system consisted of three broad phases; (i) Data recording, coding and transmission from landing sites and inland markets to the data centre (ii) A central database (iii) Query and automated response system. In summary, data was recorded once or twice a day at each of the landing sites and inland markets, and relayed by phone SMS in a coded format to a data centre based at KMFRI in Kisumu City. Here it was synthesized and appropriately packaged into a database in a special format that users (mainly fishers, fish traders, cooperatives and other consumers) could access it in real time (daily, by the hour) whenever they needed it.

To access the information a user had to send a query by SMS to the data centre from a mobile phone through a special number and get automatic response usually within 10 seconds. The system was active for 24 hours every day and could be accessed from any part of Kenya where there is a mobile phone network. Every month the project produced a summarized fish market information bulletin containing current prices and trends. This was circulated to about 1,000 stakeholders worldwide, and was also usually featured in Kenya's national newspapers. A subsidiary output is a huge database containing fish market data that can be useful for research and development studies.

**Figure 2. EFMIS system design**

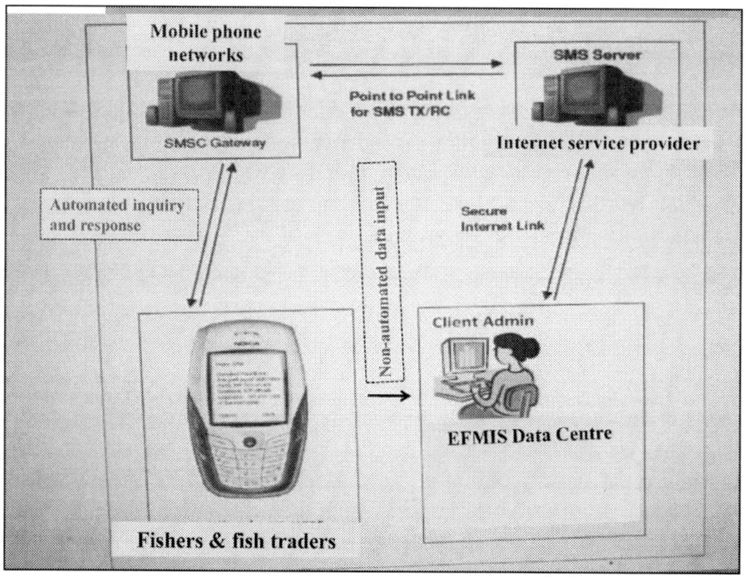

*Developing the data input phase*

The data input phase consisted of three key stages of; data recording, coding and transmission from landing sites and inland markets to the data centre in Kisumu. This segment was managed entirely by data recorders at the markets and the project staff based at the data centre using normal direct phone communication without involving external service providers.

A core principle of EFMIS was that the fisher community supplied the data which was then packaged appropriately for their own use. This dual supplier-user approach ensured that the community owned the process and its outputs. These were key conditions for sustainability of the system. Furthermore it ensured that the information disseminated was relevant to the needs of the users.

The project therefore depended on fishers to record data at the landing sites and fish traders at the inland markets. Two personnel were identified at each hub and given on the site training on how to determine weights and prices of fish, record data in a standard format, code and transmit it to the data centre. Data recorders worked on voluntary basis and did not receive any compensation for their time and effort. However, to support communication each landing site and inland market providing data was given one low-cost mobile phone and talk time per month worth about US$ 12.

### Establishing the database

A data centre was set up at KMFRI in Kisumu City which was equipped to receive data from the markets and package it appropriately into an active database. Personnel from KMFRI were trained for data centre operations, including how to retrieve, synthesize and enter data in the database. The set-up of the data centre network consisted of a phone connected to a computer enabling it to display data from markets. Data were then retrieved and transferred manually from the display into the EFMIS database. The database was based on a simple EXCEL spreadsheet structured with columns for entering each of the following variables; Market, Time; Date/month, Fish species, Quantity landed, Price, Number of trucks and basic weather.

### Developing the data output phase

The data output phase comprised of two main components: query and response. The data output phase was largely automated and managed with support of an external service provider. A special short code was leased from the SMS/internet service provider that was linked to the database through a server and dedicated entirely for use by the project. Queries sent by SMS through this number would be routed to the database. The server had software with the capability to pick out the specified data and automatically respond to the sender with information usually within 10 seconds. A user querying the system only specified the name of landing site or market and would receive all the latest information available for that market hub in terms of; the price of each of the three commercial species, quantity of fish available, number of fish trucks at a landing site and basis whether (whether wet or dry). Using this system one could access market information all the time on any day from any part of Kenya where there was phone network.

An additional information outlet was the electronic *EFMIS market bulletin* produced monthly and distributed via the internet. The bulletin presented market trends of the three commercial fish species with simple charts and figures and brief discussion of the prevailing factors. In total 21 monthly bulletins were produced and sent out individually to over 1000 people across the world Furthermore the content of the bulletin was often featured in Kenya's national press, thereby expanding the reach of the project.

## 4. RESULTS AND IMPACTS OF EFMIS

### Approach to monitoring and evaluation

The project put in place an elaborate plan for continuous internal monitoring and independent external evaluation. Objectively verifiable indicators (OVIs) for monitoring project progress and impacts were identified through a participatory process involving the fisher community and other stakeholders. At the start of project a baseline survey was conducted on the key monitoring variables, which then were monitored during the project life and progress reports were produced. An independent evaluation of project impacts was conducted towards the end of project by an outsourced consultancy firm.

### Performance indicators

The project log frame had the following OVIs for monitoring progress and those for evaluating impacts;

a) *Progress monitoring OVIs*: These indicators were designed to measure achievement of key milestones in project implementation. They included:
   (i) Number of fisheries organizations and markets participating in the project and providing data on a regular basis;
   (ii) Database established and fed regularly with market data and information;
   (iii) Number of inquiries for market information made to the database;
   (iv) Types and frequency of market data and information disseminated by various media;
   (v) Quarterly and annual reports produced in time.

b) *Impact monitoring OVIs*: These indicators were designed to measure impacts of the projects. They included:

    (i)   Fish prices at landing sites and inland markets;
    (ii)  Incomes to fishers, fish traders and processors;
    (iii) Percentage of post-harvest losses;
    (iv) Cost of marketing fish;
    (v)  Quantities of fish landed and sold at landing sites.

## *Results*

The following were the results of the key progress indicators:

*i)   Number of fisheries organizations and markets participating in the project and providing data on a regular basis*

For the purpose of this OVI, participation was viewed from the aspect of an organization providing data to the EFMIS system. The fisheries organizations participating in the program comprised of Beach Management Units, cooperatives and fish traders associations that were based at various landing sites or inland fish markets cooperating with the project. In total 165 fish markets participated in the project, comprising of 150 landing sites and 15 urban based markets. This was above the number of 150 markets that had been targeted.

*ii)   Database established and fed regularly with market data and information*

The market information database was developed in the first quarter of the project and remained fully functional to receive and respond to queries 24 hours a day, 7 days a week. Data from the landing sites and inland markets were fed into the system daily between 0800 hours and 1800 hours, although information could be obtained at any time from the database through the automated response system. At the end of pilot, the system passed, in full working status, to the new national electronic fish information project supported by the EU.

*iii) Number of queries for market information made to the database*

From zero at the beginning, a total of about 20,000 SMS queries were submitted to the database for market information and were responded to. A number of telephone call inquiries were also made to the data centre but these were not recorded in the system.

## *Impacts*

The link between market information and project impacts is built on a number of assumptions. With forehand market information (on prices, fish quantities and number of fish buyers at various markets), the beneficiaries (fishers, fish traders and cooperatives) will make informed decisions on the most cost-effective markets and avoid unnecessary transport expenses. In the same manner, knowledge of comparative fish prices at different landing sites and inland markets will give leeway for fishers to bargain for higher prices. Middlemen will not easily fix false prices as information will be available to all players.

By having information on the number of buyers and cold storage trucks at a landing site, fishers will avoid taking fish where it is unlikely to be sold and get spoilt. Bargaining time (before fish is put in cold storage) will be reduced if fishermen have information on comparative prices in other landing sites. Weather information will enable fishers make decisions avoiding inaccessible landing sites. The impact OVI targets and end-project status are as follows:

*i)   Fish prices at landing sites and inland markets*

The target set for this OVI was that the project would contribute to increased fish prices by 30 percent at end of project. There was sustained increase in average fish prices in markets covered by the project. Overall in 21 months between July 2009 and March 2011, the price of Nile perch rose by 25 percent , tilapia by 91 percent and *dagaa* by 137 percent. Clearly all the changes cannot be attributed to EFMIS alone and other external factors might have contributed to this significant increase in fish prices.

*iii)    Incomes to fishers and fish traders*

The target for this OVI was to contribute to increased incomes of fishers and fish traders by 30 percent at end of project. The gross income to fishers and fish traders is a factor of price and quantity of fish available for sale. Complete assessment of this impact has not been undertaken, however, the above significant price changes should have a strong positive influence on incomes.

*iv)    Post-harvest losses*

The target for this OVI was to bring the post-harvest fish losses down by 40 percent at end of project. A baseline survey conducted at the beginning of the project put the level of post-harvest losses for Nile perch at about 5 percent on the landing sites. Although the post-harvest situation at the end of project has not been determined, an assessment conducted in June-November 2010 on some of the landing sites reported post-harvest losses at about 4.5 percent (Abila & Werimo, 2010).

## 5. SUSTAINABILITY ISSUES

### *Cost-effectiveness of project implementation*

The budget which had been committed for running the project for one year was able to run the project for an additional year. This was attained by greater involvement of user communities, particularly in providing market information on a voluntary basis. Analysis of expenditure shows that 34.2 percent of budget was spent on talk time credit, 14.6 percent, was for holding consultative meetings with stakeholders while 10.7 percent was spent on installation and maintenance of the short-code leased for 24-hour automatic response system. The next high cost, at 10.1 percent, was on purchase of equipment, including mobile phones for the fisher organizations participating in the project. Travel by project staff to various landing sites and markets for identification and training of data collectors took 7.8 percent. The remaining significant costs were for collection and storage of market data, which took care of database operations, and internal and external monitoring and evaluation (7.4 percent). The rest of the costs were less than 5 percent of the overall and did not have significant bearing on overall project expenditure.

### *Sustainability of the system*

A key issue is the potential of the system to sustain itself beyond donor support. The expenditure structure outlined above demonstrates that much of the cost went towards establishing the system, including personnel mobilization and capital equipment, while the operational costs took less than half the budget. For sustainability, focus should be put on further reduction of the operational costs, while attention should also be on how to raise revenue from the information services.

The potential for the system to raise revenue was experimented in this pilot by charging users a small premium price above the cost of SMS sent to the data centre. The system had an inbuilt mechanism to automatically charge a user fee for an SMS query. Based on 20,000 SMS queries, the project raised a total of Ksh 200,000 (about US$ 2,500) in revenue, which was shared out in a structured way (Table 1). The revenue base is directly related to the number of queries, therefore greater use of the service would ultimately enhance revenue.

**Table 1. Revenue from SMS charges**

|     | Revenue structure | Ksh |
|-----|-------------------|-----|
| (a) | No. of SMS | 20,000 |
| (b) | Cost of SMS (Ksh) | 10 |
| (c) | Total generated by project SMS [(a)*(b)] | 200,000 |
| (d) | Cost to mobile phone companies and Government tax @ 60.3% of (c) | 120,600 |
| (e) | Balance after tax | 79,400 |
| (e) | Payment for code lease and data base maintenance services @ 50% of (d) | 39,700 |
| (f) | Revenue available for project activities @ 50% of (d) | 39,700 |

With greater sensitization and incorporation of additional information packages the number of users, therefore revenue, can substantially increase, however, it is unlikely that this source alone would suffice to sustain the system in the short-term. Besides user fees, it would be necessary that the fisher organizations play a bigger role, for instance by meeting part of the talk time costs. There is also potential to incorporate other private sector business ideas, for example, mobile money transfer services and commercial adverts targeting the fisher community as way of raising revenue.

### Implementation challenges

The main challenges faced in project implementation were:

- Despite the service being available and fully functional, it was not utilized to its capacity. The likely causes of the relatively small number queries could have been due to inadequate sensitization or the user charge fee inbuilt in the system.
- More fisher organizations and markets wanted to participate in the project faster than had been planned and budgeted for. For instance only 60 organizations had been targeted for inclusion by end of second quarter, yet up to 150 wanted to be involved, thus straining the budget especially on talk time.
- Meeting the high expectations of the different stakeholders for market information was a big challenge. Various stakeholders (fish farmers, traders, fish processors and consumers) had diverse information needs sometimes beyond what the project could provide. The information package developed fell short of the requirements of the different stakeholders. To address this situation a second package was attempted which allowed fishers to find out the availability and costs of fishing nets and other gear in different shops using their mobile phones. This, however, had not been fully developed at the end of project.
- Lack of cooperation by some stakeholders, particularly the fish processing and exporting industry. Most fishers were keen to find out factory gate fish prices paid to middlemen; however factory owners were reluctant to declare this piece of information publicly since they regarded it as their 'business secret'. It was equally difficult to get consistent information on wholesale/ retail prices in the export markets, which was of great interest to the fishing community.
- Lack of proper standardisation on fish quality and pricing units was also an important constraint; for example, fish prices could be reported as low due to poor quality rather than unfavourable market conditions.

## 6. CONCLUSION AND RECOMMENDATION

Modern ICT has great potential for improving fish trade and incomes in the sector. This is well demonstrated by the EFMIS project, which has improved the level of transparency in the fisheries industry by enhancing access to up to date market information for the fisher community. The project has managed to disseminate market information to various users using modern technology, particularly the mobile phone SMS service and internet bulletins. Though not fully validated there are indicators that the system has made a contribution to improved fish prices and incomes for fishers and fish traders, and potentially in reduction of marketing costs and post-harvest losses.

This system is relevant as it uses the mobile phone, which is the fastest growing media for communication and therefore in line with the current development trend. Despite the outlined challenges in setting and operating the system, it is a valuable tool for enhancing competitiveness of trade in small-scale fisheries in developing countries where information is a big constraint. This model can be adapted appropriately for application in other small-scale fisheries, paying attention to the lessons learnt from this pilot. The establishment of the system, however, needs to take into account the level of ICT development and put in place mechanisms for sustainability.

## 7. REFERENCES

**Abila R.O.** 2007. Assessment of Fisheries Product Values along Kenya's Export Marketing Chain. *FAO Fisheries Technical Report No. 819*. FIIU/R819. Food and Agriculture Organization of the United Nations. Rome. 262p.

**Abila, R.O.** 2008. Contribution of Lake Victoria Fisheries to the Economy of The East African Region: A Synthesis of the National Reports for Kenya, Tanzania and Uganda. Socio-economics Regional Technical Report, LVFO/IFMP.

**Abila, R.O. & Werimo, K.** (2010). Assessing Impacts of Ice Intervention on Fish Landing Beaches of Suba District, Kenya. Report submitted to *Africa Now*. Africa Now/ EU. Kisumu.

**Abila R. O., Yongo E., Lwenya C. & Omwega, R.** 2008. 'Socio-Economic Baseline Survey of Fishing Communities in Lake Victoria, Kenya'. *LVFO/IFMP Technical Report 1*. Lake Victoria Fisheries Organization. Jinja, Uganda.

**Kambewa, E., van Tilburg, A. & Abila, R.O.** 2007. 'The Plight of Small-scale Primary Producers in International Nile Perch Marketing Channels'. In: Ruben R., van Boekel, M., Van Tilburg, A. & Trienekens, J. (Eds). *Tropical Food Chains - Governance Regimes for Quality Management.* Wageningen Academic Publishers. The Netherlands. p. 309.

**Kotler, P.** 1989, "Marketing Management - Analysis, Planning, and Control", Prentice-Hall of India, New Delhi.

**LVBC.** 2007. Regional Transboundary Diagnostic Analysis of the Lake Victoria Basin. Lake Victoria Basin Commission of the East African Community. Kisumu. http://www.eac.int/lvdc.html

**LVFO.** 2008. The Fisheries Management Plan for Lake Victoria 2009–2014, Lake Victoria Fisheries Organization, Jinja. http://www.lvfo.org/

**Lwenya, C., Yongo, E.O. & Abila, R.O.** 2008. 'A Report on Fish Agents Survey - Kenya'. *LVFO/IFMP Technical Report 1.* Lake Victoria Fisheries Organization. Jinja, Uganda.

**Mann, H.M.** 1966. "Seller Concentration, Barriers to Entry and Rates of Return in Thirty Industries, 1950 - 1960". *The Review of Economics and Statistics Journal No.48,* New York.

**McNulty, P. J.** 1967. "A note on the history of perfect competition", *Journal of Political Economy,* 75, 4 part 1, August, pp. 395–399.

**Roberts, J.** 1987. Perfectly and Imperfectly Competitive Markets. *The New Palgrave: A Dictionary of Economics* 3. pp. 837–841.

# LE RESEAUTAGE ET L'ECHANGE D'INFORMATIONS DANS LE DOMAINE HALIEUTIQUE: EXPERIENCES MAROCAINES

## [NETWORKING AND EXCHANGE OF INFORMATION IN FISHERIES: MOROCCAN EXPERIENCES]

by/par

Anass Karzazi[1]

**Résumé**

De récentes expériences de réseautage dans le domaine halieutique ont vu le jour au Maroc, un des pays les plus avancés dans ce secteur en Afrique. La première, baptisée FISHPROS (www.fishpros.tk), consiste en la création fin 2008 d'un portail et d'un forum de discussion sur internet destinés au partage de tout type d'information concernant les filières halieutiques de l'amont à l'aval. Grâce à l'utilisation des techniques de positionnement et de référencement sur les moteurs de recherche ainsi que des outils de veille et de diffusion de l'information, le site a pu atteindre des résultats importants: 407 membres enregistrés, 30.000 visites depuis 112 pays dont 27 africains. La deuxième expérience consiste à la création d'une association (AMIH) créée en 2011 et regroupant les ingénieurs halieutes (IH) marocains. pour la contribution au développement du secteur halieutique national dans toutes ses dimensions: scientifique, technique, culturelle, pédagogique, environnementale et humaine. L'AMIH se base sur une organisation centrale (bureau) et régionale (3 représentations), des objectifs ambitieux et réalisables et une synergie entente entre ses membres favorisée par leur appartenance à une même école d'ingénieurs. Après l'organisation d'un premier séminaire et la création d'une base de données IH nationale, du site internet officiel et de la page facebook, l'AMIH s'est fixée un nombre d'actions pour l'année 2012 notamment l'organisation de 4 séminaires et d'une journée culturelle sur des thèmes d'actualité dans le secteur halieutique marocain et la programmation d'actions sociales et environnementales. Les postes importants qu'occupent certain IH au niveau national et régional ainsi que la répartition des IH sur tout le littoral marocain et leur pluridisciplinarité constituent des atouts qui contribueront favorablement à l'atteinte des objectifs fixés.

Les deux expériences marocaines de réseautage montrent à quel niveau l'utilisation des nouvelles technologies d'information et de communication ainsi que le rassemblement au sein d'une organisation guidée par un leadership dynamique et motivé peuvent-elles constituer des pistes pour la création et la mise en place du premier réseau africain de technologie et de sécurité des poissons (ANFTS).

*Mots clés: Réseautage, Maroc, Information, Halieute*

**Abstract**

Recent experiences of networking in the fisheries sector have emerged in Morocco, one of the most advanced countries in this sector in Africa. The first, called FISHPROS (www.fishpros.tk), was created in late 2008 and consists on internet portal and forum .to share all kind of data related to fish and seafood sector from upstream to downstream. Using techniques of positioning and referencing on web search engines as well as watch and information dissemination tools, the website could reach important results: 407 registered members, 30,000 visits from 112 countries, 27 of which were African. The second experiment consists in the founding of an association (AMIH), created in 2011 and brings together Moroccan fisheries and marine biology engineers (IH). for the goal of contribution to national fisheries sector development in all of its dimensions: Scientific, technical, cultural, educational, environmental and human. The AMIH is based on a central Board and 3 regional representations, ambitious and achievable targets and a synergy between its members favoured by the fact that they all graduated from the same Engineers School. After organizing a first seminar and the creation of a national database IH, an official website and a Facebook page, the AMIH has set a number of actions for 2012 including the organization of four seminars and a cultural day on topical issues in the Moroccan fisheries sector and the programming of social and environmental actions. Important positions occupied by some IH, both in national and regional levels, and the distribution of IH all along the Moroccan coastline and their interdisciplinarity are assets that will contribute positively to achieving the objectives.

[1] Maritime Fishery Department, Ministry of Agriculture and Maritime Fishery, PO Box 476, Haut-Agdal, Rabat, Morocco. karzazi@mpm.gov.ma

The two Moroccan networking experiences show how the use of new information and communication technologies and the gathering within an organization guided by a strong and motivated leadership can offer real examples to follow in the objective of the creation and establishment of the first African Network of Fish Technology and Safety (ANFTS).

*Key words: Networking, Morocco, Information, Fishery*

## 1. INTRODUCTION

Le réseautage et l'échange d'information constituent des outils incontournables pour le renforcement des capacités des professionnels et communautés de pêche en vue de leur développement. Dans le secteur de pêche et notamment le domaine post-capture, l'Afrique se trouve à la croisée des chemins en matière de réseautage régional et mise à profit des nouvelles technologies de l'information et de la communication. Des initiatives nationales ont toutefois vu le jour ces dernières années qui méritent d'être partagées dans la perspective de la construction d'un réseau régional. C'est le cas des expériences marocaines suivantes qui peuvent ouvrir des pistes de développement et de coopération pour les pays africains.

## 2. PRESENTATION DU SECTEUR DE LA PECHE AU MAROC

Doté de 3.500 km de côtes et de un million de km² de Zone Economique Exclusive, le Maroc a pu développer au fil des années des filières de pêche et d'industrie de transformation diversifiées et performantes.

En 2010, la production halieutique a atteint 1,13 millions de tonnes (dont 78 pour cent de pélagiques) d'une valeur de 6,6 millions de Dh[2]. La flotte nationale se compose de 450 navires hauturiers, 2.570 côtiers et 16.400 barques de pêche artisanale qui se répartissent sur les 22 ports de pêche du Royaume ainsi que les autres points de débarquements aménagés, sites et villages de pêche (DPM, 2011).

L'industrie de la transformation qui permet l'emploi de 70.000 personnes, est composée de 422 unités à terre formées par les principales filières: conserve, semi-conserve, congélation, frais et farine de poisson. En 2010, le Maroc a exporté un total de 600 000 tonnes de produits halieutiques d'une valeur de 13 milliards de Dh. Les principales destinations sont l'UE, l'Afrique, le Japon et le continent américain (DIPM, 2011).

## 3. ENJEUX ET OPPORTUNITES

Face aux grandes performances du secteur, plusieurs enjeux requièrent un intérêt spécial de la part du Département de la Pêche Maritime notamment:

- La gestion durable de toutes les pêcheries et le développement de l'aquaculture;
- Le contrôle des flux le long de la chaîne de valeur et la lutte contre la pêche illicite, non déclarée et non réglementée (INN);
- Les exigences sanitaires internationales et autres normes en changement permanent;
- La compétitivité des produits de la pêche nationaux dans les marchés mondiaux;
- La productivité de la flotte et l'exploitation de nouvelles pêcheries;
- L'efficacité de la recherche scientifique pour l'accompagnement des filières.

---

[2] 1$US = 8.4 Dh.

Les points suivants constituent, par contre, de réelles opportunités à saisir par tous les acteurs:

- La volonté de l'Etat avec le lancement par le Département de la Pêche Maritime de la *Stratégie Halieutis 2020* de développement du secteur. Basée sur 3 axes (Durabilité, Performance et Compétitivité), la stratégie compte 16 grands projets et se fixe comme objectifs, entre autres, le doublement des exportations et de l'emploi, une production de pêche de 1.6 millions de tonnes et une gestion durable de 95 pour cent des pêcheries à l'horizon 2020;
- La création de l'*Agence Nationale de Développement de l'Aquaculture* en vue de mettre sur pied le projet aquacole national;
- Le Statut Avancé avec l'Union Européenne, premier partenaire commercial du Maroc.

## 4. EXPERIENCES MAROCAINES DE RESEAUTAGE DANS LE SECTEUR DE LA PECHE

Après plusieurs années d'expériences avec les réalités du terrain, il a été constaté un grand manque d'accès aux informations sur les différents thèmes pertinents du secteur et la nécessité du *partage* et du *transfert des connaissances* pour le développement de toutes les filières halieutiques nationales. Les Nouvelles Technologies d'Information et de Communication (NTIC) constituaient une opportunité énorme qui a été saisie en conséquence, dans un premier temps, pour la création de FISHPROS, premier portail des professionnels du poisson au Maroc.

### FISHPROS (www.fishpros.tk et prochainement www.fishpros.ma)

Créé en décembre 2008 dans le but de rassembler les professionnels et autres institutions et personnes intéressées par le secteur de la pêche au Maroc pour le réseautage et le partage de l'information. Il contient deux modules:

- *Un portail*: Contenant des informations actualisées et des liens utiles sélectionnés,
- *Un forum*: Cadre d'échange divisé en 6 catégories contenant chacune plusieurs sous-forums. Différents sujets, informations et documents y sont mis en ligne régulièrement dans les domaines de la pêche et aquaculture, hygiène et salubrité des produits de la pêche, gestion des stocks, commerce et marketing, réglementation, labellisation et normes, recherche scientifique, etc.

A ce jour, le site compte 407 membres enregistrés, il a cumulé près de 30.000 visites depuis 112 pays dont 27 africains. Grâce à un travail continu d'une équipe de modération, aux techniques de veille, de référencement et de positionnement sur internet, ainsi qu'à l'utilisation des outils d'Analyse Web (Google Analytics) et de Web Marketing, le site connaît davantage de notoriété surtout parmi les professionnels avec un dédoublement de la fréquentation lors des derniers mois.

La réussite de l'expérience FISHPROS et la sensibilisation aux enjeux nationaux de la pêche, ont poussé à la création d'une autre forme de réseautage pour une participation plus active au développement du secteur. Ainsi, et avec le concours de professionnels du Département de la Pêche Maritime, l'Association Marocaine des Ingénieurs Halieutes (AMIH) a été fondée.

### ASSOCIATION MAROCAINE DES INGENIEURS HALIEUTES (AMIH)

#### *Présentation*

L'AMIH regroupe les Ingénieurs Halieutes (IH) lauréats de l'Institut Agronomique et Vétérinaire Hassan II de Rabat. Elle a pour objet de réunir les IH pour la participation au développement du monde halieutique marocain dans toutes ses dimensions: scientifique, technique, culturelle, pédagogique, environnementale et humaine.

L'AMIH œuvre entre autres pour la constitution d'un groupe de consultation scientifique et technique au profit de l'Etat et de la Profession dans les domaines liés au monde de la pêche, de l'aquaculture et de toutes les activités qui en découlent. Elle vise aussi le développement humain dans le secteur halieutique marocain et l'instauration de canaux de coopération, d'association et de communication avec les institutions gouvernementales, le secteur privé, la société civile et les associations similaires au Maroc et à l'Etranger.

### Organisation de l'AMIH

Pour atteindre ses objectifs, l'AMIH s'est organisée de la façon suivante:

- *Un bureau*: Formé par 7 membres dont un président, un secrétaire général et un trésorier.
- *Trois représentations régionales*: A savoir celles de la Méditerranée (AMIH/MED), du Centre (AMIH/CENTRE) et du Sud (AMIH/SUD). Chacune d'elles est assurée par un Représentant Régional et un Adjoint, qui veilleront à la mise en œuvre des actions régionales de l'AMIH, la remontée d'informations et la coordination locale avec les autres acteurs.
- *Trois comités:*
  - Comité Scientifique et Technique: Dont le rôle est d'assurer une veille de l'actualité scientifique et technique du secteur de la pêche au niveau national et international et coordonner entre le Bureau et les Représentations Régionales pour l'organisation des évènements et actions prévues par l'Association;
  - Comité de Communication: Organe d'information, il gère la relation de l'AMIH avec l'extérieur, les moyens de communication mis en place (Site Web, page Facebook, Bases de Données, etc.) et utilise le réseautage pour tisser des liens avec les acteurs du secteur et autres organisations d'objectifs similaires;
  - Comité Culturel: Organe responsable de la proposition et de la mise en œuvre des actions à caractère culturel.

### Réalisations

- Organisation d'un premier séminaire le 22/6/2011 à Rabat sous le thème *« Rôle de l'Ingénieur Halieute dans le développement du secteur de la pêche au Maroc »* avec la participation de hauts responsables de l'Administration (Département de la Pêche Maritime (DPM)), des Professionnels (Fédération des Pêches Maritimes et de l'Aquaculture (FPMA), Fédération des Chambres des Pêches Maritimes (FCPM), Fédération des Industries de Transformation et de Valorisation des Produits de la Pêche (FENIP)), d'Organisations Régionales (Conférence Ministérielle sur la Coopération Halieutique entre les Etats Africains Riverains de l'Océan Atlantique (COMHAFAT)), de la Formation et Recherche Scientifique (Institut National de Recherche Halieutique (INRH), IAV Hassan II). L'initiative de création de l'AMIH a été saluée par tous les intervenants qui ont exprimé leur disponibilité pour tout appui nécessaire à l'atteinte des objectifs fixes;
- Création du site internet officiel de l'AMIH *(en cours de lancement: www.amih.ma);*
- Création d'une page Facebook (www.facebook.com/halieute.ma): quotidiennement alimentée en informations et actualités sélectionnées et mises en lignes par 2 administrateurs. Statistiques actuelles: 208 «j'aime», plus de 400 publications vues par plus de 25.000 visiteurs;
- Création d'une première représentation régionale de l'AMIH;
- Constitution d'une base de données des lauréats IH depuis 1975 *(en phase de conception);*
- Recensement et constitution de la base de données des lauréats IH originaires de pays africains.

### Plan d'actions 2012/2013

- L'organisation de 4 séminaires scientifiques au cours de l'année 2012 sur les thèmes de la *« Labellisation des produits de la pêche »*, *« La valorisation et l'innovation dans les produits de la pêche »*, *« La gestion durable des pêcheries »* et *« Le développement de l'aquaculture »* avec la participation d'intervenants de renom pour une dimension plus globale des évènements;
- Création des deux autres représentations régionales de l'AMIH;
- Constitution d'une base de données des thèses des lauréats IH depuis 1975;
- Constitution d'une base de données de sites web et contacts des acteurs du secteur halieutique;
- Organisation d'une journée culturelle sur la mer et l'histoire de la pêche au Maroc;
- Réalisation d'une action sociale au profit d'une population défavorisée de familles de pêcheurs;
- Réalisation d'une action contre la pollution de l'environnement marin;
- Conception et publication d'un bulletin d'information de l'AMIH;
- Participation à la prochaine édition du Salon Halieutis;
- Publication sur le site web des thèses et publications scientifiques des membres.

*Atouts et opportunités de l'AMIH*

L'AMIH dispose d'atouts importants qui favoriseront davantage le réseautage et contribueront à l'atteinte des objectifs fixés. Il s'agit en effet de:

- L'importance des postes occupés par plusieurs IH en l'occurrence le Secrétariat Général de l'INRH, la Direction de l'Agence Nationale de Développement de l'Aquaculture ANDA, le Secrétariat Exécutif de la COMHAFAT ou aussi le Secrétariat Exécutif de la Commission Générale des Pêches pour la Méditerranée CGPM.

  De plus, plusieurs autres IH assurent des responsabilités importantes notamment au Département des Pêches Maritimes, INRH, Office National des Pêches, ANDA et IAV Hassan II et ce tant au niveau des services centraux (Rabat et Casablanca) qu'au niveau régional (Délégations);

- Pluridisciplinarité: Du fait de leur formation polyvalente, les IH sont impliqués dans des domaines clés liés au secteur halieutique (régulation, production, contrôle et inspection, recherche scientifique, assurance qualité, industrie, certification, études, assurances, banques, développement, social, commerce, etc.);

- Répartition sur tout le littoral national: Important pour l'accès à une information globale et régionale ce qui favorisera aussi les actions locales;

- Implication directe d'IH dans la conception et la mise en œuvre de projets importants dans le secteur notamment:
  - La Stratégie Halieutis de développement et de compétitivité du secteur halieutique,
  - La Stratégie de Marketing Institutionnel des Produits de la pêche,
  - La labellisation des produits de la pêche,
  - Le Système National de Traçabilité des produits de la pêche,
  - Les Pôles de Compétitivité,
  - La lutte contre la pêche (INN).

## 5. CONCLUSIONS ET PERSPECTIVES

Les deux expériences marocaines de FISHPROS et de l'AMIH constituent deux exemples différents de réseautage dans le secteur halieutique. Alors que la première puise sa force dans l'utilisation des NTIC et des techniques de veille et de partage de l'information, la deuxième se base sur l'organisation verticale et horizontale de l'association, l'entente entre ses membres favorisée par leur appartenance à une même école d'ingénieurs et les opportunités d'appui et de coopération vu les postes clés qu'occupent certains IH tant au niveau national, régional et international.

Toutefois, la réussite de telles initiatives suppose des pré-requis importants tels que l'existence d'un leadership motivé et dynamique, l'implication et la volonté de ses membres ainsi et surtout l'existence et la consolidation d'une culture de partage et de diffusion de l'information. Les experts africains du domaine post-capture pourront donc capitaliser sur ce genre d'expériences pour dynamiser la création et le fonctionnement du premier réseau africain de technologie et de sécurité des poissons (ANFTS).

## 6. RÉFÉRENCES BIBLIOGRAPHIQUES

**Département de la Pêche Maritime (DPM).** 2011. Documents statistiques et rapports d'activités internes.
**Direction des Industries de la Pêche Maritimes (DIPM).** 2011. « Répertoire des industries de valorisation des produits de la pêche ». Edition janvier 2011.

**APPENDIX/ANNEXE D**

## Opening statement by Ms Amy Quatre, Chief Executive Officer,
## Seychelles Bureau of Standards, Seychelles

Minister, Principal Secretaries, CEOs, Dr Yvette Diei-Ouadi (FAO Fish Industry Officer), Participants, Ladies and Gentleman:

Good morning

The Seychelles Bureau of Standards is delighted and honoured to host this Expert Meeting on Fish Technology, Quality Assurance and Marketing being funded by FAO.

I wish to extend a warm welcome to our beautiful shores to the delegates from the various countries. That many of you have travelled long distances serves to remind us all just how important this meeting is.

I also wish to thank FAO for pursuing this initiative as it would give the African countries the opportunity to discuss experiences and exchange information concerning research and development of fish utilization.

Seychelles is a significant exporter of fish and fishery products and the industry is an important contributor to the economy of the islands.

I recognize that the meeting sessions are principally dedicated to post-harvest fisheries technology, utilization and quality assurance.

Post-harvest losses in small-scale fisheries can be among the highest for all the commodities in the entire food production system.

Numerous actions to reduce waste already exist and the need to decrease all forms of waste and to optimize the use of fisheries resources for human food security is embedded in the FAO Code of Conduct for Responsible Fisheries as well as in the UN Fish Stock Agreement.

For many years FAO has implemented a wide range of activities including training of fish technologists in developing countries to introduce appropriate technologies for lowering fish spoilage especially for small-scale fisheries.

With that ladies and gentleman, it is my wish that this workshop is successful and that the results are adaptable and applicable by the countries in this region. It is in knowing our fisheries resources well that we can be successful in managing such resources in a sustainable manner.

I realize that you are fully dedicated to the sessions that will follow but I do hope that you will also take time to enjoy this fascinating island with its tropical settings and its friendly people.

Let me once again welcome you to Seychelles and wish you fruitful deliberations.

**APPENDIX/ANNEXE E**

**Opening statement on behalf of FAO by Mr Alejandro Anganuzzi, Executive Secretary of the Indian Ocean Tuna Commission (IOTC), Seychelles**

The Chairperson, Guest of Honour, Colleagues, Distinguished guests, Ladies and Gentlemen:

It gives a great pleasure to welcome you all to the Seychelles and to this very important expert meeting on fish technology, utilization and quality assurance. I will also take this opportunity to express FAO's and my own appreciation to you for having taken time out of your busy schedules to attend this meeting.

On behalf of FAO, I would like to thank the Government of Seychelles for having kindly accepted to host this meeting and provided the support to organize it.

The Guest of Honour, Ladies and Gentlemen:

In the 20 years between 1970 and 1990, the number of people around the world deriving a livelihood from fishing increased by 72%. Much of this increase was in developing regions, which accounted for 58% of the total production.

Most regrettably, it is in these very countries where post-harvest technologies and quality assurance systems are still in great need of improvement. Most of us here are acutely aware of the serious post-harvest fish losses that occur in artisanal fisheries in Africa. The impact of these losses on the economies of households and fishing communities is significant, resulting in reduced incomes and low household food security.

Ladies and Gentlemen:

Helping the fishermen and women fishmongers, fish processors and their communities to reduce these losses and improve the safety and quality of fish would contribute tremendously towards improving incomes and food security and reducing the level of poverty. This assistance will also help in improving access of non-fishing populations to fish and fish products and hence to achieving better levels of nutrition.

I am sure you will all agree that if this could be achieved, it will enable us to contribute towards the realization of the Millennium Development Goal on reducing food insecurity and alleviating poverty.

Achieving this, as you all know, has always been and continues to be of high priority to FAO and its partners in the fisheries sector. To better address these central issues of sustainability in fisheries, FAO has developed and disseminated guidelines and, codes of practice and has provided training, using a value chain approach, in:

- reduced post-harvest losses;
- improved value addition;
- enhanced utilization of underutilized low value fish species for human consumption;
- improved market access.

It is also becoming increasingly important to mainstream gender in development through an enhancement in development opportunities for women in fisheries. This requires both improvements in both technical and managerial skills, in addition to appropriate policy interventions that will provide a conducive framework within which the post-harvest activities operate. In fact, women play a significant role in aquaculture operations and fish handling and processing in Africa, besides their pivotal support to their families.

Adding technology related tools and providing information on the latest developments in post-harvest fisheries and promoting a fair linkage of artisanal operator to fish export industries, would be very instrumental in their accessing more lucrative local, regional and international markets.

The Guest of Honour, Ladies and Gentlemen:

The market requirements and consumer preferences, the need for ensuring fish food security, and the need to protect dwindling fisheries resources, make it imperative that the utilization of catches are improved by reducing post-harvest losses, adding value, improving resource sustainability and improving the safety and quality of fish.

These have been the major concerns in fisheries and aquaculture for the past few decades.

This meeting is therefore timely and will address the important socio-economic issues that can improve the livelihoods and food security situation of the fishing communities and the larger rural populations.

It is also expected that the meeting will review developments in international markets and in the supply of good quality products to local markets.

The sustainability of these positive developments will also be an issue in the deliberations of this meeting.

Ladies and Gentlemen:

FAO's efforts in assisting countries to address the challenges in post-harvest handling and technologies have included technical assistance and capacity building, as well as periodic thematic seminars and meetings like this one to share experience, learn from each other and develop a joint vision for post harvest fisheries in Africa.

Your contributions and recommendations to further these efforts will be most valuable.

With these few remarks, I wish you very fruitful deliberations over the next four days.

I thank you all for your kind attention and I hope you enjoy your stay in the Seychelles.

**APPENDIX/ANNEXE F**

**Opening statement by H.E. Peter Sinon, Minister for Investment, Natural Resources and Industry, Ministry of Investment, Natural Resources and Industry, Seychelles**

Principal Secretaries, Chief Executive Officers, Dr Yvette Diei (FAO Fish Industry Officer), Participants, Ladies and Gentleman:

On behalf of the Government of Seychelles and the Seychelles Bureau of Standards, Ministry of Investment, Natural Resources and Industry, it is my honour to officiate the opening of this important meeting. I welcome everyone to our islands and hope that you find a little time to enjoy what they have to offer.

In 2008, Seychelles participated for the first time in the Expert Meeting on Fish Technology, Utilization and Safety which was held in Agadir, Morocco. Seychelles was approached by FAO on the possibility of organizing the next meeting normally held on a three yearly basis. Being a major producer and exporter of fish and fishery products in Africa, and one of the world highest annual per capita consumer of fish, Seychelles viewed this event as an opportunity to further strengthen its position where fisheries is concerned on the African continent. It was therefore not surprising for, the Ministry of Environment and Natural Resources on behalf of the government in 2008 to give its commitment and full support for the event to be held in the Seychelles.

Today, the Seychelles is very proud to welcome you all on this special occasion whereby you as experts on post-harvest fisheries technology, utilization and quality assurance from all over Africa will be presenting your findings in research studies carried out in specific areas based on the needs of your respective countries. During the coming three days, you will be sharing a lot of knowledge and information from the diversity of presentation and debates, mainly on issues such as improved fresh fish handling, fish smoking, fish drying, prevention of post-harvest losses, utilization of discards and by-catch, fish marketing and trade, quality control and inspection. In light of the dwindling fish resources, especially fish from the wild, coupled with the steady increase in demands and the development in market access requirements, improving fish utilization for human consumption contributes significantly to socio economic development and food security. Post-harvest fish loss is one of the major challenges facing many small-scale fisheries in developing countries. Bad fishing methods, poor fish handling and processing as well as ineffective means of preservation and marketing cause huge losses in terms of physical, quality and market force losses. This program is designed to facilitate coordination of efforts by various institutions in research and development of fish utilization, in particular at artisanal level through the initiatives of the Products Trade and Marketing Service (FIPM) of the FAO. The initiative to establish a cooperative program amongst African institutions through a functioning network known as the African Network for Fish Technology, and Safety (ANFTS) addressing mainly technological issues associated with the small-scale post-harvest fishery has been much welcomed by FAO. However, the current globalization context requires that emerging issues both in artisanal as well as industrial fisheries be effectively addressed on a more holistic basis, integrating expertise and experience from various sources and regions.

As you may be aware, Seychelles as a small island state with a land area of 444 km² surrounded by 1.4 million km² of seas as its Exclusive Economic Zone has to depend heavily on its fisheries resources. The fisheries industry in the Seychelles is the second pillar of the economy only next to tourism. Seychelles has been exporting large quantities of canned tuna, fresh and frozen fish, fish oil mainly on the EU market, frozen tuna to Mauritius, Madagascar, Spain, etc., and fish meal to Australia, Sri Lanka and Japan. The production of fish and fishery products in the Seychelles is therefore quite diverse. However one of the government's policies on the fisheries sector is to further diversify, with special emphasis on developing more value added products. The potential for value addition using mainly by-catch from the industrial tuna vessels is big noting that about 250,000 to 300,000 tonnes of fish are landed and transshipped annually in Port Victoria and about 5% or 12,500–15,000 tonnes of that are by-catch which are currently very much underutilized. The government through the Seychelles Fishing Authority (SFA) is putting facilities at the disposal of small entrepreneurs who have shown interest to venture in this sector. They are being encouraged to seriously take the challenge with the support of governmental institutions specialized in assisting small enterprises. This will not only increase productivity and diversity of fishery products on the local and export market, but will also create job opportunities. However, in order to successfully penetrate the international market, strict standards and legislations on hygiene and quality assurance have to be met and all producers must be fully aware that there is no shortcut to this. Failing to meet international standards will make it difficult if not impossible for developing countries to trade its products internationally. This results in continued losses in earnings, wastage of valuable fisheries resources due to poor sanitary conditions during handling and processing, inadequacies in processing technology and production of low quality products.

This meeting will surely provide you participants with challenges to re-look at what is being produced in your respective countries especially at small-scale level, come up with ideas on how you can contribute to improve production, reduce wastage and maximize utilization. It will also stimulate the necessity to invest more resources in post-harvest research technologies adaptable to particular situations in different countries.

On this note let me wish you all successful deliberations and I am convinced that this meeting will achieve its objectives.

Thank you very much.